CITIES AND THE POLITICS OF DIFFERENCE

Multiculturalism and Diversity in Urban Planning

Edited by Michael A. Burayidi

Over the past decade, the growt[illegible] populations in North American cities has outpace[illegible] of the White population. Also there is a growing recognition of the multiple dimensions of diversity that include sexual orientation, income, ethnicity, religion, and disability. *Cities and the Politics of Difference* looks at the challenges and opportunities that urban planners face in addressing the needs of an increasingly diverse public.

Organized in five parts, the volume examines the practical and theoretical issues that surround this urban transformation. The chapters in part I offer a historical and critical reflection on planning practice as enshrined in the values of individualism, tolerance, and citizen rights and privileges, and on the capacity of liberal democratic institutions to accommodate immigrants and their cultures within these tenets. Part II extends the discussion of multiculturalism beyond race, culture, and ethnicity – addressing sexual minorities and planning in Native American communities – while part III looks at the proactive measures cities are implementing to make new immigrants welcome and to ensure that planning practices and governance promote tolerance and understanding of diverse groups. Parts IV and V consider ways that the design of the physical environment both constrains and enhances the needs of different groups to live, work, and play, and discuss strategies that can be taken to make urban planners more culturally competent to work with diverse communities.

Featuring in-depth essays by a distinguished international group of contributors, *Cities and the Politics of Difference* is an invaluable guide for scholars, planners, and policy-makers at all levels of government.

MICHAEL A. BURAYIDI is Irving Distinguished Professor and Chair of the Department of Urban Planning at Ball State University.

Cities and the Politics of Difference

Multiculturalism and Diversity in Urban Planning

EDITED BY MICHAEL A. BURAYIDI

UNIVERSITY OF TORONTO PRESS
Toronto Buffalo London

Toronto Buffalo London
www.utppublishing.com
Printed in the U.S.A.

ISBN 978-1-4426-4810-4 (cloth)   978-1-4426-1615-8 (paper)

♾ Printed on acid-free, 100% post-consumer recycled paper
with vegetable-based inks.

Library and Archives Canada Cataloguing in Publication

Cities and the politics of difference : multiculturalism and
diversity in urban planning / edited by Michael A. Burayidi.

Includes bibliographical references.
ISBN 978-1-4426-4810-4 (cloth). ISBN 978-1-4426-1615-8 (paper)

1. City planning – Social aspects. 2. City planning – Political
aspects. 3. Multiculturalism. 4. Cultural pluralism. 5. Social
integration. I. Burayidi, Michael A., author, editor

HT166.C58 2015   307.1′216   C2015-905450-8

University of Toronto Press acknowledges the financial assistance to its
publishing program of the Canada Council for the Arts and the Ontario
Arts Council, an agency of the Government of Ontario.

Canada Council for the Arts   Conseil des Arts du Canada

Funded by the Government of Canada   Financé par le gouvernement du Canada   Canada

Contents

Tables and Figures

Tables

Figures

Contributors

SANDEEP K. AGRAWAL is a professor and inaugural director of the Planning Program in the Department of Earth and Atmospheric Sciences at the University of Alberta. Prior to this appointment, he was graduate program director and associate director in the School of Urban and Regional Planning at Ryerson University in Toronto, where he taught for 14 years. Dr Agrawal's work focuses on Canada's ethnic communities and the effects of multiculturalism, religion, and human rights on urban structures and public policies. His research interests also include land use planning, urban design, and international planning. He is a co-author of the forthcoming book *Understanding India's New Approach to Spatial Planning and Development: A Salient Shift?* (Oxford University Press). Dr Agrawal has extensive experience in administrative law and municipal decision making. He currently serves on Alberta's Municipal Government Board.

MICHAEL A. BURAYIDI is the Irving distinguished professor and chair of the Department of Urban Planning at Ball State University. He is author of several books including *Resilient Downtowns: A New Approach to Revitalizing Small and Medium City Downtowns* (Routledge), *Urban Planning in a Multicultural Society* (Praeger), and *Downtowns: Revitalizing the Centers of Small Urban Communities* (Routledge).

PETRA L. DOAN is professor in the Department of Urban and Regional Planning at Florida State University. Her primary interests are in the area of planning for less developed areas in both rural areas of the United States and overseas in such places as Jordan, Egypt, Togo, Niger, Côte d'Ivoire, and Botswana. Research interests include integrating tourism with local development planning, developing appropriate regional development strategies, setting policy for market town development, and building capacity for local institutional development. In addition, she works on planning issues related to sexual and gender minorities in the United States.

JOHN FORESTER is a professor of City and Regional Planning at Cornell University. His research focuses on the micro-politics of planning practices with particular attention to issues of public participation, power, and conflict resolution. His best-known publications include *Planning in the Face of Power* (University of California Press, 1989), *The Deliberative Practitioner* (MIT Press, 1999), and *Dealing with Differences: Dramas of Mediating Public Disputes* (Oxford University Press, 2009).

RICHARD GALE is a lecturer in the School of Planning and Geography at Cardiff University, UK. He has a DPhil in Human Geography. He has researched ethnic and religious segregation, place and the politics of identity, religion and urban governance, ethnicity and political participation, and the response of the planning system to the needs of religious groups. He has published work on the geographical study of religion and Islam in the UK and the political engagement of young people of minority ethnic heritage. His latest published paper is "Religious Residential Segregation and Internal Migration: The British Muslim Case?" (*Environment and Planning A*, 2013).

HANNAH GILL is the assistant director, Institute for the Study of the Americas, University of North Carolina–Chapel Hill. She is an anthropologist with a specialization in Latin American/Caribbean migration studies and the author and co-author of two books, *North Carolina and the Latino Migration Experience: New Roots in the Old North State* and *Going to Carolina de Norte: Narrating Mexican Migrant Experiences*.

THOMAS L. HARPER is a professor in the Faculty of Environmental Design at the University of Calgary. His teaching and research interest is in exploring theoretical issues which impact public planners' ability to meet the challenges of a postmodern, post-industrial, pluralistic, liberal democratic society. His theoretical work (in collaboration with Dr Stan Stein) aims to develop planning theory that is relevant to contemporary planners, helping them 1) understand the rapidly changing and turbulent environments (physical, social, political, economic, legal, institutional, professional) in which they carry out their historic mission – making human environments better places to live; and 2) justify the role and purpose of public planning and plans in society. He has expressed these theoretical concerns in practice, where his primary interest has been in assisting community groups in decision making and consensus building.

STACY ANNE HARWOOD is an associate professor in the Department of Urban and Regional Planning at the University of Illinois, Urbana-Champaign. Her research, rooted in social justice, focuses on the emerging field of planning for difference and diversity. She links scholarship to planning practice by examining how practitioners deal with the mandates of participation and equity in land use planning and how planning codes and regulations differentially affect diverse populations. This interest is founded on years of participant-observation of planning in US and Latin American cities, where she has paid particular attention to the phenomenon of multicultural communities in which planning processes that endorse the ideals of justice and tolerance nevertheless often fall short.

MUKESH KUMAR is an associate professor and program director of Urban and Regional Planning at Jackson State University. He has a doctorate in Urban Studies and Public Affairs from Maxine Goodman Levin College of Urban Affairs at Cleveland State University. He has published substantially in areas of urban spatial structure and local economic development planning.

DAVID LAWS is an associate professor in the Department of Political Science at the University of Amsterdam. Before coming to Amsterdam, he worked at the Department of Urban Studies and Planning and the Sloan School of Management at MIT and with the Program on Negotiation at Harvard Law School. His research focuses on the relationship between negotiation and conflict resolution, public administration, and democratic governance. He has worked as a consultant for the New York Stock Exchange, the US Environmental Protection Agency, the US Air National Guard, the Wisconsin Department of Natural Resources, and the Ministry of Housing, Spatial Planning and the Environment.

SANG S. LEE is a PhD candidate in the Department of Urban and Regional Planning at the University of Illinois, Urbana-Champaign. Her research interests include community development and citizenship-building activities among immigrants, and efforts by local governments to make cities more welcoming for immigrants. Previous to her graduate studies, Sang Lee worked as a labour organizer in Urbana-Champaign, Illinois, and as a community organizer in Madison, Wisconsin.

KELLY MAIN is an associate professor of City and Regional Planning at California Polytechnic State University, San Luis Obispo. She holds a PhD in urban planning from the University of California at Los Angeles. Her research interests and teaching focus include community planning in culturally diverse and transnational communities, community outreach and participation methods, place attachment, place identity, and public space planning. She has conducted research in southern and central California and in Mexico. Prior to completing her doctorate and joining Cal Poly, she was a public planner in Southern California for more than 15 years.

MAI THI NGUYEN is an associate professor in the Department of City and Regional Planning at the University of North Carolina–Chapel Hill. She employs both quantitative and qualitative methods to address problems related to social and spatial inequality, urban growth phenomena, the relationship between the built and social environments, and socially vulnerable populations. She is an expert in housing policy, community development, economic development, immigration, disasters, and urban growth phenomena. Her work has been published in peer-reviewed journals, edited books, and public policy reports. Her research has been funded by the National Science Foundation, the US Department of Agriculture, the US Department of Housing and Urban Development, and the John D. and Catherine T. MacArthur Foundation. She also teaches courses in the Housing and Community Development specialization with the focus of teaching about practices and policies that create transformative community change.

MOHAMMAD A. QADEER is a professor emeritus of Urban and Regional Planning at Queen's University. He was the director of the School of Urban and Regional Planning for about 10 years and a full-time teacher for 30 years. The author of three books, numerous monographs, reports, and articles, recently he has been studying and writing about multiculturalism and cities. His pioneering article "Pluralistic Planning for Multicultural Cities: The Canadian Practice" was awarded honourable mention for best article by the American Planning Association in 1997. He regularly contributes articles on multicultural planning, ethnic enclaves and economies, and malls in *Plan Canada*, *Canadian Journal of Urban Research*, and other journals.

MICHAEL RIOS is chair of the Community Development Graduate Group and an associate professor of Community Planning and Design at the University of California at Davis. He has contributed numerous publications on the topics of placemaking, marginality, public space, the ethics of practice, and critical pedagogy. His co-edited book, *Diálogos: Placemaking in Latino Communities* (Routledge, 2012), considers how demographic changes in regions, cities, and towns both challenge and offer insight into community planning and urban design practice in an increasingly multiethnic world. He was the inaugural director of the Hamer Center for Community Design at the Pennsylvania State University (1999–2007) and president of the Association for Community Design (2003–2005). He received his PhD in Geography from the Pennsylvania State University, and Master of Architecture and Master of City Planning degrees from the University of California at Berkeley. Currently, he directs the Sacramento Diasporas Project at the UC Davis Center for Regional Change.

JAMES ROJAS is an urban planner, community activist, and artist. He is one of the few nationally recognized urban planners to examine US Latino cultural influences on urban design. He holds a Master of City Planning and a Master of Science of Architecture Studies from the Massachusetts Institute of Technology. His influential thesis on the Latino built environment has been widely cited. He founded the Latino Urban Forum, a volunteer advocacy group, dedicated to understanding and improving the built environment of Los Angeles' Latino communities. That organization has recruited urban planners, architects, artists, and public administrators to lend their knowledge and influence to innovate and address the issues of the underserved, and often underprivileged, Latino communities of Los Angeles. Most recently, he has developed the community outreach tool Place It!, a spontaneous modelling method of participation. He has facilitated over 150 interactive workshops using Place It! and built over 43 interactive urban dioramas across the country. He has collaborated with municipalities, non-profits, educational institutions, museums, and galleries to educate the public on urban planning.

SIDDHARTHA SEN is a professor and chair of the Department of City and Regional Planning and assistant dean of the School of Architecture and Planning at Morgan State University. He has written extensively on planning education at Historically Black Colleges and Universities and diversity and multiculturalism in planning education. He currently serves on the Standing Committee on Diversity of the Association of Collegiate Schools of Planning and has served on similar committees before.

SHERI L. SMITH, AICP, is an associate professor and coordinator of the Master's Program in the Urban Planning and Environmental Policy Department at Texas Southern University.

ANISHA STEEPHEN is a city planning and development professional with experience focused on successful private investments into the urban public realm through innovative partnerships. Her interests include layered development finance, affordable housing development and policy, inner-city investments, and asset building in low-wealth communities.

STANLEY M. STEIN is a professor and resident philosopher emeritus in the Faculty of Environmental Design, University of Calgary.

EMILY TALEN is a professor at Arizona State University in the School of Geographical Sciences and Urban Planning and the School of Sustainability. Her research is devoted to urbanism, urban design, and social equity. She has four books: *New Urbanism and American Planning: The Conflict of Cultures*; *Design for Diversity*; *Urban Design Reclaimed*; and *City Rules*. She edited the second edition of the *Charter of the New Urbanism*, as well as *Landscape Urbanism and Its Discontents* (with Andres Duany). She is the recipient of a Guggenheim Fellowship, and is a Fellow of the American Institute of Certified Planners.

HUW THOMAS is a reader in the School of Planning and Geography at Cardiff University, UK. He has a PhD in City and Regional Planning. Author of *Race and Planning: The UK Experience* (UCL Press, 2000), he has also edited a number of books on race equality and planning issues in Britain and Europe. He has advised professional bodies and local and national governments on planning in a multicultural society. In 2008 he guest edited a special issue of *Planning Practice and Research* on

"Race, Faith and Planning." His latest published paper is "Values and the Planning School" (*Planning Theory*, 2012).

ABBY WILES earned her master's degree from the Department of Urban Planning at Ball State University in 2013. She is the assistant planning and zoning administrator for the city of Goshen, Indiana.

NICHOLAS C. ZAFERATOS, AICP, is an associate professor of Urban Planning and Sustainable Development at Huxley College of the Environment, Western Washington University. His teaching emphasis – in urban planning, sustainable development, Native American planning, and environmental policy – complements his regional and international service learning and research interests. He is active in promoting sustainability education at WWU. Since 2005, he has directed several service learning programs in sustainable development in Washington State and internationally. His work in the Mediterranean region has received international recognition, including the 2009 Green Good Design Award (European Centre for Architecture, Art, Design and Urban Studies); the 2008 SETE National Honor Sustainability Award (Athens); and the 2006 sixth Honorific Mediterranean Sustainability Award (Spain). His professional practice in urban planning spans over 35 years and includes service as planning director and general manager with the Swinomish Indian Tribal Community and civic appointments on local government boards and commissions.

CITIES AND THE POLITICS OF DIFFERENCE

Multiculturalism and Diversity in Urban Planning

1 Cities and the Diversity Agenda in Planning

MICHAEL A. BURAYIDI

Introduction

A pervasive characteristic of cities in the twenty-first century is that they have become socially diverse and unabashedly multicultural conglomerations. In one sense, the world has become monocultural, united by a global free market system and access to information made possible through the internet and social media. At the same time, globalization and efficient communication systems have made it possible for remote ethnic groups to make their needs known to a global public. Just as the culture of North America has found its way into distant corners of the world through global corporations, so have remote ethnic cultures infiltrated North American neighbourhoods. Matters that used to be a concern for "other lands" have become pressing local issues. Wherever you are today, there is the world!

Canadians and Americans are now rubbing shoulders with foreigners on a regular basis in their neighbourhoods, at work, and on the baseball fields. The culture of the world has become local culture. Liberal democratic values in the West and the rise of civil society have enabled minority group issues to rise to the forefront of public discourse. These groups are no longer content to be passive observers; rather, they want to be direct and active participants in deliberations on matters affecting their felt needs. Invariably, these are not just social or economic issues; the issues relate to how space is used and how the built environment is manipulated to suit the needs of residents. That is the sphere of urban planners. Thus, the postmodern project that began in the mid-twentieth century not only remains incomplete but needs to be ramped up.

Although "planning for difference" is now a part of the planning discourse, and consideration of diversity and difference is acknowledged as essential to the delivery of effective planning services, this fibre of inclusion has yet to weave its way into every fabric of planning practice, nor has it been embraced by all planners, especially at the local level where it counts most. Many planners remain wedded to such ideals as "serving the public interest," "consistency," and "treating all people equally" when in fact these ideals result in a continuation of inequality and structurally hinder attempts at creating inclusive planning processes.

A study on planning and diversity conducted for the Office of the Deputy Prime Minister in Britain found that planners see their practice as "land use based" and do not consider social or cultural issues as germane to planning practice. One planning officer is quoted as saying:

> Diversity is a flawed concept. It is not necessary to seek supposed problems and representations and disadvantages. The Planning Service serves the whole community, not stratified segments of it. (Booth et al., 2004, p. 20)

Similarly, Mohammed Qadeer observed the following with respect to planning practice in Canada:

> The term multicultural planning is puzzling for planners. They are aware of the ethno-racial diversity of their clients and generally feel that they are sensitive to differences in their clients' material and aesthetic needs for community facilities, services, land uses and housing, etc. They maintain that they plan and manage by functions and not persons. Their professionalism demands a certain uniformity of treatment of all citizens ... And they are apprehensive of the accusation that they are not practicing it, particularly in the Toronto, Vancouver and Montreal metropolitan areas. (Qadeer, 2009, p. 10)

A study of local government planning practices in Australia by Dunn, Thompson, Hanna, Murphy, and Burnley (2001) revealed similar sentiments among planners regarding multicultural considerations in their practice. As one planner commented:

> We see no disadvantage in being a member of a multicultural minority group. They are particularly well provided with funding and services by

> state and Commonwealth departments, so well in fact that it is causing dissatisfaction in the working community. (p. 2488)

When Stacy Harwood reviewed land use decision-making processes in communities undergoing demographic change in Orange County, California, she found that:

> ... most of the planners interviewed not only avoided politics and the media but spoke only cautiously about issues related to ethnicity, culture or immigration. At public meetings, city officials rarely raised such topics because they were not central to the council's constituency.
>
> ... Staff reports seldom discussed the changing ethnic and cultural makeup of the community, and viewed such aspects as not relevant for policymaking. Procedures are rarely questioned. In the name of impartiality or treating everyone the same during the process, the system is not flexible enough to handle difference. (Harwood, 2005, p. 366)

So while progress has been made in bringing diversity to the attention of planners, a lot more needs to be done to embed this in their everyday practice. Most practising planners still contend that their goal is to promote the public interest without consideration for the needs of specific groups. Pursuing the public interest, they argue, ensures that people are treated equally and is congruent with liberal democratic principles. It is this line of reasoning that led Sennett to posit that "Western" urban planners are in the grip of a Protestant ethic of space. This regards the inner life as the most important and, being fearful of the pleasures, differences, and distractions of the "outside, it has had a controlling and neutralizing effect on the environment" (Sennett, 1992 as quoted in Neill, 2004, p. 15).

Barsky (2010, p. 229) reminds us that "treating everybody equally does not necessarily mean treating everybody in exactly the same way." Hence, Neill (2004, p. 220) suggests: "Planning for cultural difference must involve an awareness on the part of planners that there is not just one reading of a place and that different cultural groups relate to space in various ways."

Nonetheless, the call for diversity considerations in planning must not be a call for cultural relativism. In valuing diversity, planners must be cognizant that some cultural norms and practices are unjust and can be unfair and discriminatory. Multicultural policies need to be explicit about the set of principles that guide the social contract, by which all must abide and which form the foundations of a liberal democratic society. These are liberty, freedom of choice and association, and the unfettered right of all persons to participate in decisions governing the polity. Multicultural planning then is the continuous adjustment of planning practices to ensure parity in access to and the effective delivery of planning services to those ethnic, cultural, or demographic groups that have traditionally been marginalized or discriminated against by the planning process.

The Diversity Conundrum

In North America and other Western countries, diversity is defined in relation to the dominant European culture of the settler population. The sources of diversity in these countries are threefold: 1) Native Americans, Aborigines in Australia, or First Nations in the case of Canada, 2) descendants of former slaves and conscripted labourers, comprising primarily African Americans in the case of the US, and 3) recent immigrants, many of whom are now from non-European countries. It is the third group that is contributing to the immense diversity in Western countries, but it is the former two groups that provide a compelling moral justification for multicultural policy. After all, African Americans were forcibly brought to the New World by Europeans, and North America has been home to Native Americans for centuries before the arrival of the European settlers. The situation is of course systematically different in Europe, where diversity was driven primarily by postwar immigration. Irrespective of the source of diversity in these countries, a failure to address difference by homogenizing the diverse cultures that have different histories and origins does a disservice to the search for locally compelling responses to managing how we live together differently in the twenty-first century.

Multicultural policies in the West can be classified into two types: native multiculturalism and immigrant multiculturalism. Native multiculturalism derives from policies that are tailored to the diversity brought about by native-born minority groups, as opposed to what Kymlicka (2004) calls "immigrant multiculturalism," a term he uses in reference to policies for accommodating immigrants, particularly the recent waves of non-European immigration. Over the years, there has been a vacillation of policy responses to diversity in Western countries. For over two hundred years, all ethnic groups were expected to blend in with the dominant European culture and to lose their distinctiveness in a process of assimilation. Thus a "melting pot" doctrine was implemented in these countries right up to the twentieth century. Until the enactment of the Canadian Multiculturalism Act, Canada saw itself as comprising primarily French and English, culturally, linguistically, and politically. National policy aimed at homogenizing the different cultures in the country, not at supporting the diversity of Canada's cultural groups. For example, Aboriginal people were prohibited from practising their culture and were expected to be "civilized" into European ways. This took the form of cultural indoctrination of their children through residential schools.

Following criticism and agitation by the French in Quebec and other minority groups in Western countries, national governments began to give legitimacy to "cultural pluralism." In the US, the Civil Rights Movement initiated by African Americans to seek full inclusion in American society created a profound seismic shift in national policy. While some civil rights leaders advocated for the assimilation of African Americans into the larger American society, Black nationalists argued that African Americans should not have to give up their cultural identity in order to be part of the dominant culture and thus called for the recognition of the distinctiveness of "Black culture" through what Salins (1997) called "ethnic federalism."

Ethnic European groups such as the Ukrainians in Canada latched on to this argument and demanded recognition of their identity. Even some middle-class Whites sought to differentiate themselves from the dominant culture in the counterculture revolution of the 1960s. This group, comprising hippies and anti-war and anti-establishment protesters, rejected the dominant culture and sought a different world view and way of life displayed through social artefacts, dress code, and a subculture that was different from the mainstream. The birth of native multiculturalism was in response to these social agitations.

At the same time, the relaxation of immigration restrictions that previously favoured people from European countries resulted in the surge in immigrants from non-European origins that significantly diversified the population of Western countries. For example, it is now estimated that at least a fifth of all Canadians are foreign born. As the numbers of these new immigrants grew, it bolstered their political clout and enabled them to call for reforms, among which was recognition of their cultural identities. Instead of the "melting pot" doctrine, leaders of these groups advocated alternative visions of society along the lines of a "salad bowl," "rainbow coalition," "weaving machine," or "gorgeous mosaic," among others, to recognize the diversity of cultures in these nations. Canada led the way in adopting a multicultural policy to recognize and address the differences in culture, language, and ethnicity of the country.

At the turn of the twenty-first century, the population in Western countries has become more diverse, minorities have become a majority in some cities, and these countries continue to wrestle with questions of how best to handle the growing diversity. The planning profession is not alone in this quest. Professions as diverse as health care, education, social work, and psychology, among others, continue to grapple with this issue.

Managing Cities of Difference

Multiculturalism has been the primary approach for managing diversity since the mid-twentieth century. As a noun, multiculturalism describes the state in which people of different cultural and ethnoracial groups coexist in a given geographic area. Multicultural policy, on the other hand, is an acknowledgment of this diversity and the adoption of programs for managing cultural pluralism. In this usage, it requires policy makers to take difference into consideration in the delivery of programs and policies. Multicultural policies usually take on the following flavours, some or all of which may be discernible at the local and/or national level:

- An officially adopted policy on multiculturalism that eschews assimilation and that recognizes and supports the existence of different cultures in the country. In regard to Canada, Kymlicka (2004) termed this "the adoption of a more 'multicultural' conception of integration, one which expects that many immigrants will visibly and proudly express their ethnic identity, and which accepts an obligation on the part of public institutions (like the police, schools, media, museums, etc.) to accommodate these ethnic identities" (p. 5).
- Policies to accommodate the needs of multiethnic groups so that they are full participants in civil society. These include language programs such as the official dual languages of French and English in Canada and the provision of translation services in public meetings to ensure that the voices of diverse groups are heard in the decision-making process. In Australia this is institutionalized through the country's Translating and Interpreting Services (TIS),

a nationwide telephone service for non-English speakers, and through the language programs provided by the country's Special Broadcasting Service (SBC).

- Provision of financial support so that diverse ethnic and cultural groups can express, maintain, and celebrate their cultures, especially their language and religion. The 1988 Multiculturalism Act of Canada stated that it is the policy of the government of Canada to "recognize and promote the understanding that multiculturalism reflects the cultural and racial diversity of Canadian society and acknowledges the freedom of all members of Canadian society to preserve, enhance and share their cultural heritage" (Government of Canada, 1988).
- Proactive policies to include people of diverse backgrounds in the workplace, especially historically underrepresented groups such as African Americans and Native Americans. For example, affirmative action is used in the US to diversify the workforce and to rectify past discrimination against African Americans.
- Policies to decrease or eliminate social and economic disadvantage that is related to people's ethnicity. Here again Canada's 1988 Multiculturalism Act required all federal institutions to "promote the full and equitable participation of individuals and communities of all origins in the continuing evolution and shaping of all aspects of Canadian society and assist them in elimination of any barriers to such participation" (Government of Canada, 1988, p. 837).
- Policies against racism and discrimination based on race, ethnicity, and cultural belonging. The Civil Rights Act of 1964 in the US protects African Americans and other minorities from discrimination on these grounds.
- Involvement of minority groups in the formulation of a country's national identity. For example, the National Agenda for a Multicultural Australia in 1989 and its restatement in 1995 are applicable to all Australians, including the indigenous Aboriginal and Torres Strait Islander population. It identified three goals for Australians as "the right to cultural identity, the right to social justice and the need for economic efficiency which involved the effective development and utilisation of the talents and skills of all Australians. Balancing these rights were, however, a series of explicit obligations which included a primary commitment to Australia; an acceptance of the basic structures and principles of Australian society including the Constitution and rule of law, tolerance and equality, parliamentary

democracy, freedom of speech and religion, English as the national language and equality of the sexes; and the obligation to accept the rights of others to express their views and values" (Inglis, 1995).

In the last decade or so, however, and especially in Europe (see Gale and Thomas in chapter 5), there has been a fierce debate about the efficacy of multicultural policy in managing diversity in Western democracies. Sergei Mikheyev, a political analyst, is quoted as saying: "The policy of multiculturalism in Europe has failed. Immigrants are not integrating into the Western society; on the contrary, they do everything to lead a segregated lifestyle and establish closed communities with their own rules. They use the material luxuries that the Western countries provide, but they want to live according to their own laws and beliefs" (Kuryanova, 2013).

The twenty-first century brought with it renewed criticism of multiculturalism. Some criticized multiculturalism for contributing to the balkanization of society and the weakening of social ties. As Taylor (2012, p. 414) put it, "The underlying assumption seems to be that too much positive recognition of cultural differences will encourage a retreat into ghettos, and a refusal to accept the political ethic of liberal democracy itself." In a critique of Canadian multicultural policy, Bannerji (2009) argues that multiculturalism has been used by the state to placate the minority "other" without addressing the real issues of class conflict and power relations; "The nation state's need for an ideology that can avert a complete rupture becomes desperate, and gives rise to a multicultural ideology which both needs and creates 'others' while subverting demands for anti-racism and political equality" (p. 329). What is needed, she argues, is social justice and political equality, not multiculturalism.

Multiculturalism has also received significant criticism and aversion from civic leaders, especially in European countries that have had, to one degree or another, a national policy on multiculturalism. These criticisms were precipitated by racial tensions. In Britain, for example, racial conflicts between Asians and working-class Whites led to riots in 2001. More recently, in 2013 a young Muslim deliberately ran over a British soldier with his vehicle on a London street and proceeded to decapitate the soldier with a knife. In response to the arrests of three teens, young Muslim immigrants in France rioted in 2005. The riots that broke out in the Parisian suburb Clichy-sous-Bois quickly spread to the rest of France.

Following the British race riots in 2001, a commission was formed in the UK to investigate the cause of the unrest and to make recommendations. The *Cantle Report* concluded that minority groups lived "parallel lives" separated from the larger British society and that there was little contact between Black, minority, and other ethnic groups and the rest of the population (Cantle, 2001). Later in 2006 the Commission on Integration in its report *Our Shared Future* also identified the lack of racial and cultural contact as a national deficit. These reports and the high-profile cases involving ethnic minorities raised concerns in Europe about the efficacy of multiculturalism for managing Europe's diverse population. European prime ministers David Cameron, Nicolas Sarkozy, and Yves Leterme and German Chancellor Angela Merkel all openly declared that multiculturalism had failed. British Prime Minister Cameron publicly repudiated multiculturalism and declared that the UK needs a stronger national identity, implying that the diversity of cultures in the country has muddled what is British. The Council of Europe (2008) White Paper on Intercultural Dialogue noted that:

> old approaches to the management of cultural diversity were no longer adequate to societies in which the degree of that diversity (rather than its existence) was unprecedented and ever-growing. The responses to the questionnaire sent to member states, in particular, revealed a belief that what had until recently been a preferred policy approach, conveyed in shorthand as "multiculturalism," had been found inadequate. On the other hand, there did not seem to be a desire to return to an older emphasis on assimilation. Achieving inclusive societies needed a new approach, and intercultural dialogue was the route to follow. (Council of Europe, 2008, p. 9)

This groundswell in Europe at least for replacing multiculturalism with a policy of interculturalism is the most vociferous *coup d'état* on multiculturalism, so let's consider the merits of such a proposal.

Interculturalism has its roots in Allport's (1954) contact theory. Allport asserted that prejudice and negative attitudes towards the "other" can be decreased through intergroup contact. However, for this to be effective, the contact and interaction must occur under optimal conditions. Among these is the requirement that participants from the larger society voluntarily engage with the out-group and hold equal status in the deliberations.

Interculturalists see dialogue as the way to mend the differences between ethnocultural groups and the larger society with the goal of encouraging mutual learning and understanding between these groups. The hope is that minority ethnic groups will be provided the opportunity to interact and learn from each other, but most importantly that they will be helped to better understand the values and way of life of the host country. Parekh (2007, p. 46) calls this approach "interactive multiculturalism."

Advocates argue that interculturalism is distinct from multiculturalism. According to Salmancheema (2013), "Unlike multiculturalism, which concerns advocacy of equal respect to the various cultures in a society or the promotion of cultural diversity, interculturalism requires from all citizens an openness to be exposed to the culture of the 'other.'"

Meer and Modood (2012) identified the many ways in which interculturalism differs from multiculturalism: Interculturalism stresses the coexistence of different cultures, it de-emphasizes group belonging in favour of the whole, and it leads to the questioning of illiberal cultural practices. These attributes, they reason, contrast sharply with multiculturalism, which is said to be "essentialist, illiberal, less agency-oriented, and less concerned with unity" (Meer & Modood, 2011, p. 5).

The British Council defined intercultural dialogue as "a dynamic and challenging process that enables those engaged to explore their own and others' identities and backgrounds and their effects on attitudes, behaviours and relationships towards and within communities locally, nationally and globally" (Centre for Social Relations, 2013). Irena Guidikova, the Council of Europe's manager of the Intercultural Cities program, put it this way: "The Intercultural Cities approach addresses the issue of how to create cohesive societies that are diverse, that hold together and are resilient in the face of extremist discourse and acts" (New Europe, 2013).

In furtherance of this goal, the European Year of Intercultural Dialogue was held in 2008 to create opportunities for different cultures to learn from each other and for Europeans to develop the "knowledge and aptitudes to enable them to deal with a more open and more complex environment" (European Commission, 2007). Dialogue then is paramount to the concept of interculturalism. However, two issues are raised with respect to interculturalism. The first is how effective interculturalism is for managing diversity. The second is the extent to which it differs from multiculturalism.

The idea of dialogue and group interaction is not entirely new to urban planners. In the twentieth century, we witnessed what was considered the "communicative turn" in the profession, a paradigm not unlike the interculturalism that is now being advocated for managing diversity. In what became known variously as "communicative planning" and "collaborative planning," the role of the planner shifts from that of the rational technician seeking ends to that of a facilitator and mediator of the group process. In this position, the planner's goal is to help people of different viewpoints come to a consensus by mediating the conversation (Healey, 2003; Forester, 1989; Innes, 1995).

Huxley and Yiftachel (2000, p. 333) described this form of planning as

> communicative interchanges between interested parties, whether stakeholders or the community at large, over matters of common concern, … not necessarily confined to issues of development and land use. In some versions, such communicative action is seen as fostering community empowerment and recognition of difference, diversity, and disadvantage that has implications for the development of discursive local democracy beyond the confines of specific issues.

Healey (1992, p. 150) observed that, with collaborative planning, intersubjective deliberation "is required where 'living together but differently' in shared space and time drives us to search for ways of finding agreement on how to 'act in the world' to address our collective concerns."

These views on communicative planning are similar to Wood and Landry's (2008) description of the intercultural city:

> The intercultural city argues that we should interact more with each other because we live side by side. Only then will we foster empathy by learning more of each other and reduce the distrust between people. This way we get used to living with difference (p. 5).
>
> It seeks to consolidate different ways of living, recognizing arenas in which we must all live together and those where we can live apart. It generates structured opportunities to learn to know the "other," to explore and discover similarity and difference. (p. 9)

If interculturalism and communicative planning are seeds of the same pod, then perhaps our experience in the planning profession can be instructive to proponents of interculturalism. Heated and protracted debates ensued in the planning literature about the merits of communicative dialogue and its potential for enriching the profession and its practices. These discussions identified several flaws in communicative dialogue relating to, among others, the imbalance of power, the structure of communication, the disconnect between means and ends, and the occasional incommensurability of values.

Imbalance of Power

Communicative planning processes, like intercultural dialogue, assume that individuals enter public deliberations as disembodied and disempowered participants. In reality, "the public sphere cannot be cordoned off from the power relations of the system or the domestic sphere. Opinions formed in the public sphere or decisions taken in the appropriate forums cannot escape their embeddedness in the continuity of the social, and thus, starting from (liberal) assumptions of (legal) equality can do no more than reproduce the status quo" (Huxley and Yiftachel, 2000, p. 372). If power difference manifests itself in public deliberations, then the least powerful are at a disadvantage in exerting influence on the outcome of decisions.

The limit of interculturalism then, like its cousin of communicative planning, is that it ignores systemic and structural biases that put the less powerful in a weaker position to extract any gain from intergroup dialogue. Judith Innes, a proponent of communicative planning, concedes that "consensus building is not, in any case, the place for redistributing power" (2004, p. 12).

Another limitation of communicative planning, and by implication interculturalism, is that its evocations of dialogue may provide a way to disguise and smooth over the real issues of racism and discrimination in favour of the more celebratory but superficial instrument of group interaction. When socioeconomic inequities such as high unemployment and poverty persist, these can then be blamed on deficiencies in the ethnic and minority groups themselves and not on the structural forces to which they are prey.

All this is to say that cultural and social interaction by itself is not enough to address power imbalance. As Susan Fainstein in explaining Marx and Engels' critique of the Hegelians rightly noted, "words will not prevail if unsupported by a social force carrying with it a threat of disruption. To put this another way, the power of words depends on the power of the speakers" (2000, p. 458).

Communication Deficit

Public discourse enables the knowledge claims of individuals to be validated or invalidated through intersubjective reasoning. However, there is no guarantee that those engaged in the deliberations will comprehend each other, much less arrive at a consensus. This is particularly the case with people from diverse cultural backgrounds. Distortions in communication can mask facts and meaning, which can affect the outcome of decisions. As George Bernard Shaw aptly observed, "England and America are two countries separated by a common language." If language divides, then there is potential for communication deficit. Even if the participants understand each other, that is not to say that they will arrive at a consensus, because "To understand is not necessarily to agree" (Fay, 1987, p. 190).

It is this reasoning that led Habermas to suggest that claims made in public deliberations should be assessed on their comprehensibility, integrity, legitimacy, and truth. This is because "linguistically mediated interactions are only meaningfully understood as communicative action when they put forward validity claims that are recognized by the hearer. These are claims about truth (about the objective world), claims to rightness (about a shared social world), and claims to truthfulness (about one's own subjective world)" (Prychitko and Storr, 2007, p. 260).

The advantage of a public discourse is that the collective intelligence of the group leads to better outcomes than decisions made by the individual. If the process works well, it engenders trust and commitment to the cause. With respect to intercultural dialogue in particular, however, Meer and Modood (2011, p. 14) point out that: "What such sentiment ignores is how all forms of prescribed unity, including civic unity, usually retain a majoritarian bias that places the burden of adaptation upon the minority, and so is inconsistent with interculturalism's alleged commitment to 'mutual integration.'"

Intransigence of Values

One of the goals of intercultural dialogue is to arrive at shared values that will govern the polity. But what to do when values conflict? This leads to the irreconcilability of values and difficulties in reaching consensus. This is particularly the case with ethnocultural groups that may have similar interests but customs, traditions, and world views that are incompatible. Values are different from interests. While interests are negotiable and can be reconciled through dialogue, values are inviolable. For example, all persons regardless of their ethnicity have similar desires for shelter, health, and economic well-being. But they may differ with respect to the values that govern the achievement of these goals. For example, although all may agree on the need for improved economic welfare for their families, how this is achieved may differ. So Muslims may prefer that women perform domestic chores while men work outside of the home. These differences in values and belief systems cannot be resolved through dialogue.

Burayidi (2003) pointed to the difficulty of arriving at a compromise in GenPower's development of a power plant near Toltec Mounds, Arkansas. In this case the Quapaw Native Americans held to their beliefs that the site could not be developed because it would disturb the spirits of Native Americans. No amount of compensation could dissuade them from this view.

Another concern with intercultural dialogue is how the rules of the game are established. Are people required to use reason and be rational, or is passion accepted in such deliberations? What is the combination of reason and passion that is acceptable in such discourse? And are these rules provided a priori or are they arrived at during deliberations? If participants are given the binding rules that govern the conversation, then this obviates the need for dialogue in the first place. As Karatani (1995, p. 153) pointed out, "shared rules are the outcome of dialogue, not its point of departure."

Unjust Outcomes

Interculturalism privileges process over outcomes and assumes that open deliberative processes will lead to outcomes that are fair to all. That of course is not always the case. As Fainstein (2000, p. 457) points out, "Communicative theorists avoid dealing with the classic topic of what to do when open processes produce unjust results. They also do not consider the possibility that paternalism and bureaucratic modes of decision-making may produce desirable outcomes." Some scholars have also noted that communicative planning emphasizes consensus building rather than decision outcomes (Cameron, Grant-Smith, & Johnson, 2005).

We don't have to go too far in the planning profession to see how open processes may produce outcomes that are unjust or unfair. Since the 1960s, community participation has become an integral part of the planning process. Citizen participation provides a voice for those who will be impacted by plans and infuses mutual learning into the planning process (Friedman, 1973), creates ownership in the outcomes reached (Nelson and Wright, 1995), and ensures democratic representative decision making (Shrestha and McManus, 2008). Mahjabeen, Shrestha, and Dee (2009) reviewed the process for the development of the Sydney Metropolitan Strategy (SMS), a plan for strategic infrastructure investments and planning priorities in 43 local municipalities in the Sydney region of Australia. The authors found that the process required a thorough public participation by bringing "state government, local government and the broader community together to discuss, review and make decisions to guide the future of Sydney's economy, environment and communities" (p. 52). It included public forums, working groups, and reference panels. But while the process was commendable, the outcome left much to be desired. The authors found that "the Strategy contains few direct initiatives to address specific problems likely to be experienced by disadvantaged groups such as access to transport, health and community services on a level available to the more affluent members of the metropolis ... Hence it can reasonably be argued that input from the community groups had limited direct influence on the final decisions in relation to the SMS process" (p. 58).

The outcomes of the plan were also hijacked by influential groups such as the property Council of Australia and Landcom. Thus a proposal for a fast train linking Parramatta to Central Sydney was ignored, as was the proposal for recycling and walking "because of the Strategy's almost single focus on road and rail construction which reflects the power of government infrastructure agencies such as the Road and Traffic Authority (RTA)" (Mahjabeen et al., p. 59).

These points are not made to imply that there is no room for or benefit to dialogue. Rather they are meant to caution advocates of interculturalism about the limits to how far dialogue can go in mending fences.

Forward Multiculturalism

Contrary to its critiques, when viewed in its totality, multicultural policy is more than cultural differentiation and the assignment of group rights. The list of multicultural policies provided earlier is testimony to the breadth and depth of such policies. The primary goal of a multicultural policy is to promote societal welfare by ensuring that all persons regardless of their background have equal rights to function and contribute to the development of society. Viewed this way, multiculturalism seeks assurance that no one is marginalized because of their difference, whether this is due to their place of origin, ethnicity, age, class, sexual orientation, religion, or disability. This is achieved through policies and programs that recognize and take the diverse traits of people into consideration. Interculturalism only makes obvious what is embedded and assumed in a multicultural policy: the need for intergroup interaction.

Interculturalism is thus one strand of multiculturalism in that it operationalizes through dialogue the concept of what it takes to live amicably in a diverse world. Take as an example the case of Quebec. Unlike the rest of Canada, in the 1970s Quebec adopted an intercultural policy rather than multiculturalism for managing diversity and integrating immigrants. In a review of Quebec's interculturalism policy, Armony (2012, p. 84) concluded that "Canada's multiculturalism and Quebec's interculturalism are two variations of the same model of integration, one that favours civic inclusion rather than assimilation, plays down public displays of patriotism, values diversity in itself, and judges immigrants' contribution to society as mostly positive."

Gutenberg's (2013) description of interculturalism sheds further light on the similarities between the two concepts:

> Intercultural integration encourages the view that diversity is an asset and fights discrimination with the active support of public authorities, business, civil society and the media. The intercultural city is anchored by the principles and standards of democracy and human rights and dismisses cultural relativism, its sponsors committed to the importance of a rights-based approach to diversity management.

A policy of interculturalism by itself is deficient for managing diversity because it focuses on process and not outcomes. Dialogue in and of itself without attention to ameliorating the conditions of immigrants will only exacerbate racial tensions and the conflict that emanates therefrom. Furthermore, the deficiency of intercultural policy is that it does not address the underlying causes, such as those behind the European ethnic riots, that brought it about. Isolation and lack of interaction of ethnic groups with the larger society was certainly one of the reasons for the riots, but it wasn't the major cause. The riots in the UK were to a large extent the result of higher unemployment rates among young immigrants and their perceptions of discrimination. Interculturalism won't resolve the economic and spatial isolation of these groups.

What is needed therefore is not an abandonment of multiculturalism but the reframing of multicultural policy to make it more robust for current needs. To do so, multicultural policy needs to be explicit about the set of principles that guide the social contract to which immigrants and all others must abide and which form the foundations of a liberal democratic society. Basic to this contract is acceptance of individual rights and freedoms, equality, and democratic citizenship, among others. In addition, the concept of diversity must be broadened to include the variety of differences in society such as age, class, sexual orientation, and disability, the attention to which is crucial to the achievement of a just city.

The purpose of planning with difference is thus to enhance public welfare with the understanding that this goal cannot be attained if the different elements that make up the "public" are not taken into consideration. "Since planning serves the public interest and the public is diverse, then planners have to listen to all the different voices, not just those who are adept and articulate in presenting their case" (Office of the Deputy Prime Minister, 2005, p. 18).

There is evidence that the equal treatment of persons in planning regulations, policies, and programs without consideration for difference results in disparate impacts, which can be mitigated by attention to difference. This issue is made evident in Qadeer's observation of housing provision: "The availability of subsidized housing affects all those in need, regardless of the culture, but the policy to allocate units only to nuclear families effectively bars the multigenerational or relatively large families of immigrants" (Qadeer, 2009, p. 12).

Reflections on Planning in a Multicultural Society

A multicultural society brings opportunities to mine ideas from diverse minds for the benefit of society, but it also poses challenges to cities and the planning profession. The purpose of this book is both reflective and prescriptive. It is reflective in looking at planning practice and its past (in)adequacies in addressing the needs of a diverse public and how these practices will serve an ever-growing diverse population in the new century. It is also prescriptive in providing guidance to planners on the path that the profession needs to take to better serve a pluralist society. The essays in part I (Multicultural Praxis) take a deeper and critical view of multicultural planning within the context of a liberal democratic society. The chapters provide a critical reflection of planning practice as enshrined in the values of individualism, tolerance, and citizen rights and privileges and examine the capacity of liberal democratic institutions to accommodate cultural diversity within these tenets.

In chapter 2, T.L. Harper and S.M. Stein provide conceptual clarity on the key issues of citizenship, reciprocity, responsibility, obligation, and tolerance. They argue that *only* liberal democratic societies have the institutions to allow for cultural diversity. Often overlooked, they argue, is that the respect for (tolerance of) diversity by a liberal democratic society rests on an overlapping consensus supporting the public values and institutions of the society. The authors argue that Anglo-American multicultural societies should not assume the resilience of their liberal democratic institutions in the face of continual in-migration but must ensure their durability by insisting that immigrants respect these core beliefs and by limiting the exercise of those values that violate core liberal democratic ideals.

In chapter 3, Mohammad Qadeer notes that multiculturalism strikes at the root of the presumed uniformity of all groups and persons through the unitary public doctrine. He is of the view that multiculturalism has been incorporated (with varying success) both in urban planning practices and in the structures of North American cities. The chapter uses case studies to show that the multicultural ethos has been incorporated into the planning of cities in North America and then argues that a strategy of reasonable accommodation should be the basis of multicultural planning. It concludes with a policy index to guide the institutionalization of this strategy.

David Laws and John Forester explore the challenges of multiculturalism in planning as they surface as matters of "street-level democratization." In chapter 4, they draw from testimony and reflective accounts arising from a conflict in Amsterdam, Holland, for perspective. The authors used practice-focused interviews with community workers, housing association staff, elected politicians, and others involved in the dispute to show how such micro-scale work can have immediately more macro implications and how what seems to be personal interaction can be a testing ground for democratic politics writ large in a global city. Essential to such interactions are not only capacities of recognition and abilities to engage local, particular conflicts creatively but also capacities for responsive and even transformative improvisation.

Richard Gale and Huw Thomas in chapter 5 discuss the evolution of multicultural planning in Britain, noting that its implementation has ebbed and flowed, influenced by a mix of local political circumstances, the attitudes and level of understanding of professional planners, and action and pressure from minority groups. The authors use examples to illustrate some promising developments which are taking place at what might appear to be a generally inauspicious time in British planning. Following recent social unrest, British national policy has shifted towards ensuring a collective identity rather than the promotion of multiculturalism, a policy that is blamed for contributing to a fractured society by emphasizing identity politics. And, because British planners see their role as technical experts concerned primarily with promoting order in development, they are loath to enter the debate on social issues and therefore view race equality and multiculturalism as tangential to their work.

Part II extends the discussion of multiculturalism beyond race, culture, and ethnicity. The contributors consider planning issues affecting sexual minorities and Native Americans, which have hitherto not been given sufficient attention in planning discourses. In chapter 6, Petra L. Doan observes that although progress has been made in multicultural planning by drawing planners' attention to a wider set of perspectives especially as they relate to race and ethnicity, a similar importance has not been given to the recognition of non-conformist individuals such as sexual minorities. Case material from the redevelopment of Atlanta's Midtown neighbourhood is used to illustrate such failure. Doan challenges planners to consider the needs of LGBT-identified individuals, those living in queer-identified areas as well as those who live beyond queer spaces. The chapter then extrapolates lessons for planning practice to incorporate the needs of the entire spectrum of non-normative populations.

Nicholas C. Zaferatos follows this line of reasoning in chapter 7. He makes the case that until quite recently, the planning profession failed to pay sufficient attention to the particular needs of tribal communities. As a result, it may have inadvertently contributed to impeding the advancement of planning in those communities, which continue to rank among the most disadvantaged in American society. As a starting point for supporting the work of planning in Native American reservations, Zaferatos unveils the complex context of tribal planning to inform practice in these communities. The chapter examines the context under which tribal planning occurs, with the practical aim of introducing new approaches to make more decipherable the conditions that impede tribal development. As a pathway towards furthering a tribe's self-determination, the author contends that effective planning should be grounded in a tribe's own historical experience and its political capacity as the starting point from which to form effective planning strategies.

Part III provides an account of the proactive measures cities are implementing to make new immigrants feel welcome and to ensure that planning practices and governance promote tolerance and understanding of diversity. Michael A. Burayidi and Abby Wiles in chapter 8 review planning documents in the 25 largest majority-minority cities in the US with a goal of identifying the modifications that such cities have instituted in the wake of a growing minority population. Their findings of planning practices in these cities provide guidance to other cities contemplating changes to their planning instruments to meet their circumstances.

Mai Thi Nguyen, Hannah Gill, and Anisha Steephen observe in chapter 9 that in the US there is currently no federal immigrant settlement policy. They contend that traditional models of immigrant "assimilation" were unidirectional, focusing on how immigrants adapt to the receiving society. With time, and after several generations, through improved language skills and upward socioeconomic mobility, immigrants are expected to assimilate into their host society. The authors note that more contemporary scholarship suggests that integration is a two-way process whereby the receiving society plays an active role in accelerating the resettlement of immigrants. This has resulted in a myriad of immigrant integration policies and practices by local governments and civil society organizations across the country. However, while such strategies abound, little is known about which models work and which practices are most promising.

The authors assess the landscape of immigrant integration practices around the country to develop a better understanding of how local governments and administrators are facilitating the resettlement of immigrants. An evaluation is provided of such strategies in three communities in North Carolina. The lessons learned from these examples provide guidance to local government officials and administrators contemplating the adoption of immigrant integration strategies.

Stacy Anne Harwood and Sang S. Lee in chapter 10, like Nguyen, Gill, and Steephen in the previous chapter, review the emergence of immigrant-friendly practices at the local level and assess the role of planners in planning for immigrant-friendly communities. They ask what "immigrant-friendly" means and then evaluate the various strategies in use to see which ones are effective and why.

All planning activities have a manifestation in space. Part IV advances strategies in the design of the physical environment that meet the needs of a diverse population even as they compete for the use of space. Emily Talen explains in chapter 11 that sustaining neighbourhood-level diversity requires a level of tolerance and inclusiveness that has never been easily accomplished. No matter how enduring the power of a shared American dream, no matter how compelling our historical sense of the moral rightness of American pluralism, we are always going to have to work hard to make diversity in local contexts endure. From the point of view of planning and urban design, American pluralism has been confounded by a physical context that fails to accommodate it. Talen shows ways in which design can be leveraged to provide a much more supportive framework for diversity.

In chapter 12, Kelly Main and James Rojas note that current planning outreach and participatory practices that rely heavily on verbal communication reinforce the growing gap between those who speak the technical language of a planner and those who do not. Moreover, cultural conflicts between communities over the appropriate use of city space can be heightened by language differences that make communication and understanding during outreach extremely challenging. The authors suggest that public outreach/participation in multicultural communities can benefit from employing the visual arts and approaches to understanding and communicating about place that de-emphasize language, both to enhance cultural expression and to expand cross-cultural communication and problem solving regarding planning issues. The authors illustrate this with the use of two methods: photovoice and Place It!.

In chapter 13, Sandeep K. Agrawal discusses the development of two religious clusters which emerged in Canada due to incremental zoning changes over time. Using key informant interviews as the method and Allport's contact hypothesis as the guiding theoretical framework, he assesses the effects of proximity and contact on intergroup relations. The goal is to show whether land use zoning that promotes physical proximity can help create a space for religious interactions between the members of different faiths and thus lead to attitudinal shifts concerning the religious "other."

The chapters in part V suggest ways to make urban planners more culturally competent in working with diverse communities through both formal training and informal interactions with these groups. Michael Rios admits from the onset in chapter 14 that there are no easy technical solutions to the problems of cultural difference and intergroup relations. The author uses examples to highlight the importance of place as the space where cultural citizenship is produced, as marginalized populations struggle to build community and gain social and political standing. The chapter draws attention to the difference that culture makes for marginalized communities, and, specifically, how the convergence of place and culture instigates negotiations of belonging, authorship, and power to establish what groups can expect of one another. These agreements, or "cultural contracts," measure the degree to which cultural values and commitments are exchanged between groups – including professionals and the publics they purport to serve. The implication for planning education is a greater focus on cultural competency as measured by the level of cross-cultural communication between individuals and among different social groups to determine why place matters, for whom, and with what results.

In chapter 15, Siddhartha Sen, Mukesh Kumar, and Sheri L. Smith discuss planning pedagogy at three historically Black colleges and universities to show that multicultural education is an integral part of the schools' curriculum. Using Banks's five dimensions of multicultural education, they discuss the critical learning skills these schools impart to their students. These skills, as they relate to multicultural planning, include content integration, knowledge construction, prejudice reduction, equity pedagogy, and an empowering school culture. They conclude that historically Black colleges and universities need support so that they can continue to assist in the education of planners with a multicultural focus.

The book concludes with Burayidi's chapter on moving the diversity and multicultural planning agenda forward. He teases out the key suggestions and lessons from the chapters in the book on planning with diversity and how to retool planning practices to better serve a diverse poly-ethnic public. Three suggestions are provided in this respect: acknowledge diversity, plan with diversity as the norm, and support and enhance the development of socially cohesive cities.

REFERENCES

Allport, G.W. (1954). *The nature of prejudice*. Cambridge, MA: Addison Wesley.

Armony, V. (2012). Multiculturalism, interculturalism and the effects of a weak ethnos. *Canadian Diversity*, *9*(2), 82–84. Retrieved July 23, 2013 from http://www.ieim.uqam.ca/IMG/pdf/armonyV_2012_canadian_diversity.pdf.

Bannerji, H. (2009). On the dark side of the nation: Politics of multiculturalism and the state of "Canada." In S. Mookerje, S. Szeman, and G. Faurschou (Eds.), *Canadian cultural studies: A reader* (pp. 327–343). Durham, NC: Duke University Press.

Barsky, A.E. (2010). *Ethics and values in social work: An integrated approach for a comprehensive curriculum*. Oxford: Oxford University Press.

Booth, C., Batty, E., Gilroy, R., Dargan, L., Thomas, H., Harris, N., & Imrie, R. (2004). *Diversity and planning: Research report on planning policies and practice*. London: ODPM.

Bouchard, G., & Taylor, C. (2008). *Building the future: A time for reconciliation*. Montreal: Commission de Consultation sur les Pratiques d'Accommodement Reliées aux Différences Culturelles.

Burayidi, M.A. (2003). The multicultural city as planners' enigma. *Planning Theory & Practice*, *4*(3), 259–273. http://dx.doi.org/10.1080/1464935032000118634

Cameron, J., Grant-Smith, D., & Johnson, A. (2005) Formative evaluation of improving collaborative planning: A case study at the regional scale. *Australian Planner*, *42*(4), 22–29.

Cantle, T. (2001). *Community Cohesion: A report of the Independent Review Team*. London: Home Office.

Centre for Social Relations. (2013). Intercultural Dialogue. Coventry, West Midlands, UK. Retrieved May 18, 2015 from http://www.cohesioninstitute.org.uk/Resources/Toolkits/InterculturalDialogue/InterculturalDialogue.

Commission on Integration and Cohesion. (2007). *Our shared future*. Wetherby, UK: Commission on Integration and Cohesion.

Council of Europe (2008). *White paper on intercultural dialogue: Living together as equals in dignity*. Strasbourg. Retrieved September 5, 2014 from http://www.coe.int/t/dg4/intercultural/source/white%20paper_final_revised_en.pdf.

Dunn, K., Thompson, S., Hanna, B., Murphy, P., & Burnley, I. (2001). Multicultural policy within local government in Australia. *Urban Studies (Edinburgh, Scotland)*, *38*(13), 2477–2494. http://dx.doi.org/10.1080/00420980120094623

European Commission 2007. *European Year of Intercultural Dialogue* [online]. European Commission. Available from http://ec.europa.eu/culture.

Fainstein, S. (2000). New directions in planning theory. *Urban Affairs Review*, *35*(4), 451–478.

Fay, B. (1987). *Critical social science*. Cambridge: Polity Press.

Forester, J. (1989). *Planning in the face of power*. Berkeley: University of California Press.

Friedman, J. (1973). *Retracking America: A theory of transactive planning*. Garden City, NY: Anchor Press/Doubleday.

Government of Canada. (1988). Canadian Multiculturalism Act, Bill C-93–1988. Ottawa, Ontario. Canada. Retrieved September 5, 2014 from http://laws-lois.justice.gc.ca/eng/acts/C-18.7/FullText.html.

Harwood, S. (2005). Struggling to embrace difference in land-use decision-making in multicultural communities. *Planning Practice and Research, 20*(4), 355–371. http://dx.doi.org/10.1080/02697450600766746

Healey, P. (1992, Apr.). Planning through debate: The communicative turn in planning theory. *Town Planning Review, 63*(2), 143–162.

Healey, P. (2003). The communicative turn in planning theory and its implications for spatial strategy formation. In S. Campbell & S.S. Fainstein (Eds.), *Readings in planning theory* (pp. 237–255). Blackwell.

Huxley, M., & Yiftachel, O. (2000). New paradigm or old myopia? Unsettling the communicative turn in planning theory. *Journal of Planning Education and Research, 19*(4), 333–342

Inglis, C. (1995). Multiculturalism: A policy response to diversity, Paper prepared on the occasion of the "1995 Global Cultural Diversity Conference," April 26–28, 1995, and the "MOST Pacific Sub-Regional Consultation," April 28–29, 1995, both in Sydney, Australia. Retrieved June 24, 2014 from http://www.unesco.org/most/pp4.htm.

Innes, J.E. (1995). Planning theory's emerging paradigm: Communicative action and interactive practice. *Journal of Planning Education and Research, 14*(3), 183–189. http://dx.doi.org/10.1177/0739456X9501400307

Karatani, K. (1995). *Architecture as metaphor: Language, number, money* (S. Kohso, Trans.). Cambridge, MA: MIT Press.

Kuryanova, L. (2013). Policy of multiculturalism in Europe has failed – expert. Retrieved June 24, 2014 from http://voiceofrussia.com/2013_05_07/Policy-of-multiculturalism-in-Europe-has-failed-expert/.

Kymlicka, W. (2004). Marketing Canadian pluralism in the international arena. *International Journal, 59*(4), 829–852.

Mahjabeen, Z., Shrestha, K., & Dee, J. (2009). Rethinking community participation in urban planning: The role of disadvantaged groups in Sydney Metropolitan Strategy. *Australian Journal of Regional Studies, 15*(1), 45–63.

Meer, N., & Modood, T. (2012). How does interculturalsim contrast from multiculturalism? *Journal of Intercultural Studies, 33*(2), 175–196.

Neill, W.J.V. (2004). *Urban planning and cultural identity*. London: Routledge. http://dx.doi.org/10.4324/9780203402245

Nelson, N., & Wright, S. (1995) Participation and power. In N. Nelson & S. Wright (Eds.), *Power and participatory development: Theory and practice* (pp. 1–18). London: Intermediate Technology Publications.
New Europe (2013). Can interculturalism succeed where multiculturalism has failed? Analysis, p. 7. Retrieved May 18, 2015 from http://www.scribd.com/doc/122462623/New-Europe-Print-Edition-Issue-1015#scribd.
Office of the Deputy Prime Minister. (2005). *Diversity and equality in planning: A good practice guide*. London: ODPM.
Parekh, B. (2007). Multiculturalism. In J. Baggini and J. Strangroom (Eds.), *What more philosophers think* (pp. 45–56). London: Continuum.
Prychitko, D.L., & Storr, V.H. (2007). Communicative action and the radical constitution: The Habermasian challenge to Hayek, Mises and their descendants. *Cambridge Journal of Economics*, *31*(2), 255–274. http://dx.doi.org/10.1093/cje/bel017
Qadeer, M.A. (2009). What is this thing called multicultural planning? *Plan Canada: Special Edition: Welcoming Communities: Planning for Diverse Populations*, 10–13.
Salins, P. (1997). *Assimilation, American style*. New York: Basic Books.
Salmancheema (2013). From multiculturalism to interculturalism: A British perspective. Retrieved July 14, 2013, from https://www.opendemocracy.net/ourkingdom/ali-rattansi/from-multiculturalism-to-interculturalism-%E2%80%93-reply-to-british-political-elite.
Shrestha, K.K., & McManus, P. (2008). The politics of community participation in natural resource management: lessons from community forestry in Nepal. *Australian Forestry*, *71*(2), 135–146. http://dx.doi.org/10.1080/00049158.2008.10676280
Taylor, C. (2012). Interculturalism or multiculturalism? *Philosophy and Social Criticism*, *38*(4–5), 413–423. http://dx.doi.org/10.1177/0191453711435656
Wood, P., & Landry, C. (2007). *The intercultural city: Planning for diversity advantage*. London: Cromwell Press.

PART 1

Multicultural Praxis

2 The Centrality of Liberal Democratic Values in a Multicultural Society

THOMAS L. HARPER AND STANLEY M. STEIN

Introduction

One of the aims of this book is to provide "guidance to planners" in liberal democratic societies (particularly the US and Canada) regarding "the path that the profession needs to take to better serve a pluralist society" (chapter 1).

In considering how to respond to current contexts, we must recognize that 1) many local planners are unaware of their responsibility to accommodate cultural differences; 2) other planners emphatically reject measures that treat people differently, as being contrary to professional obligations (see Burayidi, chapter 1, and Qadeer, chapter 3); and 3) the normative idea of "multiculturalism" has recently been challenged politically, particularly with the growing perception of terrorism and civil unrest as threats.[1] This relatively recent shift has been well documented by Bilge (2008).

Given these realities, those who strive to make planning more inclusive of people with different cultural origins must know how to justify and vigorously defend their policies and plans. Our aim is to enhance the impact of their arguments by 1) identifying some mischaracterizations of multiculturalism, and underlying confusions regarding *key concepts* in the debate; 2) explaining why *liberal democratic values* both *underlie and require* multiculturalism; 3) outlining liberal democratic values and examining some of their *implications*; 4) illustrating how clarity about these values helps to address some *current issues* concerning multiculturalism; and 5) suggesting how planners and policy makers can be *more effective* in a multicultural society, and encouraging them to *advocate* for liberal democratic values. It should be noted that our focus is on planning and policy involving voluntary in-migrants from other different cultures, rather than on those who migrated from Europe, those whose history includes involuntary migration (i.e., slaves), or those whose residence pre-dated European explorers (i.e., aboriginals). This focus reflects issues most current in Canada and Europe at the time of writing.

Two clarifications also seem necessary: 1) The term "liberal" is used here in quite a broad sense, which applies to the public realm – i.e., Rawls's "political liberalism" (1993);[2] and 2) in advocating liberal democratic values, we do not intend to provide an apologia for neoliberalism, mixed capitalism, or democracy, nor for the practices of societies which claim, or have claimed, to hold these values. The values discussed here provide a normative basis for reflective critique – an important responsibility of public planners (Forester, 1989).

Confusions Underlying Debates about Multiculturalism

Debates about multiculturalism are plagued by semantic confusions, which muddle the issues unnecessarily. We first seek to clarify those confusions.

i) The Meaning of Multiculturalism

Dictionary definitions of "multiculturalism" are usually purely descriptive. In the sense of having members from different cultures, numerous societies (past and present) are undeniably "multicultural." In contrast, this discussion is concerned with the *normative* meaning of "multiculturalism," which is often lacking in clarity. Kymlicka found that recent critiques *mischaracterize* multiculturalism "as the uncritical celebration of diversity at the expense of addressing grave societal problems." His use of the term refers to a "set of [normative] ideas about the legal and political accommodation of ethnic diversity [which] emerged in the West as a vehicle for replacing older forms of ethnic and racial hierarchy with new relations of democratic citizenship," where these ideas include some form of accommodation to different cultures (Kymlicka, 2012, p. 1). We will use this definition as a starting point, recognizing that it requires the delineation of the set of ideas, and criteria for what should be accommodated. We are not intending to deny evidence that this normative ideal has never been fully, or even well, implemented (Bannerji, 2000).

ii) Framing the Choice as: National/Cultural Values vs Multiculturalism

One mischaracterization is to present the values of a nation or culture as being *opposed* to multiculturalism. This frames the choice of values and institutions as a choice between competing cultures. This is evident in the Netherlands' 2011 announcement of their new integration bill:

> The government shares the social dissatisfaction over the multicultural society model and plans to shift priority to the values of the Dutch people. In the new integration system, the *values of the Dutch society* play a central role. With this change, the government steps away from the model of a multicultural society. (*EU Times*, 2011; emphasis added)

This statement raises many questions. What parts of the "model of a multicultural society" are going to be abandoned? Which values are "Dutch"? Are liberal democratic values which have been part of the Dutch heritage going to be given up?

The same confusion can be seen when the (then) government of the Canadian province of Quebec escalated an intense public debate over the accommodation of minorities with

> a proposed *"Charter of Quebec Values"* ... [which] would include that public employees in schools, hospitals and other government offices will be barred from wearing religious clothing in the workplace ... and ban religious symbols such as kippas, hijabs and turbans and ostentatious crosses for public-sector workers. (*Maclean's*, 2013a)

Many questions of justification arise. What makes *your* (Quebec) values so special? What values are included as "Québécois"? Do they include the values of citizens who espouse Nazi or Marxist or fundamentalist religious values? Do they include only the values held by the elite or by the majority? Do they include substantive values (e.g., drinking Molson's "Canadian" beer, watching or playing curling, Canadian football, and hockey, eating poutine, and wearing toques) as well as procedural[3] ones? The Quebec minister responsible for democratic institutions claimed that "It's a good balance between respect for individual rights and the respect of Quebecers' common values" (*Maclean's*, 2013a). Does this mean that respect for individual rights is not part of Quebecers' common values?

iii) Framing the Choice as: How Much Tolerance and Accommodation?

Another mischaracterization is to debate in terms of the appropriate amount of tolerance or accommodation needed. In the 2007 Quebec election "debate over the accommodation of minorities," the leader of an opposition party asserted that "Quebec had gone too far in catering to newcomers" (*Maclean's*, 2013a). How much accommodation is "too far"? How much is enough? Too much? What is the evaluative metric? What behaviours should or should not be accommodated?

Although he takes an opposing view, current Canadian Liberal party leader Justin Trudeau frames the choice in a similar way: he believes that no amount of tolerance is enough. Eulogizing his father, former Prime Minister Pierre Trudeau, in 2000, he advocates going *beyond* tolerance, because:

> … mere tolerance is not enough. We need genuine and deep respect for each and every human being notwithstanding their thoughts, their *values*, their *beliefs*, their *origins*. (Khan, 2013; emphasis added)

Trudeau's comment is ambiguous. Does he intend that we should respect each *individual person*, or that we should respect *all* of their thoughts, *values*, beliefs? Should we also respect *all behaviours* flowing from their beliefs? What if their beliefs include wife and child beating, clitorectomy, "honour" killing?

The columnist admiringly quoting Trudeau believes that going beyond tolerance should involve "allophilia" – "positive intergroup dynamics that supersede tolerance" by "recognizing and embracing differences" (Pittinsky, 2012). For inspiration, she looks to the Toronto-based Inspirit Foundation, which promotes "initiatives related to pluralism among young Canadians of different spiritual, religious and secular backgrounds" by supporting projects that promote "a society where we actively engage with each other's differences and where all these *differences*, which include our ethnicity, culture and beliefs, are *celebrated*" (Khan, 2013; emphasis added).[4] Are *all* differences in culture and belief worthy of celebration? If not, what are the criteria for deciding which ones should be celebrated?

iv) Framing the Choice as: Secularism vs Religious Values

A third mischaracterization is to view the "secularism" of a liberal democratic society not as a description of public neutrality with regard to religion but as a kind of competing cultural (anti-religious) perspective which "Western" cultures seek to impose on other cultures. In a journalistic discussion of what should be required of immigrants and their descendants, the *Economist* reported:

> The Nordic countries, long seen as strong supporters of multiculturalism, now seem divided on immigration policy issues. There has been recent uncertainty, even in Sweden, where support for multiculturalism has historically been highest. While they regard their open-armed approach to asylum-seekers as an expression of what is best in their culture, they now ask: "Is it more enlightened to *impose secular values* on devout Muslims or to *dilute liberal values* in the name of multiculturalism?" (*Economist*, 2013, 8; emphasis added)

Should any values be "imposed"? What are the "secular values" which have been imposed? Which liberal values should be "diluted"? What are the criteria for "dilution"?

A Basis for Addressing the Confusions

i) The Liberal Democratic Context

In addressing mischaracterizations, confusions, and questions about multiculturalism, it is useful to recognize that debates about multiculturalism arise *only* in the context of a particular kind of society – a liberal democratic one. By liberal democratic society, we mean one which holds as central two core values: a) *equality* of persons – each entitled to inalienable individual rights, and to fair, impartial treatment; b) *liberty* – freedom of choice (or autonomy) to live a good life, a life the person deems worth living. These values, generally traced back to Kant (1993 [1785]) reflect a belief that the free and equal, rational and reasonable, individual person (Rawls, 2001, p. 8) is the basic unit of society, the ultimate object of moral concern, and the primary source of value.

The debates and questions raised in the previous sections actually *presuppose* such a society. They would never even arise in other political systems, e.g., theocratic, totalitarian, fascist, collectivist, communist, traditional hierarchical, or tribal societies. A "multicultural society" (in the normative sense) *must* be a liberal democratic society, because *only* liberal democratic societies have the *values and institutions* to allow for the tolerance, acceptance, or accommodation of cultural diversity.

Thus, liberal democratic values and institutions are a *necessary condition* for a "multicultural society."[5] To claim that a measure can advance multiculturalism but violate liberal democratic values would be a complete contradiction.

Well before the two core values above were widely affirmed, many normative political theorists (e.g., Hobbes, 1651; Rousseau, 1762) used the idea of an implicit "social contract" to justify (and limit) the powers of the state to govern a society. Most contractarians, including some twentieth-century ones (Gauthier, 1978; Buchanan & Tullock, 1962), appeal to individual self-interest as a rationale for demarcating the powers of government.

More recently, this approach has been taken by demographer-planner Dowell Myers (2007), who persuasively argues that the aging "baby boomer" generation in the US should support education and integration of immigrants and their children (particularly Hispanics) to ensure enough workers and taxpayers to support the "boomers" in their declining years. Although his proposed new social contract does allude to the moral responsibility of working generations to provide for dependent generations (young and old), the primary appeal seems to be to self-interest.

American political philosopher John Rawls (1993, 2001) takes a different contractarian approach. He begins by assuming, not self-interest, but a society whose members are *committed* to the two core moral values of equality and liberty, and that holds a conception of "justice as fairness" in the public social realm. This seems like a most reasonable starting point, given the recent history, constitution or tradition, and legislation of the US, Canada, the UK, Australia, New Zealand, and many western European societies.

Rawls recognizes that pluralistic liberal societies (e.g., Canada, the US, or Britain) need to be able to accommodate very diverse viewpoints and values. He traces this ability historically to the Anglo-European experience with clashes between different incompatible (but equally reasonable) substantive conceptions of the "good" (which he calls "thick theories of the good"). Originally, these different views were religious (Roman Catholic vs Protestant). They developed a conception of society which respected and allowed very different thick theories of the good within a framework of shared public political values. This "thin theory of the good," which includes basic democratic political moral values like fairness and tolerance, does not substantively define "the good."

ii) Private and Public Values

In order to justify the conception of a "political liberal" democratic society, Rawls (1993) imagines a group of people who come from a variety of different cultural backgrounds, with diverse philosophical and religious perspectives. They are committed to the two core values of liberty and equality, and want to live together as a peaceful society, dealing with their disagreements in a *non-violent* manner. Rawls seeks to articulate the principles and social institutions which would enable such a society.

In developing a consensus regarding basic social institutions, it is essential that they put aside or bracket certain controversial (though not unimportant) issues, particularly differences between thick theories. The reason they must be put aside is that they are both *reasonable* and *irreconcilable* – there is no universal forum (with objective criteria which all reasonable persons would accept) where these debates could be resolved. These controversial issues must be placed in the private realm, and excluded from the public realm.

What a liberal democratic society requires is a public conception of the good that is as *wide* as possible, one that accommodates as many different (conflicting) private views of the meaning of life as possible. Thus the aim in the *public* realm of pluralistic liberal democracies is to develop a moral basis of communal existence, a conception of society (a "thin" theory of the good) that tolerates many different thick theories of the good within a framework of shared basic democratic public political values.

Public and private realms can be distinguished pragmatically by their very different aims. In the *private* realm, our aim is to develop a conception that guides our lives, giving us an account of how things "hang together, in the largest sense" (Sellars, 1963, p. 1), and a framework for structuring and guiding our lives, one which gives meaning to them. Individuals have different affections, devotions, and loyalties, which are central to their self-conceptions and identities, and which organize and shape their lives. Without these private conceptions, their lives would be disjointed. Religious and philosophical perspectives provide such frameworks. These perspectives play a valuable role in providing a life-guiding framework, giving meaning and coherence to the lives of those who accept them. In addition, shared private conceptions help members of a subculture make sense of their collective experience (Galloway-Cosijn, 1999). As Peter Winch points out, culture-based beliefs and practices give participants a "sense of the significance of human life" (1970, p. 105). Individuals' conception of the good and their private identities may change over time, without any change in their public identities as free and equal, reasonable and rational citizens of a liberal democratic society.

Another way of expressing the contrast: the *private* realm is defined as "that which one does as an individual pursuing what one takes to be the truth, such as theology, moral theorizing, metaphysics" (Hampton, 1989, p. 806). In the *public* realm, we are concerned, not with whether an idea is true or false (arriving at the "truth of the matter"), but with whether the idea can command consensual support as providing a reasonable basis for public policy in a democratic society.

iii) Social Contract

Rawls further imagines the principles of a "social contract" being devised by free and equal, reasonable and rational persons, behind a "veil of ignorance." The idea of the veil is that the hypothetical contractors know nothing about themselves which would distort their judgments or tempt them to be unfair. They should not know anything about their own attributes (race, gender, talents, disabilities, capacities, or assets), or about their own position in society; they should know only what is *relevant* to making *fair* judgments about the principles of justice.

The idea is that when making normative judgments about the suitability of general social institutional arrangements, we should *put aside* what we know about our *own position*. Rawls's hypothetical social contract is intended to give a perspicuous representation of our ordinary moral intuitions about society and governance, as explicated and generated by him via a process which he calls "Wide Reflective Equilibrium" (WRE).[6]

He argues that two principles would be generated by the contractors using his process, relating to liberty and equality. The first principle of *Liberty*: "Each person has the same indefeasible claim to a fully adequate scheme of equal basic rights and liberties, which scheme is compatible with the same scheme of liberties for all" (Rawls, 2001, p. 42). The first principle is given priority.[7] This priority means that the basic rights and liberties are constitutionally fixed, off the agenda for political debate, beyond any utilitarian or consequentialist calculus. It ensures mutual recognition of perpetual equality for all, regardless of their comprehensive doctrine. Our discussion will centre on this principle, because most of the debates about multiculturalism seem to focus on rights, rather than on equality of consequentialist outcomes.

iv) Core Liberal Democratic Values

The set of values which best achieves these purposes, and can be shared by everyone committed to the two core values, is the one which allows (or tolerates) the widest variety of different conceptions of the good (i.e., it protects human rights, individual freedom). This "liberal democratic" value set has been identified with "the open society" by Popper (1945), and with "political liberalism" by Rawls (1993; 2001). (We will use the terms "liberal democratic," "political liberal," and "open" as roughly synonymous adjectives in our discussion.)

Other core values are derived from the first two core values: (a) equality of persons and (b) liberty – freedom of choice (or autonomy). They include: (c) tolerance – acceptance of a plurality of conflicting conceptions of the good life, provided they do not violate (a) and (b); (d) societal structure as a fair system of cooperation between free and equal persons; (e) the rule of law – the impartial application of law to all persons; (f) a primary government role in protection of liberty; and (g) restrictions on the coercive powers of government.[8]

These core liberal democratic values guide the creation of social (political, legal, economic) institutions which are designed to secure "justice as fairness." A central aim is to ensure basic liberties and to provide certain "primary goods,"[9] including basic rights and liberties, which Rawls (2001, p. 58) argues a *just* liberal democratic society must provide to all of its citizens.[10] These primary goods are necessary in order for the person to have the capacity to revise their ideas of the good life (that they inherit from their culture or subculture), to consider alternative ideas in formulating their own conception, and to participate in public life as a rational and reasonable citizen. Only in this "political liberal" context is there an argument supporting respect for (and accommodation of) individual differences (so long as their behaviour is not disrespectful of the rights of others).

The essential idea for a multicultural society is the Rawlsian "political liberal" *separation* between public and private realms. This is the idea that allows for accommodation of the broadest diversity of values. A liberal democratic society must maintain this distinction in order to accommodate diversity. In contrast, societies without this separation (e.g., a theocracy) *enforce* the same set of values on everyone.

v) An Overlapping Consensus

Core liberal democratic public values (and the principles and social institutions which implement them) require support by an overlapping consensus of all citizens who hold the conception of society just outlined. Rawls's term "overlapping consensus" indicates that the support comes from citizens with very different private "thick theories" of the good. The institutions specify procedures for adjudicating specific substantive decisions where a consensus may be impossible (e.g., many land use disputes). The more diverse the society, and the more diverse the differing thick theories (conceptions of the good life), the more the focus[11] is on the provision of an institutional framework (political, legal, economic) that satisfies a procedural concern for full, free, uncoerced, undistorted communication and reason in democratic public decision making.[12]

In practice, the boundary between public and private realms will be complex and contested. Different thick theories may have different views as to what is legitimately private (Mandelbaum, 2000). Advocates of thick theories will tend to urge that more of their private convictions be addressed in the public realm. For example, some will believe that abortion is a private matter; others, that it should be a public concern. The boundaries will be the subject of ongoing political debate.

vi) Rights and Duties

The previous sections have focused on the rights of each citizen. Now we want to stress that the rights, basic liberties, and public goods provided to its citizens by a liberal democratic society come with reciprocal obligations. Each person has a duty to respect the same rights for all other persons. The Canadian Multiculturalism Act clearly specifies that all Canadians, "whether by birth or by choice, enjoy equal status, are entitled to the same rights, powers and privileges and are subject to the same obligations, duties and liabilities" (Government of Canada, 1985). Thus each person in the receiving culture has a duty to respect the rights of immigrants as persons, and to be tolerant of all differences which do not violate liberal democratic values. Immigrants have a right to this respect, but they also have concomitant duties to accept, respect, and abide by the core liberal democratic values, principles, and institutions which underlie their rights.

In encouraging in-migration, some liberal democratic societies seem to have over-emphasized the "tolerance" aspect, and under-emphasized the reciprocal obligation. It may not have been made clear to in-migrants that this duty requires changes in the practice of those in-migrant *private* values which violate *public* liberal democratic values. In-migrants should not expect to re-create their own closed culture of origin within an open multicultural society. They should not expect the receiving society to be open to them if they isolate themselves with their own language, education, communication media, and values. It is *not a*cceptable for them to be intolerant of liberal democratic values, while expecting the liberal democratic societies to be tolerant of theirs.

What often seems to have been overlooked is that the respect for (tolerance of) value diversity by a liberal democratic society rests on the overlapping consensus supporting the public values and institutions of the society. Tolerating and accommodating practices and customs which violate core liberal democratic values amounts to undermining and destroying the very basis of tolerance (i.e., the public values) in the name of tolerance. This makes no sense.

It should be understood that openness to expressions of cultural diversity is an expression of the character of an open democratic liberal society, making it distinctive from all other kinds of societies. Ideally, all aspects of a liberal society should be open to dialogue: science, philosophy, politics, religion, etc. New views and ideas are always welcome, but those expressing their views and ideas are expected to be equally open to other ideas.

vii) Multiculturalism and the Liberal Democratic Society

We have already discussed why liberal democratic values and institutions are a necessary condition for a "multicultural society" – one which respects each person's right (regardless of culture or subculture) to seek their own ends. But respect ends where anyone wants to deny another person's right to express her views or to act in any way (which does not deny the same rights of others). Violent practices like wife and child beating, clitorectomy, subjugation of women and children, "honour" killing, and murdering those who criticize your culture (e.g., the murder of Van Gogh in the Netherlands [Buruma, 2006]) are morally repugnant violations of the victims' rights. Those who assert their right to such practices appeal to liberal democratic values, while at the same time denying them.

Liberal democratic values are not just another set of cultural values; they are the specific values which support freedom, equality, and autonomy. Without these core liberal democratic public values,[13] a society cannot be multicultural. Thus, a society which is diverse (multicultural in the descriptive sense) but not multicultural (in the normative sense) cannot claim to be liberal, democratic, open, or free.

It is also important to recognize that a liberal democratic society is *inherently* multicultural. Multiculturalism is not so much a value as it is simply a description of the kind of society that most Western societies claim (and aspire) to be – one which is focused on a respect for, and concern for the human rights of, each person,[14] regardless of any personal attributes.

Some Implications of Liberal Democratic Values

i) Tolerance

The meanings of terms like "tolerate," "accommodate," and "secular" start to become clearer when they are set within a liberal democratic framework. What a liberal democratic society should tolerate is any and all private values, beliefs, and behaviours which do not violate the rights of others to live the kind of life they freely choose. The cultural or religious origin of the values is morally irrelevant, although the political appeal of liberal democratic values may be enhanced by pointing out the history of the values.

One thing the open liberal society must be *intolerant* of: the dogmatism of any kind of fundamentalism[15] – whether it be Protestant, Roman Catholic, or Orthodox Christianity, Islam, Hindu, atheist, feminist, environmentalist, cultural traditionalist, Marxst, Nazi, Fascist, etc. Fundamentalist beliefs can be tolerated as thick theories of the good, but only as long as they do not violate, or advocate violating, the rights of others. To participate in an open liberal society, fundamentalists *must give up* the idea that (what they see as) the absolute universal truth of their dogma justifies *imposing* it on others. Many fundamentalists fail to recognize this necessity.

ii) Democracy

The most important element in a liberal democratic society is not free elections. Imagine that, in response to popular demands for "freedom," a country holds a free election,[16] and the party elected is one which advocates a theocratic state. Once in power, this government proceeds to blatantly violate the rights of people with different religious beliefs. Is this society a liberal democratic one? Should other nations endorse this government as morally legitimate because it was elected? Emphatically NO! Elections are meaningful only to the extent that liberal democratic institutions (e.g., independent courts, uncensored news media, an effective constitution protecting liberal democratic values) are functioning.

iii) The (Non-)Resilience of Liberal Social Institutions

Anglo-American and European multicultural societies seem to have assumed the resilience of their liberal democratic institutions in the face of continual in-migration of people from different cultures. The underlying assumption may have been that the inherent attractiveness (or the obvious correctness) of these liberal values and institutions means that all in-migrants (or at least their descendants) will adopt them. Recent experiences of "home-grown" terrorism in a number of societies have demonstrated that this assumption is not always true.

One of the dangers of any liberal democratic society is that the overlapping consensus supporting liberal values may erode. As Fukuyama points out:

> … nothing in the formal institutions themselves guarantees that the society … will continue to enjoy the right sort of cultural values and norms under the pressures of technological, economic and social change. Just the opposite: the individualism, pluralism, and tolerance that are built into the formal institutions tend to encourage cultural diversity, and therefore have the potential to undermine moral values inherited from the past. (1999, 59)

A liberal democratic society needs to actively promulgate the essential liberal democratic values in the educational and political realm, and should actively communicate the elements of their political consensus to newcomers (Stein & Harper, 2000). This concern is at the root of calls for the US to return to its assimilationist ideal of the "melting pot" rather than move towards the opposite Canadian ideal of the "cultural mosaic" (Gibbon, 1938; Porter, 1965). As discussed earlier, it is a mistake to debate moving along a single continuum. Rather, we should distinguish which aspects of our culture are essential. Liberal democratic societies do not need to completely assimilate in-migrants, but they do need to *integrate* them into the *public* realm. Although Salins (1997) is popularly associated with assimilation, what his "American style" version advocates seems closer to integration.

If newcomers fail to experience "conversion" to these public values, eventually liberal democratic institutions will give way as the overlapping consensus supporting them is eroded by a growing minority who oppose them. When this loss of consensus is perceived to be happening, it may well generate a nationalistic backlash. Politicians, civil servants, and citizens all need to strive to make these institutions function well, to actively advocate and to defend them.

Clarification of Some Current Issues

Many of the debates about multiculturalism become clearer when we recognize (1) the inherent link between the liberal democratic society and multiculturalism, and (2) the crucial distinction between private and public values. This also eliminates at least some of the apparent dilemmas which arise when liberal democratic societies deal with immigrants from different cultures of origin.

Now we can address the questions we posed at the beginning. What is special about the values of a liberal democratic society is not their national history, but their rightness in protecting human rights. Traditional customs and symbols of national, ethnic, or religious identity are fine, unless and until they violate liberal democratic values. We should respect and celebrate each person, but not all their values and behaviours. We must object whenever, wherever, and by whomever liberal democratic values are violated. Liberal democratic values must be maintained, and not diluted. A liberal democratic society is secular in its neutrality to religious values, but it can accommodate any religion or value set which agrees to respect the rights of others.

In this section, we illustrate how going back to core liberal democratic principles helps to illuminate some commonly debated questions regarding multiculturalism. Because we are using brief summaries of news media accounts, our discussion will be simplified and not detailed enough to capture all the relevant nuances for the particular examples. It should be remembered that the application of any values or principles always requires interpretation in cultural contexts. Any regulation, convention, or practice which no longer expresses the intended value or principle needs to be changed.

i) Sikh Turban and Dagger

The Sikh religion dictates that males wear turban and dagger. (1) The turban: RCMP ("Mountie") tradition and regulations dictate the headgear to be worn with the uniform. Canadian authorities correctly decided, after some public debate, that wearing a turban does not interfere with an officer's ability to carry out their duties, so there was no reason why a person could not wear other headgear for religious reasons (CBC, 1990). (2) The dagger (kirpan): As a weapon, a dagger could be a danger to the rights of others. When a student appealed his school's refusal to allow him to carry it, a Canadian court devised a compromise. Sewing the dagger into a sheath rendered the threat negligible (CBC, 2006). These resolutions seem like ideals of "accommodation" – both the dominant and the minority culture made reasonable adaptations by carefully examining their principles as they applied to the contemporary context.

ii) Islamic Body Covering

Some Islamic cultures dictate that women should wear clothing which conceals some of their bodies, ranging from the hair (scarf) to half the face (niqab) to the entire body (burka or chador). Do any of these practices contradict liberal democratic values? The key here is "openness." Liberal democratic societies are "open societies" (Popper 1945; e.g., in court, an accused person has the right to face their accuser.) Communication is essential to participation in an open society. This is impossible if a person's face is covered. Citizens must be identifiable to each other. Thus, *face covering* is generally *inconsistent* with liberal democratic values.

What about other body coverings? It seems that liberal democratic values are generally *not violated* by other coverings. Practical concerns will be situation specific. In a recent incident, a soccer federation refused, on safety grounds, to allow a young woman to play wearing a head-scarf. This generated a great deal of public protest. Eventually a scarf was designed that eliminated the danger (of choking), and the provincial association lifted its ban (CBC, 2013). A reasonable accommodation seems to have been reached.

iii) Parental Rights

However, there are other subtle considerations in determining whether wearing a garment violates the rights of the wearer. Is wearing it entirely voluntary? Is it forced upon them by family? By subcultural norms? Further, do parents have the right to impose a cultural custom on their minor-aged children? The rights of parents are a "grey area" between public and private values, and will likely continue to be so. It seems reasonable that setting the boundary requires situational assessment of the possible harm to the child. At the extreme end of the continuum, the highest Canadian court held that the child's right to life (via a blood transfusion) overrides both the parents' religious beliefs and the child's own convictions (CBC, 2009).

iv) Religious Symbols

In considering the question of banning religious symbols, there is an ambiguity regarding the meaning of "secular." It can have: (a) A *procedural* meaning – the liberal democratic concept that individuals have the right to select their own substantive conception of the good. The procedural view focuses on (to use Rawls's term) "the right." The wearing of religious symbols will generally not be regulated. For public servants, dress regulation should relate only to whether the performance of their job is impeded. If not, they should be free to passively express their private beliefs via their attire. This is the interpretation of a liberal democratic society that fully supports multiculturalism. (b) A *substantive* meaning – usually particular to a nation, culture, or religion. A substantive view adds some conception of "the good." When secularity is interpreted in a substantive way, then it may be argued that public servants are the "public face" of the government, and their appearance should symbolize the religious neutrality of the state. Exhibition of religious symbols by public servants is then deemed to be a violation of the religious neutrality of the state. This seems to be the view of governments which have banned (or propose to ban) religious symbols worn by public servants. This interpretation of "secularism" could be seen as competing with other thick theories of the good, rather than being neutral among all such conceptions.

One substantive interpretation is an "anti-religious" view – that a society is opposed to any such beliefs, and limits some or all public expression of religious values. According to a Canadian columnist, the contemporary French term *laïcité* is not fully anti-clerical, but apparently still implies "a subtle hope that religion will ultimately dwindle – that it will not so much be defeated, in the long run, as left behind" (Cosh, 2013). The controversy over the proposed Charter of Quebec Values has been described in a news report as a "War on Religion" (*Maclean's*, 2013b). A substantive interpretation of secularism might arguably be consistent with a monolithic liberal democratic society (if there were such a thing), but certainly not with a multicultural one. If the society has been multicultural, then the imposition of a dress code would certainly violate the rights of existing citizens who were affected.

In the case of France, Netherlands, and Quebec, those in favour of banning the passive expression of private religious beliefs (via attire) might argue that the very survival of their culture is at stake. In this sense, their referral to "French," "Dutch," or "Québécois" values (rather than liberal democratic values) is the correct terminology. But they should recognize that a society which is multicultural descriptively but not normatively must abandon its historic claim to be liberal democratic, open, or free. Their prohibitions are a violation of the affected wearers' rights, as Amnesty International warned Quebec (*National Post*, 2013).

The case for regulating students' attire seems even more dubious, since they are not representatives of the state. Unless it interferes with educational activities, it is difficult to make any case that the wearing of religious symbols to school violates someone's rights.

v) Education

A primary purpose of education in a liberal democratic society is to prepare citizens, so they can be reasonable, rational, and committed to its public values and institutions. Education about these values and institutions, including citizens' rights and duties, should definitely be part of every child's education. It is the duty of government to ensure that it is, via curriculum requirements that are applied to all types of schools.

vi) Immigration Processes

The immigration processes of liberal democratic societies should emphasize that the same public values which enable us to accommodate a diversity of private values also limit the exercise of some of those private values – whenever they violate core liberal democratic values (e.g., with regard to the rights of women and children). It is crucial that immigrants not arrive with the impression that their entire culture of origin can be transplanted unaltered into a liberal democratic society. They must not be given any reason to believe that no accommodation will be required on their part.

Prospective immigrants should receive full information about the priority of liberal democratic values over their private values when they clash. Knowledge of these values and their implications should be a condition for immigration. Another condition (perhaps with exceptions for elderly family members) should be competency in a dominant or official language, so that they have the potential to fully participate in the public life of the receiving society.

Liberal democratic societies should also be cautious about receiving people from radically different cultures. It can be more difficult for them to accept (or even understand) liberal democratic values, and to integrate into the public realm (Scruton, 2002).

vii) Changes in the Dominant Culture

Should aspects of the dominant culture change to accommodate minorities? Should holidays with a religious origin (e.g., Christmas or Easter) be eradicated or changed into purely secular celebrations? Should symbols with a religious connotation (e.g., Christmas trees) be eliminated from public places?

It is clearly a violation of rights to force anyone to participate in religious observances. But it is not clear that exposure to symbols of the religion(s) underlying the dominant culture (i.e., Christianity, Judaism) is really a threat to other religious beliefs. The liberal democratic society is "secular" in not endorsing or enforcing religious practice, but not in being "anti-religious," e.g., banning religious symbols from public display.

vii) Equality before the Law

The UN convention on racial discrimination recognizes the liberal democratic ideal that "all human beings are equal before the law and are entitled to equal protection of the law" (United Nations General Assembly, 1965). The Canadian Multiculturalism Act aims to "ensure that all individuals receive equal treatment and equal protection under the law, while respecting and valuing their diversity" (Government of Canada, 1985). Equality before the law means that the *same laws* apply to *each* and every person, regardless of culture of origin. A liberal democratic society cannot have different laws for different people, unless it is bound by prior historical commitments.[17] However, religious traditions or cultural customs can be used by mutual consent to resolve interpersonal disputes; then they can function in the same way as binding arbitration.

More Effective Planning and Policy Making

A primary goal of accommodation should be to engage all minorities as full participants in the public realm. This goal requires a sensitive understanding of their cultures of origin.

i) Local Planning

In Canada, there is no effective linkage between federal immigration policies and local planning; yet local conditions are a key factor in successfully integrating immigrants into our society. The availability and affordability of appropriate housing, recreation, library services, education, child care, and public transit are important to settling new families (Cappe, 2011). Urban planners should be alert to regulations and policies which inadvertently disadvantage recent low-income immigrants (e.g., inadequate or expensive public transit or limitations on the supply of rental housing), or which inhibit cultural expression (e.g., maximum unit sizes too small for multi-generation families, minimum unit sizes too large for small vendors, limitations on street vendors).

ii) Engagement and Cultural Sensitivity

In order to engage people from other cultures of origin in collaborative planning processes, we should seek a "participant" (Winch, 1958) or "communicative" (Habermas, 1984) or "inside" (Walzer, 1987) understanding of their culture of origin. We should avoid condemning customs without sufficient understanding of the meaning of the customs to the participants. Planners in a multicultural society need to have this cultural sensitivity.

iii) Dialogue

We should also be open to the opportunities to learn from other cultures, without assuming that our dominant/receiving culture is superior in all respects. We should take a critical view of our own culture, acknowledging that past and present failures to live up to our own liberal democratic values have had oppressive consequences, e.g., for women and minorities (Bannerji, 2000). Where injustices are ongoing, we should acknowledge and seek to change them.

As much as possible, we should avoid having to *impose* liberal democratic values on immigrants. We should first seek *dialogue* with minorities whenever we see a practice that is clearly oppressive. Even though it may be difficult, we should try to communicate an understanding of our open liberal democratic culture, and to appreciate our society's tolerance and accommodation of the many aspects of immigrants' cultures which do not violate liberal democratic values. In this limited sense, we need to integrate immigrants into the public realm of our society. In an open society, the ideal is a respectful dialogue about differences.

iv) Collaborative Engagement

To be effective, engagement of people from other cultures in collaborative public planning processes must be proactive. Planners should understand the relevant cultural community well enough to adapt their planning processes where necessary. For example, women in some cultures do not speak if men are present. Engaging women with this cultural background may require that a woman planner meet with women separately. Interaction with women from our culture can begin to educate them regarding their human rights.

iv) Avoidance of Labelling

One thing that could encourage respectful treatment of everyone is to avoid (where possible) the use of names or labels which separate people into groups (Harper & Stein, 2012, c. 11; Jamal, Stein, & Harper, 2003). As Bertrand Russell (1905) pointed out, nominalization (labelling with nouns) leads to trouble. The use of labels encourages an implicit assumption that all persons in group "X" share the same attributes (concepts, values, beliefs, customs, practices, behaviours). But individuals have different ideas, values, and customs, even within the same group.[18] As much as possible, it is better just to describe (using adverbs and adjectives) people's beliefs and actions.

vi) Celebration of Difference

Communitarians criticize Rawls's thin theory of the good as inadequate to sustain a society (Sandel, 1996). They believe a healthy society must have shared substantive values ("civic virtues"). In a society that has a less diverse culture, the shared consensus may well include substantive values. But in a pluralistic society that includes many diverse cultures, shared values will likely be limited to the procedural ones essential to political liberalism. Rawls (1985) argues:

> a democratic society is not and cannot be a community, where by community I mean a body of persons united in affirming the same comprehensive … doctrine. The fact of reasonable pluralism makes this impossible. This is the fact of profound and irreconcilable differences in citizens' reasonable comprehensive religious and philosophical conceptions of the world. (p. 3) [vi]

Its claim to citizens' allegiance rests on the fact that it provides a climate that allows many communities, with their very different thick theories of the good, to flourish.

There may be some merit in the suggestion mentioned earlier that we should promote the idea of "allophilia" – which goes beyond tolerance or acceptance to an appreciation (even love) of cultural differences (Khan, 2013). Educational programs can tap the latent potential of individuals to develop strong bonds with those outside their own "group" (Pittinsky, 2012). This could provide additional social "glue" to hold a society together. One aim of Canadian governments' promotion of their policies in this area has been to make *multiculturalism itself* a positive *unifying element* of Canadian identity.

vii) Celebration of Commonality

Liberal democratic societies do need to be cautious in encouraging (or subsidizing) the maintenance of cultures of origin. While diversity clearly enriches the dominant culture, an overemphasis on celebration of difference could also fragment the society, or give the impression that *all* practices of cultures of origin are acceptable. In celebrating differences, equal emphasis should be placed on celebrating the sameness (common humanity and shared public values) of all citizens, and on celebrating the uniqueness and rightness of our liberal democratic values, which provide the framework for multiculturalism. It is great if a thriving liberal society can attain a sense of shared citizenship which goes beyond the consensus on core values (including multiculturalism) to other commonalities and symbols of national identity[19] – food or drink, flags, national holidays, clothing, sports, dance, natural beauty, etc.

Conclusion

Here are four important "take-aways" from our discussion: 1) Plans, policies, and programs which aim to incorporate and accommodate immigrant cultures into a liberal democratic society require justification. Planners can more effectively *provide justifications* if they understand, and can articulate, how multiculturalism reflects and requires core liberal democratic values, and then frame public debates in clearer terms. 2) Multiculturalism is not so much a value as it is an *expression* of liberal democratic values. 3) Planners should seek to involve immigrants as full participants in collaborative public planning. Their efforts can be much more effective if they have a sensitive *communicative understanding* of the cultures of origin with which they are working. 4) Liberal democratic values are *not inherently resilient*. Arguments, policies, and plans which misinterpret or misapply these values undermine the underlying overlapping consensus, and could ultimately destroy it. Planners should work to integrate immigrants into the public realm. They should *communicate*, *advocate*, and *defend* these values vigorously at every opportunity.

NOTES

1 Security (desecuritization of ethnic relations) is one of five factors which Kymlicka (2012) found would enhance the acceptance of multiculturalism. When these facilitating conditions are absent, multiculturalism is more likely to be seen as a high-risk option.
2 Our definition of liberalism is intended to encompass a variety of political philosophies – Berlin, Bernstein, Dworkin, Habermas, Hooke, Kymlicka, Nielsen, Nozick, Rawls, Rorty, Walzer – and planning theorists – Forester, Healey, Hoch, Innes, Krumholz, Mandelbaum, Throgmorton, Friedmann, Sandercock. This broadly "political liberal" perspective generally underlies the policies of nearly every mainstream political party in North America and western Europe – e.g., Republican, Conservative, Democrat, Liberal, New Democrat, Labour. Also see note 14.

3 As we are using the terms, "procedural" refers to the structure and rules that govern rule making or decision making; "substantive" refers to the content of the rules or decisions.
4 This view resembles interculturalism (discussed in chapter 1).
5 It is possible that someone might be able to conceive of another kind of society being multicultural in the normative sense.
6 The intention of a Wide Reflective Equilibrium (WRE) process (Rawls, 1971, 2001) is to objectively devise the structure or rules for particular types of situations, independent of our own interests or position in society, within a particular set of value commitments (Stein & Harper, 2000). In the moral realm, a WRE has been defined as a coherent set of beliefs which includes the following components: (1) a set of considered moral judgments (which may be intuitive); (2) a set of normative substantive and/or procedural ethical principles; (3) a set of background theories which show that the set of normative ethical principles are more acceptable than alternative normative ethical principles. These background *theories* may be both (a) normative – incorporating ethical notions (different from those in the normative ethical principles held e.g. fairness, impartiality), and (b) empirical theories which seek to explain and predict observed behaviour (Daniels, 1985).
7 The second principle of *Equality*: "Social and economic inequalities are to satisfy two conditions: first, they are to be attached to offices and positions open to all under conditions of fair equality of opportunity, and second, they must be to the greatest benefit of the least-advantaged members of society (the difference principle)" (Rawls, 2001, p. 42). In addition, Rawls argues for a *Priority:* "the first principle is prior to the second; also, in the second principle, fair equality of opportunity is prior to the difference principle" (p. 43). In contrast to the first principle, implementation of the second principle rests partly on empirical judgments, and will unavoidably be the subject of political debate.
8 Our outline of these values is adapted from Rawls (1985).
9 Rawls's primary goods are (1) basic rights and liberties; (2) freedom of movement and free choice of occupation; (3) powers of offices and positions of authority and responsibility open to all; (4) income and wealth necessary for personal autonomy; (5) social bases of self-respect. The basic rights and liberties should be constitutionally protected. All others can be legislated – they are open to political debate.
10 Our interpretation of "citizen" here is very broad – encompassing all legal residents: citizens, landed immigrants, refugees.
11 Rawls describes this focus as the "priority of the right" (the thin theory) "over the good" (the thick theories) (1971, 31).

12 As many critics of Habermas have pointed out, this aim can never be fully met. Pragmatically, we view it as a useful heuristic: the results of trying to move towards it are positive.

13 Acceptance of "political liberalism" in the public realm does not require adopting liberalism (or secularism or any other "Western values") as a *private* system of belief. This is why such a range of different thinkers can be characterized as "liberal."

14 Habermas (1984), Forester (1989), and many others have urged planners to identify deviations from this heuristic ideal, and to work to bring their society closer to it.

15 We define "fundamentalism" as the view that one's beliefs are universally true and absolute, generally accompanied by rigidity with regard to any change in interpretation.

16 We are assuming no rigged voting, suppression of opposition parties, a free press, etc.

17 Constitution and law may give some groups special rights, which may or may not be ethically justifiable. For example, in Canada: French language rights, Roman Catholic education rights, and aboriginal treaty rights.

18 For example: within Islam, Sunnis, Shi'ites, Wahhabis, Ismalis, etc. are quite different.

19 Problems arise with unifying symbols of national identity when they are used to demonize the "other," whether internal or external. Nations have frequently used a common enemy (real or imagined) to promote national identity.

REFERENCES

Bannerji, H. (2000). *Dark side of the nation: Essays on multiculturalism, nationalism, and gender*. Toronto: Canadian Scholars Press.

Bilge, S. (2008). Between gender and cultural equity. In F. Engin (Ed.), *Recasting the social in citizenship* (c. 5). Toronto: University of Toronto Press.

Buchanan, M., & Tullock, G. (1962). *Calculus of consent*. Ann Arbor: University of Michigan Press.

Buruma, I. (2006). *Murder in Amsterdam: The death of Theo van Gogh and the limits of tolerance*. London: Penguin Press.

CBC. (1990). Sikh mounties permitted to wear turbans. Retrieved September 2013 from http://www.cbc.ca/archives/categories/society/crime-justice/mounties-on-duty-a-history-of-the-rcmp/sikh-mounties-permitted-to-wear-turbans.html.

CBC. (2006). Ban on Sikh kirpan overturned by Supreme Court. Retrieved September 2013 from http://www.cbc.ca/news/canada/ban-on-sikh-kirpan-overturned-by-supreme-court-1.618238.

CBC. (2009). Girl's forced blood transfusion didn't violate rights: Top court. Retrieved September 2013 from http://www.cbc.ca/news/canada/girl-s-forced-blood-transfusion-didn-t-violate-rights-top-court-1.858660.

CBC. (2013). Quebec Soccer Federation reverses turban ban. Retrieved September 2013 from http://www.cbc.ca/news/canada/montreal/quebec-soccer-federation-reverses-turban-ban-1.1319350.

Cappe, M. 2011. Starting on solid ground: The municipal role in immigrant settlement. Ottawa: Federation of Canadian Municipalities. Retrieved August 2014 from http://www.fcm.ca/home/issues/housing/issue-resources.htm.

Cosh, C. (2013, Aug. 31). The real reason to fear Quebec's charter of values. *Maclean's*.

Daniels, N. (1985). Two approaches to theory acceptance in ethics. In D. Comp & D. Zimmerman (Eds.), *Morality, reason and truth* (pp. 81–102). Totowa, NJ: Rowman and Allanhold.

Economist. (2013, Feb. 2). Immigrants: The ins and the outs, 7–8. Retrieved July 4, 2013 from http://www.economist.com/news/special-report/21570836-immigration-and-growing-inequality-are-making-nordics-less-homogeneous-ins-and.

EU Times. (2011, June 27). The Netherlands to abandon multiculturalism. Retrieved July 1, 2013 from http://www.eutimes.net/2011/06/the-netherlands-to-abandon-multiculturalism/.

Forester, J. (1989). Planning in the face of power. Oakland: University of California Press.

Fukuyama, F. (1999). The great disruption. *Atlantic Monthly, 283*(5), 55–80.

Galloway-Cosijn, A. (1999). A theoretical framework to facilitate cross-cultural communication in the Aboriginal planning context. Environmental design master's degree project. Calgary: University of Calgary.

Gauthier, D. (1978). The social contract as ideology. *Philosophy & Public Affairs, 6*(2), 130–164.

Gibbon, J. (1938). *The Canadian mosaic*. Toronto: McClelland and Stewart.

Government of Canada. (1982). Charter of Rights and Freedoms.

Government of Canada. (1985). Canadian Multiculturalism Act. Retrieved August 1, 2013 from http://laws-lois.justice.gc.ca/eng/acts/C-18.7/page-1.html.

Habermas, J. (1984). *The theory of communicative action*. Boston: Beacon Press.

Hampton, J. (1989). Should political philosophy be done without metaphsyics? *Ethics, 99*(4), 791–814. http://dx.doi.org/10.1086/293122

Harper, T.L., & Stein, S.M. (2012). (Orig. 2006). *Dialogical planning in a fragmented society: Critically liberal, pragmatic and incremental.* New Brunswick, NJ: Transaction Publishers.

Hobbes, T. (1651). Leviathan, or the matter, forme, and power of a commonwealth, ecclesiasticall and civil.

Jamal, T.B., Stein, S.M., & Harper, T.L. (2003). Beyond labels: Pragmatic planning in multistakeholder tourism-environmental conflicts. *Journal of Planning Education and Research, 22*(2), 164–177. http://dx.doi.org/10.1177/0739456X02238445

Kant, I. (1993) (Orig. 1785). Transl. J.W. Ellington. *Grounding for the Metaphysics of Morals* (p. 30). 3rd ed. Hackett.

Khan, S. (2013, Feb. 28). Allophilia: Beyond tolerance lies true respect. *Globe and Mail.* Retrieved July 3, 2013 from http://www.theglobeandmail.com/globe-debate/allophilia-beyond-tolerance-lies-true-respect/article9130024/.

Kymlicka, W. (2012). *Multiculturalism: Success, failure, and the future.* Washington, DC: Migration Policy Institute.

Maclean's. (2013a, Aug. 22). How the Parti Québécois got to where it is now on minority rights. Retrieved August 24, 2013 from http://www.macleans.ca/general/how-the-parti-quebecois-got-to-where-it-is-now-on-minority-rights/.

Maclean's. (2013b, Sept. 20). Quebec's war on religion, p. 18.

Mandelbaum, S. (2000). *Open moral communities.* Cambridge, MA: MIT Press.

Myers, D. (2007). *Immigrants and boomers.* New York: Russell Sage.

National Post. (2013). Amnesty International warns Quebec values charter would violate "fundamental rights." Retrieved September 2013 from http://news.nationalpost.com/news/canada/canadian-politics/amnesty-international-warns-quebec-values-charter-would-violate-fundamental-rights .

Pittinsky, T.L. (2012). *Us plus them: Tapping the positive power of difference.* Cambridge, MA: Harvard Business School Press.

Popper, K. Orig. 1945. *The open society and its enemies.* 6th ed. 1966. Princeton University Press.

Porter, J. (1965). *The vertical mosaic: An analysis of social class and power in Canada.* Toronto: University of Toronto Press.

Rawls, J. (1971). *A theory of justice.* Cambridge, MA: Harvard University Press.

Rawls, J. (1985). Justice as fairness: Political not metaphysical. *Philosophy & Public Affairs, 14*(3), 223–251.

Rawls, J. (1993). *Political liberalism.* New York: Columbia University Press.

Rawls, J. (2001). *Justice as fairness: A restatement*. Cambridge, MA: Harvard University Press.

Rousseau, J.J. (1762). *The social contract or principles of political right*.

Russell, B. (1905). On denoting. *Mind, 14*(4), 479–493. http://dx.doi.org/10.1093/mind/XIV.4.479

Salins, P. (1997). *Assimilation American style*. Oxford: Oxford University Press.

Sandel, M.J. (1996). *Democracy's discontent: America in search of a public philosophy*. Cambridge, MA: Harvard University Press.

Scruton, R. (2002). *The West and the rest: Globalization and terrorist threat*. Wilmington, DE: ISI Books.

Sellars, W. (1963). Philosophy and the scientific image of man. In W. Sellars (Ed.), *Science, Perception and Reality* (c.1). London: Routledge.

Stein, S.M., & Harper, T.L. (2000). The paradox of planning in a multicultural society: A pragmatic reconciliation. In M. Burayidi (Ed.), *Urban planning in a multi-cultural society* (c. 5). Westport, CN: Greenwood Press.

United Nations General Assembly. (1965). International Convention on the Elimination of All Forms of Racial Discrimination, 21 December. Treaty Series, vol. 660, p. 195. Retrieved August 28, 2013 from http://www.refworld.org/docid/3ae6b3940.html.

Walzer, M. (1987). *Interpretation and social criticism*. Cambridge, MA: Harvard University Press.

Winch, P. (1958). *The idea of a social science*. New York: Humanities Press.

Winch, P. (1970).(Orig. 1964). Understanding a primitive society. In B. Wilson (Ed.), *Rationality* (pp. 78–111). Oxford: Blackwell.

3 The Incorporation of Multicultural Ethos in Urban Planning

MOHAMMAD A. QADEER

The Meaning of Multicultural Planning

Multicultural planning is an ambiguous term. It is commonly spoken about but used sparingly in writings.[1] The ambiguity arises about its meaning and scope. What does it mean for planning to be multicultural? The answer lies in how planning addresses the cultural diversity of the cities or societies that are the objects of its operations. It is not that planning should take on some special quality called multicultural. The focus is on how urban planning can and should become inclusive and pluralistic in response to the ethnocultural diversity of the people it serves. And about this meaning there is almost a consensus. It is unsurprising that most of the writers on multicultural planning use phrases linking planning to the cultural diversity of cities to define the scope of their investigations, e.g., *Urban Planning in a Multicultural Society* (Burayidi, 2000), *Pluralistic Planning for Multicultural Cities* (Qadeer, 1997), *Planning in the Ethno-culturally Diverse City* (Pestieau & Wallace, 2003, *Planning and Diversity in the City* (Fincher & Iveson, 2008).

The critical problem meant to be addressed by multicultural planning is urban planning's responsiveness to the ethnocultural diversity of people living in cities and their entitlement to the rights and resources of the city as equal citizens. Multicultural planning is driven by two values, diversity and equality (Reeves, 2005, pp. 4–6). Diversity for planning purposes suggests two propositions: 1) culture and identity matter in the sense that they lay the bases of difference in people's ways of meeting their needs both as individuals and groups; 2) the recognition of these differences necessitates the forging of multiple or pluralistic modes of satisfying public needs. These propositions strike at the root of the uniformity of measures and standards as the criterion of good planning. They militate against the universality of needs and modes of satisfying them.

For example, the notion of family may differ between immigrant and native-born Americans. The former may include parents and relatives, while the latter may consider only parents as the cohabiting members of a family. In the same vein, gay couples or single mothers with children are legitimate families, though some local housing and zoning codes may not recognize them as such. Multicultural planning demands that the differences in family/household composition should be recognized and policies should be flexible to correspond to these differences.

Equality is the value that energizes diversity and makes it a driving force of multicultural planning. As a value, equality is the right and entitlement to equal status in law and equal access to public benefits, resources, and treatment, without discrimination on the basis of race, colour, age, gender, disability, culture, or religion. Equality rights are enshrined in the Canadian Charter of Rights and Freedoms (1982), in the 14th Amendment to the US Constitution (1868), and in the US Civil Rights Act (1964).

Equality is the ideal or goal, but equity is the term that gives it practical form by incorporating it in a program, policy, or institution. It turns the principle into practice. Susan Fainstein in *Just City* prefers equity to equality on pragmatic grounds as a criterion for urban policies (Fainstein, 2010, p. 36). It isn't about treating everybody the same, but about treating everybody appropriately to meet their needs, without the necessity of favouring those already better off (ibid). Equity could also mean modifying a policy or program to equalize outcomes or to favour the disadvantaged. It is affirmative and accommodative in intent. Employment equity, for example, means proactive recruitment of under-represented groups. It establishes preferences for recruiting persons of some designated identities to ensure fairness.

In urban planning, equity is the bridge between equality and diversity. The objective is to equalize outcomes not inputs. For example, many cities in the US and Canada provide interpretation and translation services in ethnic languages to facilitate the participation of immigrants in planning decision-making processes. These services give non-English speakers equal opportunity to participate. Another example is the inclusive housing policies in some jurisdictions that require some quota of affordable housing in new developments of apartments or subdivisions.

The challenge of realizing social equity is much broader than the domain of urban planning. Thomas Piketty has shown in a comprehensive study of the history of economic growth in the past century that equity remains an elusive goal, because capital takes a greater share of the growth in national income (Piketty, 2014). The current phase of corporate capitalism and the rise of neoliberal ideologies have institutionalized economic and social inequities into the economic and social systems. Urban planning's jurisdiction extends only to a small part of these systems. Its contributions to equity are limited to framing policies that at best give priority to the needs of the disadvantaged in the physical and social development of regions, cities, and neighbourhoods, and that too only in some designated sectors. Multiculturalism injects the considerations of race, ethnicity, and immigrant status, in addition to class, into the criteria for determining the distribution of resources and services.

My objective in this chapter is to critically review the evolution of the discourse on multicultural planning and contrast it with current planning practices of responding to ethnocultural diversity in US and Canadian cities. The underlying question is whether there is disparity between the discourse and planning practices, and if so, in what direction? This analysis will point out that a strategy of reasonable accommodation is the most practical approach to multicultural planning. It concludes by spelling out a policy index to help institutionalize this strategy.

Evolution of the Theoretical Discourse about Multicultural Planning

Pluralism as an approach has a long history in urban planning. It goes back to Paul Davidoff's seminal article "Advocacy and Pluralism in Planning," which argued against unitary plans and advanced the idea that multiple proposals be prepared to promote the interests of different groups, particularly non-White minorities. These plural plans ought to be advocated in the public arena to hammer out equitable policies (Davidoff, 1965). Advocacy planning on behalf of disadvantaged groups has been a significant part of the urban planning practice in North America since then. Although its primary focus has been to promote the rights of African Americans and advocate for their needs, with the post-1965 wave of immigrants in both the US and Canada, Latinos, Chinese, Russians, Caribbean natives, and others have come to demand equity and recognition of their rights to the city and its services. This has heightened the politics of difference and expanded the agenda of pluralistic planning, particularly in practice.

The theoretical discourse on multicultural planning began in earnest in the 1990s with the emergence of sizeable populations of immigrants and their native-born second generations, whose sense of citizenship and awareness of rights is stronger. The discourse has evolved through roughly three thematic but overlapping phases in the last two decades as follows:

1 Advocacy of minority rights and critiques of the urban planning approach and methods. Though continuing to reverberate in the literature up to the present, these themes were front and centre in earlier writings. Urban planning is criticized for its modernist, universalistic, and scientific-deductive paradigm of value-neutral professionalism, which presumably keeps out voices of minorities. Planning has to be driven by the divergence of people's needs and interests and guided by communicative and participative modes of understanding and policy making (Burayidi, 2000, Sandercock, 2003). The idea that "it is morally and ethically incumbent on planners to treat different groups differently" was the crux of advocacy for multicultural planning (Burayidi, 2000, p. 2). Another theme in the early literature was the pointing out of the systematic bias in urban planning against immigrants and minorities, rising to accusations of racism in some cases. This bias has been attributed to the Anglo-European norms and preferences that are coded into the precepts of urban planning, turning them into principles of planning (Reeves, 2005; Milroy & Wallace, 2002).
2 Sensitivity to cultural differences and empowering of ethnoracial minorities. This theme emerged as a corrective to the criticism levelled at urban planning. It has been a logical consequence of the critiques of current planning practices and has been offered as a solution to the cultural insensitivity and myopia of the planning profession (Milroy & Wallace, 2002; Sandercock, 2003). This theme has spawned soaring rhetoric on making urban planning a transformative practice, highlighting justice and equity as its premier goals (Viswanathan, 2010). In some accounts, urban planning is said to be steeped in a colonial outlook towards racial minorities, aboriginals, and immigrants (Viswanathan, 2009). Leonie Sandercock challenged planners to be audacious, daring to break rules

and take risks of involving the public in decision making (not just consultation) and to develop creative capacities through visionary leadership and participatory action research (Sandercock, 2004a, pp. 136–137). Patsy Healey observed that planners "are under pressure not merely to absorb new sensitivities and demands into their thought worlds and practices, but to transform their own 'culture'" (Healey, 2003, p. 245). The sum total of this phase of planning theory is that urban planners have to change their modes of thinking by becoming aware of their biases and assumptions and learn to listen to the diverse voices of people in making plans and policies. It envisages a participatory and communicative mode of urban planning. It also makes planners' values and outlook the critical factors in ensuring equity and recognition of ethnocultural groups.

3 Mainstreaming diversity, incorporating pluralism, and institutional learning. By the middle of the first decade of the 2000s, the discourse of multicultural planning began to shift away from critiques and exhortations of planners per se to ideas about accommodating diversity and incorporating it into the mainstream. This change occurred subtly, prompted by the imperatives of integrating immigrants coming in such large numbers and impacting the demography of North American cities. Independently in urban theory, diversity came to be looked upon as an asset in the global economy and a source of creativity and innovation in cities (Florida, 2005). Theorists of multicultural planning began calling for institutional change to cultivate diversity as one of the important qualities of a good city. The discourse began to shift beyond the exhortations for planners to be sensitive to cultural differences, towards reconstruction of planning principles and common institutions (Qadeer, 2009). Leonie Sandercock, for example, argued for an intercultural approach, one that promotes mutual engagement among different groups in contrast to separating people on the basis of differences (Sandercock, 2004a, 2011). Her seven-point multicultural manifesto ranges from recommendations for increased spending to political and policy support for multiculturalism at all levels of government, creation of institutions for cross-cultural negotiations, and developing new notions of citizenship, among others (Sandercock, 2004b, pp. 156–157). This emphasis of interlinking different groups and integrating them in a civic community has been packaged in the concept of the intercultural city (Wood & Landry, 2008, pp. 296–297), which has not yet caught on in North America, but

the integration of different groups in a cohesive city has become the new expected outcome of multiculturalism. In one sense, the argument has come full circle back to the idea of pluralistic planning within a framework of negotiated urban policies.

Assessment of the Multicultural Planning Theories

Multicultural planning thought has raised consciousness about the cultural rights of ethnic groups, racial minorities, and immigrants in the field of urban planning. Yet it could not have been insulated from the changing perspective of multiculturalism, which evolved from the celebration and promotion of ethnoracial identities to the recognition of cultural rights and the forging of inclusive modes of meeting the public needs of those groups, and recently has tilted towards balancing cultural differences with social cohesion (Kymlicka, 1998, 2007; Fleras, 2009). The strong advocacy of the social and cultural rights of minorities has been tempered by the European experiences of divisiveness and exclusion expressed in the form of race riots, immigrant ghettoization, and episodes of terrorism (Rath, 2011). There is an emerging consensus around the need to balance social integration and minorities' rights to build shared citizenship. Will Kymlicka, the pre-eminent theorist of multiculturalism, maintains that "multiculturalism combines robust forms of nation building with a robust form of minority rights" (Kymlicka, 2007, p. 83). My point in referring to the changing perspective of multiculturalism is that the conceptual ground under multicultural planning thought has been shifting. It could not have maintained its single-minded advocacy of giving primacy to cultural differences in public policies.

Multicultural planning theorists have focused almost exclusively on the need to bring diverse voices of minorities into the planning process. They have advanced the need for minorities to be at the decision-making table. Facilitation of the ethnocultural groups' participation in planning decision-making processes is regarded as a way of empowering them and making planning more inclusive. This emphasis on the process, reinforced by the preference for postmodern modes of analysis with little regard to the substance of urban planning, has been a long-standing bias in planning theory. It has limitations. As Ruth Fincher and Kurt Iveson observe, "planning frameworks must enable planners to make calculations about 'what should be done,' not just 'how it is done'" (Fincher & Iveson, 2008, p. 5).

This multicultural planning thought has an expansive and unrealistic view of the scope of planning. It presumes urban planning to be synonymous with city governance and management. Its criticism of planners has an aura of disappointment in their failures to lead social change. These ideas do not take into account the institutional restraints of urban planning.

Urban planning is a bounded institution. It has a defined functional jurisdiction ranging over matters of land use, transportation, housing, environment, and services. Even its social planning is concerned with community organization and programming for the provision of social services, particularly for disadvantaged groups. As an institution, urban planning is embedded in the constitutional sphere of public powers, laws, property rights, markets, and individual and group liberties and rights. Its decision-making processes are multilayered, in which mayors, councils, courts, state/provincial authorities, and citizen groups and communities, as well as developers, are participants. Planners per se are one among many actors, and laws and traditions limit their authority. This has to be kept in mind in theorizing about planners' embrace of multicultural planning.

Multicultural planning theories are normative and prescriptive. They almost always lay down what "ought" to be done, without giving much attention to what is being done in urban planning practice. The empirical evidence of the state of multiculturalism in cities has had little bearing on multicultural planning thought. There is plenty of evidence suggesting that urban planning systems have responded sensitively to the ethnocultural diversity of urban residents. Let us turn to this discussion.

The Practice of Multicultural Planning

My hypothesis is that urban planning practice has forged ahead of the theories in responding to ethnocultural diversity. Major cities of North America are vivid evidence of this proposition: Chinatowns, Latin barrios, or Russian neighbourhoods, ethnic economies and niches, Asian malls, street signs in ethnic languages, radio and TV stations filling the airwaves with the sights and sounds of distant lands, colourful fairs, parades, and nationality days, multicultural curricula in schools, soccer and cricket fields in city parks, translation and interpretation facilities, immigrant settlement services, and that most striking expression of multiculturalism, the proliferation of ethnic cuisines and restaurants, are testimony to flourishing ethnocultural diversity whether in

Toronto, New York, Los Angeles, Miami, Calgary, Windsor, or Minneapolis. Practising planners recount these developments as expressions of their recognition of diversity. After all, many say these developments would not have come about without the responsiveness of the planning system to such proposals.

Aren't the thriving multicultural communities, institutions, and practices in North American cities the result of private initiatives enacted through markets, with little input from urban planners? The answer is that urban planning in both public and private forms is involved in approving, consolidating, and servicing ethnic enclaves, mosques, malls, and other similar facilities, even if they emerge incrementally through private initiatives, which is usual in most forms of urban development, as various case studies of multicultural planning show (Harwood, 2005; Qadeer & Chaudhry, 2000; Agrawal, 2009; Zhuang, 2013). Second, urban planning as an instrument of regional and local policy making is a partner with the respective operational agencies in developing strategies for the development and coordination of schools, public transport, public housing, local economic development, amenities, parks, and recreation facilities. Race, ethnicity, and poverty are implicated in these strategies. So the thriving multicultural forms of urban development could not have come about without urban planning's responsiveness to diversity.

Of course, there is more specific evidence of the accommodation of the demands of diversity in both the process and policies of urban planning. That evidence is in the form of measures taken in particular cities to involve minorities and accommodate their interests.

Planning Process and Involvement of Minorities

Urban planning as an institution pioneered citizens' involvement in public decision making in Canada and the US almost 50 years ago. This practice arose from the realization that without taking minorities' interests into account, planning policies produce deleterious outcomes by displacing Blacks and the poor, as in the much-touted Urban Renewal program (1954–74). Since then the participation of citizens through community meetings, statutory public hearings, empowering of community boards made up of residents' representatives, and other forms of people's involvement has become an integral part of the planning process. That is why urban planning is described as a communicative and collaborative activity (Innes, 1998; Healey, 1997).

Almost all levels of planning decisions in the US and Canada, be those long-range policies or applications for a minor variance from zoning standards at a site, are required to have community involvement. It is in this decision-making structure that issues of ethnoracial diversity play out. There are often special provisions for facilitating the participation of minorities and ethnic groups by specifically targeting such groups, offering translation and interpretation services in order to seek their input. Giving attention to diversity in public discussions is a part of the Code of Professional Conduct of the Canadian Institute of Planners (CIP).[2]

How widespread these practices are for the ethnoracially diverse publics can be observed from the results of a mailed questionnaire survey of 23 US and 19 Canadian municipalities based on the policy index of multicultural planning (to be discussed later).[3] A majority (52–57 per cent) of the responding municipalities reported having policies of providing minority language facilities, ensuring their representation in planning committees/task forces, and involving them in decision-making processes (Qadeer & Agrawal, 2011, pp. 142–143). Among the responding municipalities were Los Angeles, San Francisco, Chicago, Houston, Toronto, Vancouver, Montreal, Calgary, and many others small and big.

More concrete evidence about the inclusion of minorities in planning processes has to be gleaned from case studies of individual cities. Toronto, for example, is obligated under provincial legislation to hold public meetings for policy decisions, secondary plans, and any site plan or zoning changes. Depending upon the ethnic composition of an area, it offers translation and interpretation services for the non-English-speaking participants and citizens. Furthermore, it funds community organizations engaged in anti-racism programs and ethnic/immigrant services. These programs have fostered a large number of advocacy groups who speak for their respective communities. The same is, by and large, the case in other cities of Canada.

Minorities' representation in the processes of planning decision making in US cities is equally robust. Despite the fact that the US is a country where assimilation is emphasized and the "melting pot" is held to be the national condition, in contrast to Canada's official policy of multiculturalism, minorities have been deliberately involved in urban planning processes. American cities are particularly sensitive to the issues of racial equity and representation and actively seek involvement of Blacks, Latinos, and other minorities. In preparing a general plan, Los Angeles held 60 meetings and involved 3,000 citizens of diverse backgrounds. It provided interpreters in meetings and translations of planning documents from English to Spanish, Japanese, Korean, Chinese, and Persian (City of Los Angeles, 2011). New York City has a standing policy that all city agencies, including the planning commission, provide interpretation and translation services for non-English speakers.

Similar examples can be made for other cities, but the point being underscored is that, institutionally, urban planning processes in both countries are open to minorities, who are welcomed into discussions of policies, plans, and actions. The vigorous public debates about planning policies in almost every city and the occasional flare-ups of community confrontation are proof of the fact that planning processes are not lacking in providing platforms for minority voices. The process may not always be satisfying to one or the other group, but that is democracy. NIMBYism, community politics, and necessary trade-offs among multiple objectives of planning are some of the factors that affect the outcomes of planning processes. Minorities' interests are part of this equation.

Does participation in planning decision making lead to equitable outcomes? Of course there is a weak correlation between participation of minorities and equity. Well-organized and politically strong groups have greater influence on outcomes than small and disadvantaged groups. And this applies as much to the competition among minorities as to that between the dominant majority and minorities. Furthermore, the power structure and corporate interests have many ways of swinging decisions in their favour. Yet if a "with and without" test is applied to the utility of minorities' participation in planning process then their being able to articulate their interests and agitate for them in some cases yields relatively more equitable outcomes than if they were not heard or included in the decision-making process.

That having been said, it must be pointed out that planning processes, despite their communicative orientation, do not rise above the disparities of power, resources, and political influence. In the context of planning for diversity, I would say that the theorists' holding equitable outcomes to be dependent on minorities' inclusion in planning decision making offers a limited promise of advancing equity. The practitioner faced with competing political pulls and divergent objectives has to forge substantive ideas that are satisfying to different interests, including minorities in varying degrees.

The Making of Multicultural Cities: Embedding Diversity

Turning to the substance of city development, one can readily see that the major US and Canadian cities have been transformed in the last 20 to 30 years through immigration. Culturally diverse forms, functions, and activities have permeated all of their social, economic, and spatial institutions. They have come to have "the world in a city," a phrase which is the title of two books, one for Toronto and the other for New York (Anisef & Lanphier, 2003; Berger, 2007).

As Burayidi and Wiles discuss in chapter 8, 58 of the top 100 US cities are now majority-minority in population (Frey, 2011). Major Canadian cities have either turned or are at the cusp of turning majority-minority. What it means in stark terms is that Whites are less than 50 per cent of the population in many cities. The demography of cities is changing and with it the dominant culture. This is the result of the current and unending wave of immigration in both countries, prompted by the falling birth rates of the native population, the pressing need for youthful labour to replace the aging workforce, and the actuarial demands of Social Security and pensions. New Americans and Canadians, and their native-born second/third generations, are fusing their cultures into North American cities.

Ethnic enclaves are the most striking spatial manifestation of cultural diversity in cities. Neighbourhoods where a particular ethnoracial group dominates residentially, commercially, and institutionally have emerged in Toronto, Vancouver, New York, Los Angeles, and elsewhere, in suburbs as well as in the city centres. Toronto's long-established Italian and Jewish enclaves are complemented by Chinese enclaves in the northeastern sector of Scarborough and the town of Markham and by South Asian neighbourhoods in the northwestern suburbs of Brampton and Mississauga (Qadeer, Agrawal, & Lovell, 2010). New York has Chinese enclaves in Forest Hills and Sunset Park, Colombian and South Asian neighbourhoods in Jackson Heights, Russians on Brighton Beach – all in outer boroughs, beyond Manhattan's Chinatown, Spanish East Harlem, and Dominican Washington Heights (Lobo & Salvo, 2013; Berger, 2007). Los Angeles City's barrio, East Los Angeles, along with Koreatown, Chinatown, Watts, and Little Tehran, are the expressions of cultural diversity within the city. One can recount similar developments in Miami, Chicago, and even small places like Utica, New York. The point is that new ethnoracial minorities are realigning the urban landscape. Ethnic enclaves are not planned by city planning departments. They are the product of ethnic households and businesses choosing to locate near each other in order to carve out territorial communities where their cultural, social, and religious institutions can be re-created. They should not be viewed as ghettos (Marcuse, 2005). Urban planning has not initiated such developments, but it has supported them in various ways. These developments are what practising planners point out as outcomes of their multicultural planning.

Ethnic enclaves often turn into bases of ethnic economies, though the latter are not limited to territorial concentration of businesses. Commercial developments in the form of ethnic malls and plazas are now a feature of many metropolitan areas. The Toronto area has five Chinatowns and 66 Chinese shopping centres (Wang & Zhong, 2013).

How ethnicity transforms the economic and social structure of a region is illustrated by the case of San Gabriel Valley in Los Angeles County. Immigrants from Taiwan, China, and Korea have changed not only the demography but also the economy and social organization of the cities and towns in the valley. Suburban municipalities of Monterey Park, Arcadia, Alhambra, Walnut, and others in the valley were bastions of the old-style Americana until almost the 1970s and were swept by nativist sentiments of resisting "Chinese takeover." By the 1990s, the tide had turned. Asians, particularly Taiwanese investors, poured hundreds of millions of dollars into real estate, housing, and businesses. Projects like Atlantic Times Square in Monterey Park, developed by Kam Sang Company, which includes 210 condo units, a 14-screen movie theatre, and 200,000 square feet of retail space offering choices for shopping, dining, and entertainment in Mediterranean style, have not only transformed Monterey Park but have also affected the whole of the San Gabriel Valley (Zhou, Chin, & Kim, 2013). In 2010, Asians were a majority of the population in many of these suburban cities.

These ethnoburbs have become a link in the American global connections with Asia-Pacific. All this change has been undoubtedly market driven and fuelled by immigration, but it could not have happened without urban planning's responsiveness in the form of revised economic and social policies, accommodation of cultural differences, and restructuring of zoning and design standards. Incrementally and reactively, urban planning has accommodated the change. It may not have led the change, but it did not resist it. This is how multicultural planning has advanced in practice.

Multicultural Institutions and Services

Ethnic and religious diversity leads to the formation of new institutions and the realignment of services, modifying existing provisions and promoting new practices. How has urban planning responded to these demands? I will review trends and cases in major cities on the responsiveness of urban planning. One defining characteristic of the current long wave of immigration in both the US and Canada is that it is racially non-White and religiously non-Western and non-Judeo-Christian to a large extent. This injects many new institutions, practices, and services into local societies.

One obvious example is the development of non-Christian places of worship, such as Islamic mosques, Hindu mandirs, and Sikh gurudwaras in cities. The history of development of these institutions follows a mutual learning path. The first proposal for a mosque or temple in a city comes across planning regulations meant for the development of churches. It precipitates a process of policy revision, involving community consultations, sometimes confrontations, as well as formulation of new standards and planning norms; e.g., substituting the generic term *places of worship* instead of *churches* in a by-law, or, as in one case in the Toronto area, removing the architectural requirement of having a church-like clock tower (Qadeer & Chaudhry, 2000, p. 19). Once a precedent of building a mosque or temple has been set, the path is cleared for others. The local planning departments learn about the requirements of different types of places of worship, and the faith communities come to understand the objectives of planning policies. This is the process of mutual learning.

Of course, this learning does not mean that proposals for such places will move smoothly through the planning processes. There are objections, oppositions, even racist remarks in community meetings, but planning departments acting as mediators attempt to steer community discussions towards matters of land use and their impacts, thus keeping a lid on political controversies engendered around such proposals. Over time, pluralistic standards for the development of places of worship emerge. It should also be noted, as pointed out by Gale and Thomas in chapter 5, that many places of worship, including new ones, are established in stores and structures where zoning allows such uses. Those developments occur with few procedural requirements. This is what has also happened in Toronto, New York, Houston, and Vancouver.

The results of this mutual learning are imprinted on the urban landscape of North American cities. Metropolitan Toronto has 100 mosques and numerous Islamic prayer rooms in colleges, schools, and even offices. It has 70 mandirs, 19 gurudwaras, and a host of ethnic churches. New York City has 75 mosques, 30 in Brooklyn alone, 30 mandirs, and numerous Zen temples. Houston has 90 mosques and many mandirs. Even churches take on ethnic identities such as Korean, Chinese, or African. The story is the same in other cities.

As the congregations of a faith increase, its places of worship begin to be divided by denominations and sects, generating demands for new forms of development. Then they start to get built in industrial zones or seek large sites on the peripheries of cities, posing challenges for planning policies attempting to contain sprawl and promote compact cities. Thus new challenges of reconciling broader planning goals with minority rights are precipitated. Planning policies for places of worship continue to evolve, but they have a distinct local flavour.

In the case of the development of mosques, which sometimes arouse oppositional sentiments, urban planning's role is reassuring. The US has about 2,000 mosques and Canada about 200. These numbers speak for urban planning's record of responsiveness in processing such proposals. A case in point is the proposal to develop an Islamic community centre and mosque by rebuilding 51 Park Place, around the corner from the Ground Zero of the former World Trade Center at the tip of Manhattan. It conformed to the zoning and other regulations, but its proposal was turned into a national controversy by conservative groups on the plea that an Islamic centre on this site would be an insult to the victims of the terrorist attacks of 9/11.

New York's Lower Manhattan Community Planning Board voted 29 to 1 with 10 abstentions to approve the development of the Islamic centre, despite strong public protests. A subsequent attempt to get 51 Park Place declared a historic building to block the changes was also turned down by the city's Historic Landmark Preservation Commission. The mayor of New York unequivocally declared that the city is open to all faiths. Today, the Islamic centre is functioning, in the existing building, and the sponsor of rebuilding it as a new centre has withdrawn his application.

I do not mean to suggest that all mosque proposals have positive outcomes. Local politics, planning policies, and the nature of the proposal affect the results. Yet mosques continue to be developed, some with long-drawn struggles and modifications, others smoothly. In the same city one proposal may meet fierce opposition, and another a few miles away may sail through with little difficulty, as it happened in Staten Island, New York City. The case study of mosques points out that diversity is alive and well, and urban planning as an institution has been, by and large, fair though buffeted by local politics and community participatory processes.

It must also be pointed out that long-drawn-out processes of approving proposals for mosques and other institutions should not always be interpreted as evidence of planning systems' bias or discrimination. The planning process and its participatory requirements allow all kinds of objections to be raised for any development, even for those of the mainstream. Urban planning is a localized process, which means it may be discriminatory in its policies in one jurisdiction or at one time, but entirely opposite in the neighbouring municipality and at another time. There should not be a rush to call it discrimination whenever a minority's proposal runs into opposition. The fairness with which minorities' goals are reflected in substantive plans is the test of urban planning's responsiveness.

Denominational schools, ethnic language and cultural centres, ethnic radio/TV channels, and community organizations are found in almost all cities where ethnic groups have a sizeable presence. Immigrant settlement services are provided by many cities, including Toronto, New York, and Los Angeles. Neighbourhood centres as meeting grounds for immigrants are funded by cities and states. New York and Toronto have developed cricket fields, bocce lawns, and soccer stadiums as part of park policies.

Most cities now promote ethnic parades, fairs, and national days by permitting marches and suspending parking and traffic regulations on such occasions. Ethnic leaders, sportsmen, and literary/scientific figures are honoured by the naming of streets and public spaces after them. Housing developments catering to a particular ethnoracial group have been built by both the private and public sectors. All in all, diversity has been woven into the urban structure of North American cities.

The city planning departments responding to the survey of 42 US and Canadian cities, referred to above, have identified a very high level of accommodative actions on items, such as: ethnic heritage preservation – 71 per cent; ethnic signage and street names – 64 per cent; immigrant special services – 57 per cent; ethnic diversity as a plan goal – 57 per cent; promoting ethnic art and culture – 71 per cent; and accommodating ethnic sports – 54 per cent (Qadeer & Agrawal, 2011, p. 145).

How Has Urban Planning Practice Fared in Recognizing and Incorporating Diversity?

The openness to people and their ways of life has come about by the granting of freedoms and liberties of civil rights. These rights are formally guaranteed by the national constitutions and specifically in the US by the Civil Rights Act (1964), which outlawed discrimination on the basis of race, religion, sex, colour, or national origin, and in Canada by the Canadian Bill of Rights (1960) and the Canadian Charter of Rights and Freedoms (1982).

Urban planning operates within these values and rights. Institutionally, it is committed to advancing the public interest, and, by professional tradition and ethics, it leans towards fairness for all groups and individuals. Since the 1960s, urban planning has espoused values of equally serving minorities and disadvantaged groups, even if its actions may not entirely live up to these goals.

Canada is officially a multicultural country with multiculturalism enshrined in its Charter. Though the US subscribes to the ideology of assimilation and espouses the "melting pot" as the cultural metaphor, its civil rights and history also make it a country of multiple identities and cultural mosaic. Toronto, Vancouver, and increasingly Montreal in Canada are recognized multicultural cities. New York is globally known as a multiracial, multiethnic, and cosmopolitan city (Foner, 2013). Los Angeles has been long described as a multicultural city (Davis, 2000; Waldinger & Bozorgmehr, 1996). Chicago, Miami, Boston, Houston, and Phoenix are not far behind. Overall, urban planning has helped sustain this multiculturalism.

The following observations sum up the scope and limits of urban planning in responding to ethnocultural differences.

1 Urban planning has responded to ethnocultural diversity incrementally and reactively. Prompted by private initiatives and community pressures, it has reacted by modifying policies and regulations.
2 The recognition of cultural differences is almost universal, but local politics, resources, and communities' situations affect the tailoring of policies to specific groups. Thus there is considerable local variation and contingent decision making in accommodating the demands of different groups.
3 Multicultural planning has not evolved a vision or overarching model of a multicultural city. This may be one of its shortfalls. Yet this shortfall has not impeded pluralism in urban planning.

4 Community initiatives and private entrepreneurship are strong driving forces in promoting culturally specific services and institutions. Different groups will differ in injecting their cultural identities into the urban structure. Some will be more successful than others, because they may be large in numbers, form a big voting bloc, be economically resourceful, or be politically organized.
5 Cumulatively, urban planning has incorporated cultural sensitivity in its operations.
6 Urban planning's functional jurisdiction is limited to land uses, transportation, housing, environmental sustainability, services, and local economic development. Even in these functional areas, its authority and role vary from development to regulatory, coordinative, and advisory. Its role overlaps with that of other public and private agencies. It does not have a direct influence on policies of income distribution, taxation, and social welfare. The distributional justice for minorities or fairness to immigrants in jobs and housing are matters over which urban planning has limited influence. It cannot serve as the primary platform for social transformation, as some theorists assume. It can follow these ideals in its limited sphere.
7 Ethnoracial and religious communities are divided in their interests by class, nationality, regional affiliation, and denomination. Their needs and demands are not uniform. Urban planning has to mediate among intra-ethnic differences. A highly differentiated sensitivity to cultural difference is required in urban planning.[4]
8 Urban planning has a long tradition of citizens' involvement in policy making. Some of the racist confrontations and nasty stereotyping occur in public meetings for planning decisions. Planners have the delicate task of managing these negotiations. As Burayidi pointed out in the introductory chapter, more participatory planning may not necessarily improve the responsiveness to cultural differences, contrary to the theorists' assertions.
9 Overall, the hypothesis that planning practice has overshadowed the theoretical discourse in multicultural planning is upheld.

That urban planners have accommodated diversity in their professional work is attested by other observers. Domenic Vitiello, writing about immigrants in American cities, observes, "planners and community development practitioners have incrementally addressed many problems immigrants face" (Vitiello, 2009, p. 251). Susan Fainstein envisions that "justice in the urban context encompasses equity, democracy and diversity" (Fainstein, 2010, p. 5). She found New York measuring well on the criteria of ensuring diversity (ibid., 7). On the meter of immigrants' discontent in North American cities, urban planning does not even register. Except for the site-specific controversies of land development about ethnic malls, mega-houses, or mosques/mandirs that occasionally flare up, there seldom is any mention in ethnic media of the planning issues. There are frequent accounts of police racial profiling, job discrimination, devaluing of foreign credentials, and income disparities, but hardly any mention of urban planning's inequities.

The practice of urban planning suggests another model of multicultural planning, which will be discussed below. Before elaborating on that model, the concept of multiculturalism needs to be clarified.

Clarifying Multiculturalism: Diversity and Common Ground

The core idea of multiculturalism is that the culture of a society or community is a composite of many subcultures of distinct beliefs, behaviours, and values and that those are recognized in their authenticity in the public sphere. The multiplicity of cultures in a shared and defined space is the basis of multiculturalism. The space is an unstated but well-understood dimension of multiculturalism, because it refers to the recognition of ethnocultural identities within a nation, city, or institution (e.g., education). After all, the world as a whole is multicultural and multiracial, but that is not the scale at which we normally speak of multiculturalism. Shared and defined space, including its infrastructure, has a bearing on the scope and limits of multiculturalism. I will come to that soon.

The civil and human rights revolution of the post–Second World War era helped to launch multiculturalism (Kymlicka, 2007, p. 31). It extends the right to act upon individual as well as group identities in the private domain of home, religion, and community and their recognition in the public domain of laws, politics, economy, and civic order (Rex, 1996, p. 18). Multiethnicity without rights is not multiculturalism, as is, for example, the case in the Gulf States, Saudi Arabia, and Israel, and to some extent Japan and other countries. Multiculturalism thrives in the liberal-democratic countries of Europe, North America, Australia, New Zealand, and India. It combines demographic diversity with civil rights (see Harper and Stein, chapter 2).

Canada, the first country to pass a national multicultural law (1988), includes in multiculturalism "the freedom of all members of Canadian society to preserve, enhance and share their cultural heritage ... to ensure all individuals receive equal treatment and equal protection under the law while respecting their diversity" (Canadian Multiculturalism Act 1988, clauses 3a, 3e). Freedom to preserve cultural heritage with equality under the law is the defining element of multiculturalism.

Multiculturalism is a contested idea. It is particularly picked apart for its potential to divide a society into separate groups and weaken social cohesion. Britain, Germany, and the Netherland have retreated from the ideology of multiculturalism for presumably impeding the integration of immigrants, promoting illiberal values, and laying the ground for parallel societies (Rath, 2011; also see Gale and Thomas's discussion in chapter 5). These conditions, if true, arise from exclusionary policies, anti-immigrant politics, and the denial of equality rights rather than from multiculturalism.

The cultures of multiculturalism are subcultures. They are partial, not extending to all institutions. They are encased in several other levels of cultures, i.e., national, regional, and city. They are limited in scope and do not extend to constitutional, criminal, and civil laws, which are applicable to everybody, as are political structures, economy, technology, good government, environment, and national ideology. Multiculturalism requires that these institutions recognize and incorporate the interests of minorities without compromising national interests. Multicultural theorists argue about the scope of cultural rights, but they all concur that multiculturalism balances diversity rights with social cohesion and what Will Kymlicka calls societal culture (Kymlicka, 1998, p. 27). The existence of societal culture and citizenship rights and responsibilities serves as the overarching culture for interlinking subcultures. Multiculturalism emphasizes social integration, cultivating a sense of belonging while recognizing differences.

Multiculturalism is a two-sided coin. On the one side are the community subcultures largely limited to a few areas of social and political life, and on the other is what I call common ground that knits together diverse interests and identities into common values, laws, economy, politics, narratives, symbols of shared citizenship, and national/regional institutions (Qadeer, 2007, p. 91). The common ground also includes norms of everyday behaviour such as tolerance, trust, lining up to get into a bus, punctuality, etc. There is little scope for diversity in traffic laws, public health measures, or values such as gender equality, peace, and public order.

A city is itself a part of the common ground, both as a shared space and as a package of collective goods, which makes urban living possible.

These goods generally are for collective consumption. For example, rat control in my house depends on the cleanliness of my neighbours' homes and vice versa.

Collective goods and services come with associated behavioural imperatives embedded in civic norms and values. Urban living requires common rules, norms, and mutual trust. In all these respects, the cultural differences in cities are continually negotiated with the shared interests and collective welfare. The point is that a city is a community with shared norms, values, and behaviours, which constitute civic culture. These interlinkages and interdependencies make a city itself a part of the common ground, which is an integral part of multiculturalism.

Thus multicultural planning has two distinct tasks: 1) to recognize cultural rights and infuse pluralism in community structures and urban policies, 2) to reconstruct common ground by incorporating interests of minorities. A comprehensive view of multicultural planning has to balance the demands of recognizing cultural rights with the need to integrate ethnoracial groups in urban and societal institutions. The process of integration is a two-way adjustment process. Ethnic minorities have to imbibe foundational values, language(s), and narratives of the society, and the society-at-large has to incorporate their interests. This view of multicultural planning spells out an approach, which has been followed in planning practice.

The Strategy of Reasonable Accommodation and Harmonization of Multiple Interests

Ethnic enclaves, economies, and mosques have come about through a process of reconciling and adjusting the forms and purposes of these developments with planning policies and regulations. This process is an expression of the strategy of reasonable accommodation, which is a legislated concept in labour law, disabilities legislation, and human rights codes of the US and Canada (Abu-Laban & Abu-Laban, 2007, p. 30).

Recently the province of Quebec's official commission on accommodating cultural differences, popularly known as the Bouchard and Taylor commission, recommended adoption of the strategy of reasonable accommodation as the approach to multiculturalism. It defines reasonable accommodation as an adjustment of norms, practices, and policies on cultural grounds if those accommodations do not cause "unreasonable costs, a disruption of an organization's or establishment's operations, the infringements of other people's rights or the undermining of security and public order" (Bouchard & Taylor, 2008, p. 19). The same criteria are used in disabilities legislation, labour practices, and universities and schools. Reasonable accommodation is a two-way process. It allows a policy to be modified to accommodate differences of a multicultural population, but it requires that such adjustments do not compromise the purpose or function of a rule or objective and do not impose excessive costs, and particularly in urban planning that they are also not iniquitous to others.

Urban planning allows minor variances from zoning rules and standards, which is a practice of reasonable accommodation. Although reasonable accommodations usually occur on a case-by-case basis, they are cumulative, leading to the revision of policies and institutional restructuring. This is what has happened in practice. Multicultural planning has to formalize this strategy in the form of a principle stated firmly in comprehensive plans.

The criticism of reasonable accommodation comes both from the right and left perspectives. The former look upon it as a dilution of public law and unity, and the latter regard it as a compromise of minorities' rights (Grey, 2007). Yet it has proven to be an effective instrument for balancing community goals and minorities' rights. The accommodations allowed to the physically handicapped are a good illustration of managing differences and retaining institutional purposes. The same argument can be extended to the management of cultural differences.

In urban planning, accommodations lead to mutual learning and harmonization of differences by the revision of a policy or standard. Harmonization is a step towards making policies inclusive and thereby reconstructing common ground.

Ethnic malls are a case in point. In Toronto, the first Chinese mall, Dragon Centre (1983), required the modification of parking standards and the dispensation of the requirement that a supermarket anchor a neighbourhood mall. Soon after, the mall policy was revised to dispense with the anchor store requirement and to accept condominium ownership of commercial units. This harmonization of policies has evolved in parallel with the evolution of Chinese malls, as they have become larger, multifunctional, and diverse in the types of stores (Qadeer, 1998, p. 13; Wang & Zhong, 2013, p. 20). A similar harmonization of local planning policies regarding ethnic businesses has taken place in Monterey Park and the cities of San Gabriel Valley, Los Angeles County, and in other cities. The development of mosques in North America has followed a similar path of reasonable accommodation leading to harmonization of the policies for the development of places of worship as a generic category including, but not limited to, churches in major cities.

Policy Index of Multicultural Planning

The strategy of reasonable accommodation and the process of harmonizing policies have been organized in a 19-item index of policies covering the issues of cultural difference that usually arise in the course of urban planning. The index represents a workable model of multicultural planning. It includes both the process and substance of urban planning.

The index does not suggest a parallel system, another species, of urban planning but aims at making planning systematically responsive to the needs of ethnocultural communities. This policy index has been used to frame the questionnaire for the planning departments' survey reported above. It has been tested on 42 municipalities and was independently applied for assessing the multicultural responsiveness of Toronto-area municipalities (Newman, 2008). The survey of 42 municipalities revealed that on average, among large cities, Canadian cities adopted 15.4 policies and the US cities 12.6. For medium-size cities, the numbers were expectedly smaller, 11.6 and 6.5 respectively (Qadeer & Agrawal, 2011, figure 2, p. 144).

Reconstruction of Common Ground

The second component of multicultural planning is the incorporation of minorities' interests in a common ground, which is always evolving. There are many layers of this task, mostly matters of ideologies, economic and social institutions, laws and governance, human rights, values, norms, and citizenship. For example, the norms and attitudes towards race, ethnicity, and gender have evolved towards greater equality in both Canada and the US, so much so that some observers talk about post-racial and post-ethnic societies (Hollinger, 2000). The common ground changes by legislation, policies, social movements, and by evolving national values and attitudes. A current example of the change underway in national values is the increasing acceptance of gay rights, including the right to marry.

The common ground of cities' space, services, and organization with which urban planning is directly concerned also evolves with the national ethos. More specifically it incorporates new goals of racial and ethnic equality, environmentalism, sustainability, healthy living, equitable services, and fairness in land use, etc. Multicultural planning requires the

Table 3.1. Policy Index of Multicultural Planning

1. Providing minority-language facilities, translations, and interpretation in public consultations.
2. Including minority representatives in planning committees and task forces as well as diversifying staff.
3. Including ethnic/minority community organizations in the planning decision-making processes.
4. Routinely analysing ethnic and racial variables in planning analysis.
5. Studies of ethnic enclaves and neighbourhoods in transition.
6. Recognition of ethnic diversity as a planning goal in official/comprehensive plans.
7. Citywide policies for culture-specific institutions in plans, e.g., places of worship, ethnic seniors' homes, cultural institutions, funeral homes, fairs, etc.
8. Policies/design guidelines for sustaining ethnic neighbourhoods.
9. Policies/strategies for ethnic commercial areas, malls, and business improvement areas.
10. Incorporating culture/religion as an acceptable reason for site-specific accommodations/minor variances.
11. Accommodation of ethnic signage, street names, and symbols.
12. Policies for ethnic-specific service needs.
13. Policies for immigrants' special service needs.
14. Policies/projects for ethnic heritage preservation.
15. Guidelines for housing to suit diverse groups.
16. Development strategies taking account of intercultural needs.
17. Promoting and systemizing ethnic entrepreneurship for economic development.
18. Policies/strategies for ethnic art and cultural services.
19. Accommodating ethnic sports (e.g. cricket, bocce, etc.) in playfield design and programming.

Source: Qadeer, 2009, p. 13.

infusion of minorities' perspectives into these goals. Correspondingly it defines the limits of minorities' demands to conform to common ground.

The thrust of reconstructing the city as a common ground is to normalize diversity and integrate people in shared spaces, services, and institutions. It is done by building more public places in diverse cultural idioms, offering the choice of cricket fields and baseball diamonds, ethnic or aboriginal heritage places and museums in addition to the mainstream art galleries, provision of mixed housing, enforcement of labour rights in ethnic economies, and multicultural events, fairs, and music concerts. These are examples of how diversity can be turned into shared assets for everybody. It is a matter of what Ruth Fincher and Kurt Iveson call "convivial encounters among 'strangers', thereby facilitating opportunities for people to identify with one another in the course of shared activities and labours" (2008, p. 160). It promotes convergence of values and norms towards a shared vision of a good city. Fainstein's idea of a just city offers a pragmatic decision rule for reconstructing the city as common ground, "in which public investment and regulation would produce equitable outcomes rather than support those already well-off" (2010, p. 3). Another rule that can help reconstruct the urban common ground is to embed human rights into urban planning legislation, so that all policies and projects have to conform to these human rights.

Multicultural planning has not led to the development of a coherent vision of a multicultural city. There are motherhood statements of promoting diversity among the goals of comprehensive plans of cities like Toronto and New York, but almost no policies linking this goal to the proposed measures. The city as a common ground is evolving in the usual incremental process of cumulative accommodations. It needs to be anchored in a vision that serves as a guiding principle of balancing diversity with shared goals of equity, sustainability, liveability, and efficiency in cities.

Conclusion

Multicultural planning is an approach within the field of urban planning, emphasizing infusion of diverse forms and functions in urban institutions. New cultural and religious spaces, services, and identities are grafted into the existing urban structures. Pluralism of urban policies is the essence of multicultural planning. Yet multicultural planning is not a new type of planning but a new sensitivity and responsiveness.

The theory or thought of multicultural planning has lagged behind its practice or action, or at least it has not provided adequate guidance to resolve issues of cities of diversity. It is largely normative, process-oriented, expansive in its view of the scope of urban planning, and indifferent to the empirical facts about the state of diversity in North American cities. It is certainly high-minded in advocating for minorities' interests.

The multicultural planning practice operates within the institutional mandate of urban planning, which ranges from the development and management of space, services, local economy, and environmental resources to transportation and housing. Urban planning has absorbed the lessons of the 1960s and now is a participatory decision-making process involving citizens and various stakeholders. Thus what theorists advocate is already an established part of planning practice.

Urban planning practice has addressed the demands of ethnic communities incrementally as they arose. It has followed a process of mutual learning where planning agencies modify their policies to be sensitive and inclusive and the ethnocultural communities learn to appreciate the planning objectives. So wide ranging is this process that major North American cities are vibrant places with multicultural institutions, facilities, functions, symbols, and practices. This vibrancy is an indicator of the responsiveness of urban planning. This chapter provides concrete evidence of the emergence of ethnic enclaves, economies, places of worship, fairs and parades, and multicultural narratives. I do not mean to suggest that it has been without inter-ethnic confrontations, NIMBYism, charges of racism, or resistance of the established order. But the communicative and participatory modes of urban planning have provided a platform for the expression of such sentiments.

Multicultural planning has to be balanced with the shared goals of a city as a whole. Goals such as sustainability, healthy living, energy conservation, adaptation to climate change, and economic growth also apply to minorities. They affect their welfare. Reconciling minorities' interests with these overarching goals is a part of multicultural planning.

Multiculturalism is a two-sided coin. It has to promote the right of ethnocultural minorities to live by their values and norms and to preserve their heritage in their personal life as equal members of society as long as these do not conflict with the norms of the society. Correspondingly, multiculturalism has to contribute to the restructuring of the common ground (shared laws, institutions, and values of collective life) by infusing minorities' interests into it. A good city is also in the interest of minorities. This task has not been addressed. The integration of minorities and balancing their identities with the common ground are the challenges for research and theorizing.

The chapter has highlighted the strategy of reasonable accommodation and presented a policy index as a guide for multicultural planning. The index is also a statement of what multicultural planning is and could be.

Class differences apply as much to ethnoracial communities as to the mainstream. Multiculturalism is primarily concerned with cultural recognition and expression of identities in the public arena as equal citizens. The balance between the mainstream (historical majority) and minorities (immigrants) is shifting. Cities in North America are turning into majority-minority places. Minorities are gaining voting power and demographically becoming majorities. The social and political equation is changing. Multiculturalism is going to be the rule. The focus of urban development will shift from majority mainstream to a conglomeration of minority cultures. The intercultural relations and integration of diverse groups are the urban questions of the twenty-first century.

NOTES

1 For example, in one of the earlier collections of articles on this topic, Michael Burayidi's book, *Urban Planning in a Multicultural Society*, only two articles out of the 15 use multicultural planning explicitly in their titles.
2 The Canadian Institute of Planners' code of ethics requires members to "Practice in a manner that respects the diversity, needs, values and aspirations of the public and encourages the discussions of these matters." Retrieved from ontarioplanners.ca/knowledge-centre/professional-code-of-practice. Similar provisions are part of the American Institute of Certified Planners' code of professional ethics: "American Institute of Certified Planners' code of ethics requires that planners take a special responsibility to plan for the needs of disadvantaged groups and persons and to promote racial and economic integration." AICP code of personal and professional conduct. Retrieved August 10, 2015 from http://www.planning.org.
3 This was a purposive (non-random) survey of the city planning departments in the American and Canadian cities. A questionnaire based on the 19 items of the Policy Index was mailed out with the request that a senior staff member knowledgeable about the city's policies and practices dealing with issues of diversity be assigned to fill in the questionnaire. The response rate was 38.5 per cent. For details of this survey see Qadeer and Agrawal, 2011.
4 The checklist type of ethnicity is not adequate for programming purposes. There are internal divisions on interests and goals among broad categories of ethnic groups. For example, in the Toronto area, many Chinese opposed the building of more Chinese malls in Markham (Preston & Lo, 2009, p. 73). In ethnic media, many Muslims have expressed the opinion that Muslim communities need to build seniors' homes, schools, and other community institutions and not invest all their resources in building more mosques.

REFERENCES

Abu-Laban, Y., & Abu-Laban, B. (2007). Reasonable accommodation in a global village. *Policy Options, 26*(8), 30–33.

Agrawal, S. (2009). New ethnic places of worship and planning challenges. *Plan Canada: Special Edition: Welcoming Communities: Planning for Diverse Populations*, 64–67.

Anisef, P., & Lanphier, M. (Eds.). (2003). *The world in a city*. Toronto: University of Toronto Press.

Berger, J. (2007). *The world in a city*. New York: Ballantine Books.

Bouchard, G., & Taylor, C. (2008). *Building the future: A time for reconciliation*. Quebec: Commission de Consultation sur les Pratiques d'Accommodement Reliés aux Différences Culturelles.

Burayidi, M. (2000). Urban planning as a multicultural canon. In M. Burayidi (Ed.), *Urban planning in a multicultural society* (pp. 3–14). Westport: Praeger.

Canadian Multiculturalism Act. (1988). C. 31, assented to July 21, 1988. http://laws-lois.justice.gc.ca/eng/acts/C-18.7/page-1.html

City of Los Angeles. Los Angeles Comprehensive Plan. Retrieved October 25, 2011 from http://cityplan.lacity.org/cwd/gnlplan/tranlot/TE/T1introduction.htm#citizen.

Davidoff, P. (1965). Advocacy and pluralism in planning. *Journal of the American Institute of Planners*, *31*(4), 331–338.

Davis, M. (2000). *Magical urbanism*. London: Verso.

Fainstein, S. (2010). *The just city*. Ithaca: Cornell University Press.

Fincher, R., & Iveson, K. (2008). *Planning and diversity in the city*. Basingstoke: Palgrave Macmillan.

Fleras, A. (2009). *The politics of multiculturalism*. New York: Palgrave. http://dx.doi.org/10.1057/9780230100121

Florida, R. (2005). *The flight of the creative class*. New York: Collins.

Foner, N. (2013). Introduction: Immigrants in New York City in the new millennium. In N. Foner (Ed.), *One out of three* (pp. 1–34). New York: Columbia University Press.

Frey, W. (2011). Melting pot cities and suburbs: Racial and ethnic change in metropolitan America in the 2000s. *Metropolitan policy program at Brookings*. http://www.brookings.edu?papers/2011/0504_census_frey.Apex.1.

Grey, J. (2007). The paradox of reasonable accommodation. *Policy Options*, *28*(3), 34–36.

Harwood, S. (2005). Struggling to embrace difference in land-use decision-making in multicultural communities. *Planning Practice and Research*, *20*(4), 355–371. http://dx.doi.org/10.1080/02697450600766746

Healey, P. (1997). *Collaborative planning*. Vancouver: University of British Columbia Press.

Healey, P. (2003). Editorial. *Planning Theory & Practice*, *4*(3), 245–247. http://dx.doi.org/10.1080/1464935032000118616

Hollinger, D. (2000). *Postethnic America*. New York: Basic Books.

Innes, J. (1998). Information in communicative planning. *Journal of the American Planning Association*, *64*(1), 52–63. http://dx.doi.org/10.1080/01944369808975956

Kymlicka, W. (1998). *Finding our way*. Toronto: Oxford University Press.

Kymlicka, W. (2007). *Multicultural Odysseys*. New York: Oxford University Press.
Lobo, A., & Salvo, J. (2013). A portrait of New York's immigrant mélange. In N. Foner (Ed.), *One out of three* (pp. 35–64). New York: Columbia University Press.
Marcuse, P. (2005). Enclaves yes, ghettos no: Segregation and the state. In David Varady (Ed.), *Desegregating the city* (pp. 15–30). Albany: State University of New York Press.
Milroy, B., & Wallace, M. (2002). *Ethnoracial diversity and planning practices in the Greater Toronto Area*. Toronto: Centre of Excellence for Research in Immigration and Settlement, University of Toronto.
Newman, G. (2008). Responding to ethno-cultural diversity through planning policy and practice: an analysis of five select municipalities in the Toronto CMA. In *A Master's Report to the School of Urban and Regional Planning*. Kingston, ON: Queen's University.
Pestieau, K., & Wallace, M. (2003). Challenges and opportunities for planning in the ethno-culturally diverse city: A collection of papers – introduction. *Planning Theory & Practice*, *4*(3), 233–245.
Piketty, T. (2014). *Capital in the twenty-first century*. Cambridge, MA: Harvard University Press.
Preston, V., & Lo, L. (2009). Ethnic enclaves in multicultural cities: New retailing patterns and new planning dilemmas. *Plan Canada: Special Edition: Welcoming Communities: Planning for Diverse Populations*, 72–74.
Qadeer, M. (1997). Pluralistic planning for multicultural cities: The Canadian practice. *Journal of the American Planning Association*, *63*(4), 481–494. http://dx.doi.org/10.1080/01944369708975941
Qadeer, M. (1998). Ethnic malls and plazas: Chinese commercial developments in Scarborough, Ontario. *CERIS Working Paper*. Toronto: Joint Centre of Excellence for Research on Immigration and Settlement, University of Toronto.
Qadeer, M. (2007). The Charter and multiculturalism. *Policy Options*, *28*(2), 89–93.
Qadeer, M. (2009). What is this thing called multicultural planning? *Plan Canada: Special Edition: Welcoming Communities: Planning for Diverse Populations*, 10–13.
Qadeer, M., Agrawal, S., & Lovell, A. (2010). Evolution of ethnic enclaves in the Toronto metropolitan area, 2001–2006. *Journal of International Migration and Integration*, *11*(3), 315–339. http://dx.doi.org/10.1007/s12134-010-0142-8
Qadeer, M., & Agrawal, S. (2011). The practice of multicultural planning in American and Canadian cities. *Canadian Journal of Urban Research*, *20*(1), Supplement, 132–156.

Qadeer, M., & Chaudhry, M. (2000). The planning system and the development of mosques in the Greater Toronto Area. *Plan Canada*, *40*(2), 17–21.

Rath, J. (2011). Debating multiculturalism: Europe's reactions in context. *Harvard International Review*, *6*(January). http://www.janrath.com/wp-content/uploads/@Rath_Debating_Multiculturalism_Harvard_Int_Rev_2011.pdf

Reeves, D. (2005). *Planning for diversity*. London: Routledge.

Rex, J. (1996). *Ethnic minorities in the modern nation state*. Houndmills: Macmillan Press. http://dx.doi.org/10.1057/9780230375604

Sandercock, L. (2003). *Cosmopolis II: Mongrel cities*. London: Continuum.

Sandercock, L. (2004a). Towards a planning imagination for the 21st century. *Journal of the American Planning Association*, *70*(2), 133–141. http://dx.doi.org/10.1080/01944360408976368

Sandercock, L. (2004b). Sustaining Canada's multicultural cities. *Our Diverse Cities*, no. 1 (Spring), 153–157.

Sandercock, L. (2011). Commentary: Where do theories come from? *Canadian Journal of Urban Research. Canadian Planning and Policy*, special issue, 157–159.

Viswanathan, L. (2009). Postcolonial planning and ethno-racial diversity in Toronto: Locating equity in contemporary planning context. *Canadian Journal of Urban Research*, *18*(1), 162–182.

Viswanathan, L. (2010). Integrated, equitable, and transformative: A hopeful future for planning. *Plan Canada*, *50*(3), 33–35.

Vitiello, D. (2009). The migrant metropolis and American planning. *Journal of the American Planning Association*, *75*(2), 245–255. http://dx.doi.org/10.1080/01944360902724496

Waldinger, R., & Bozorgmehr, M. (1996). *The making of a multicultural metropolis*. In R. Waldinger and M. Bozorgmehr (Eds.), *Ethnic Los Angeles* (pp. 3–38). New York: Russell Sage Foundation.

Wang, S., & Zhong, J. (2013). Delineating ethnoburbs in Metropolitan Toronto. *CERIS Working Paper No.100*. Toronto: CERIS–The Ontario Metropolis Centre.

Wood, P., and Landry, C. (2008). *The intercultural city*. London: Earthscan.

Zhou, M., Chin, M., & Kim, R. (2013). The transformation of Chinese American communities; New York vs. Los Angeles. In D. Halle & A. Beveridge (Eds.), *New York and Los Angeles* (pp. 358–382). New York: Oxford University Press.

Zhuang, Z. (2013). Rethinking multicultural planning: An empirical study of ethnic retailing. *Canadian Journal of Urban Research*, *22*(2), 90–116.

4 The Pragmatic Politics of Multicultural Democracy

DAVID LAWS AND JOHN FORESTER

This chapter explores a shift in the practical meaning of "multiculturalism" as an element of public policy. We focus on the case of the Netherlands, specifically on one case in the City of Amsterdam where political events – including the first political murders that had occurred there since the sixteenth century – led many residents to reconsider long-standing commitments to a multicultural society. Rethinking the role of the state in society also prompted a re-evaluation of the role of citizens.

These developments raised important questions in everyday settings. They arose initially as apparently small, if practical, questions about public order, spatial development, the administration of schools, and other domains of "street-level bureaucracy" (Lipsky, 2010; Laws & Forester, 2015). In such street-level contexts, diverse citizens could engage these issues in a pragmatic approach that shifts attention from clashes between doctrinal belief systems to practical questions: where, for example, to put the stop signs, or what to do about the latest incident in the neighbourhood (Forester, 2009, 2013). Before examining how one such incident unfolded on the streets of Amsterdam, we provide a bit of background.

I. Background: Multicultural Fears and Tensions Erupt in a City of Tolerance

In the late twentieth century in Holland, several settled issues relating to multiculturalism became quite unsettled. What would relationships be like between urban residents with an ethnic Dutch background and residents descended from ethnic minorities? The latter included descendants of Moroccan and Turkish guest workers who came to the Netherlands beginning in the 1960s and stayed. They form a relatively small minority in the country overall, but a prominent presence in the larger cities in the west.[1] Surinamese and Antilleans who emigrated from former colonies make up another substantial presence in Dutch cities.

Tensions between these groups suggested to some analysts that the policies that had let Protestants and Catholics live together were not keeping pace with changing circumstances (cf. Gale & Thomas, 2015). An essay entitled "The Multicultural Drama" in the *NRC Handelsblad*, one of the Netherlands' leading newspapers, brought the debate into sharp focus. The author, Professor Paul Scheffer, claimed that "the culture of tolerance is coming up against its limit" and argued that the time had come to "honestly tackle" what he called the country's "lazy multiculturalism" (Scheffer, 2000).

Scheffer differentiated between tolerance and pragmatism. The latter had characterized the approach that had historically worked in the Netherlands, he argued, but the old forms of pragmatism were no longer working.

The tensions prompting these discussions were dramatized in two episodes, both involving murders. The central character in the first was Pim Fortuyn, who rose quickly to political prominence just after the turn of the twenty-first century. Fortuyn was controversial – openly gay, a Marxist sociologist and academic turned politician – known for his direct and controversial remarks on Islam and on Dutch policies and for his disregard for multiculturalism. Fortuyn inspired a populist movement that peaked in Rotterdam, where *Leefbaar Rotterdam* (Liveable Rotterdam) won over a third of the city's district council elections in March 2002, making it the largest party in the council and pushing the Labour party out of power for the first time since the Second World War. Fortuyn also inspired a national *party Lijst Pim Fortuyn* (List Pim Fortuyn or LPF) that won 26 seats (17 per cent) in the lower house in the 2002 elections. LPF became part of the ruling coalition with the Christian Democrats and others, but the cabinet collapsed within the year, and in the elections that followed LPF won only 8 seats (which declined to 0 in the next national election).

Fortuyn was assassinated just before these national elections by Volkert van der Graaf, an animal rights activist and not, as many had feared when news of the assassination first became public, a Muslim. At his trial, Van der Graaf "said his goal was to stop Mr. Fortuyn exploiting Muslims as 'scapegoats' and targeting 'the weak parts of society to score points' to try to gain political power. He said: 'I confess to the shooting. He was an ever-growing danger who would affect many people in society. I saw it as a danger. I hoped that I could solve it myself.'"[2]

A little over two years later, on November 2, 2004, Theo van Gogh, a filmmaker, columnist, and author, was killed as he bicycled to work in the morning in the east of Amsterdam. Van Gogh was shot and then stabbed by Mohammed Bouyeri, a Dutch-Moroccan Muslim. Bouyeri's statements and the details of the murder linked the killing to Van Gogh's role in the production of a film called *Submission*, which he made with Ayaan Hirsi Ali, a Somali refugee who was then a member of Parliament. The film became controversial for its criticism of the treatment of women in Islam (on conflicts of marginalized identities, cf. Doan's discussion in chapter 6). Both Van Gogh and Hirsi Ali had received death threats after Dutch TV broadcast the film.

Street interviews conducted by the BBC captured the atmosphere in the Netherlands following Van Gogh's murder.[3] Eric Hulscher, 39, a taxi driver, remarked:

> The atmosphere is grim after the murder. You feel strange, especially having seen that attacks [on mosques and an Islamic school] have now taken place. It will only get worse, all the talk about integration is a lot of nonsense. There's been talk of "us" and "them" and that's absolutely right.

Some believed Van Gogh's murder was part of a larger plot. Citizens with a Moroccan or Turkish background described feeling at risk as a threatening identity was pressed upon them. In another BBC street interview, Radouan Veldmeijer, 21, a cook, captured the ambiguities and feelings of vulnerability in this new situation:

> I'm half Moroccan and the atmosphere now in the Netherlands is terrible. I was adopted, and since I was seven months old I have been brought up by Dutch parents, so for me it's doubly difficult. I'm not a Muslim but since the murder I've been sworn at in the street by skinhead types.

Van Gogh's murder made tensions around multicultural policies and practices even more explicit. Rita Verdonk, the minister of integration, was widely quoted: "for too long we have had a multicultural society [where] everyone would simply find each other. We were too naïve in thinking people would exist in society together." Jozias van Aartsen of the VVD (Liberal party) took Verdonk's dramatic framing further: "Jihad has come to the Netherlands and a small group of jihad terrorists is attacking the principles of our country. They don't want to change our society, they want to destroy it."

Representatives of the Muslim community were quick to denounce the murder. Ahmed Aboutaleb, then an alderman in Amsterdam before becoming mayor of Rotterdam, took a strong stand that Moroccans needed to accept the "core" of Dutch society:

> For people who do not wish to share the fundamental core values of our society [there] is no room in a society like the Dutch one … Everyone who does not share these values is best advised to make his or her conclusions and leave. It just can't be that someone demands all of us to respect his viewpoints and at the same time refuses to respect the viewpoints of others.

Gerd Wilders, who would later lead his own populist party, used Van Gogh's murder to express in harsh terms the misgivings many seemed to share:

> It's not a coincidence that the unfortunate slaughter of Mr. van Gogh happened in the streets of Amsterdam and not anywhere else. For too long we've been tolerant of the intolerant. We've had a policy for years that everything should be tolerated, that anything is possible … Only three years ago journalists on public television recorded Imams in The Netherlands saying things on the record about how women could be beaten, homosexuals should be killed and the friends of democracy are the sons of Satan. Our secret service has already known for two years that the recruitment for jihad in mosques and prisons were no longer incidents but a structural phenomenon. (Wilders in Hurewitz, 2005)

These events provided the backdrop for the following case. Tensions between groups were high and still in flux. People were asking, "What is going on here?" and local politics, the sphere of action to which we turn, would be central to answering that question in practical and policy terms.

II. A Local Drama of Multicultural Relationships: Football and Fighting, Violence and Leadership

We turn now to an extended story about conflict and local politics in the borough West in the City of Amsterdam. West has roughly 130,000 residents from over 130 distinct national and ethnic backgrounds. The borough is large, dense, and diverse, containing over 25 distinct neighbourhoods. These include the Kolenkit, identified as the worst neighbourhood in the Netherlands just before our story begins, and more affluent neighbourhoods in Old West. Young professionals and other new groups were moving into neighbourhoods like Westerpark, where our drama unfolds. Long-standing social housing policies administered through "housing corporations" moderated these trends and kept the neighbourhood diverse.

As our story begins, a football tournament has just concluded in Westerpark. The tournament had been set up by residents and the local government to try to build relationships across group boundaries. This effort reflected typical urban policy initiatives in Amsterdam and in the Netherlands more generally. The idea that social relationships could be influenced through policy and physical design has a long history. The residents' engagement in part acknowledged the limits of the state's ability to direct this process.

Martien Kuitenbrouwer, the borough president, described the events that set the stage for the drama:

> The football tournament is named after this little area, an ecological neighborhood, a socially close-knit community and an area where people are very proud to live and like to do things together. This football tournament was set up because some people started to notice that the mixing between Moroccans, Antillean people, and Dutch people wasn't as straightforward as people thought or wanted it to be. So they thought: "We're going to play together, it's going to be fun."[4]
>
> The football tournament[s had been] a big success. [They had] gotten bigger and bigger each year and more teams participated. Four years ago, the tournament was organized by one of the autochtoon[5] Dutch inhabitants and the Moroccan boys were able to play but were not really part of the organization. What happened was that some of them kept on disturbing the game. These were Moroccan and Antillean guys of different groups, they were not together.
>
> My view on the situation is that they didn't feel part of it, although they were invited to play. Because they didn't feel part, they interrupted the game and when the teams were playing they kept cycling over the field and such – which was really annoying. Then one of the referees and the wife of the referee started to tell these boys off and said: "You have to go, this is annoying."
>
> These boys were provocative [and] ... they kept saying: "You don't want us to be part of this, and we are not allowed to play with you."

> Then the wife of the referee took one of the guys and took him by the arm and pulled him from the field. This was a boy of 15 or 16 years old. He left, there were also smaller guys there, and they got their mothers, and then they came back. And then there was this whole group of Antillean mothers who started fighting. The referee got badly hurt and had to flee.[6]

The fight disrupted the very relationships that the tournament had tried to restore. The local government had supported the tournament to bring the community together across ethnic and cultural lines. The tournament's initial success had confirmed this approach. But now that policy frame was challenged by the escalation unfolding on the pitch. This triggered uncertainty among local officials: How should they respond?

> What happened then was that the [local officials] didn't really know how to react; it took a while. They were a bit hesitant. They tried to intervene and tried to set up a meeting in which all parties ... were put together in a room – [to] confront [the] victims and all [of the] people involved in [the] conflict together.
>
> ... [T]he Stadsdeel [officials] just didn't want this to happen again. But it took quite a while (6 weeks) before [we] had actually organized that, and everybody said: "It's too late; I don't want to be a part of this." One of the boys, who was one of the main actors in the conflict, was sent to a youth detention centre because he also had a lot of other things on his file. So it was not quite satisfactory how that was dealt with. It faded out and the neighbour who was hit moved away partly because of this conflict.

This was no isolated incident. There were other episodes of low-level violence. On Saint Martin's day, as children went from house to house to sing for candy, several children were attacked and robbed of their sweets. Some believed that the same group of young men was responsible and police were called in to patrol. Kuitenbrouwer recounted:

> [It was] not a very nice atmosphere. It wasn't a big thing, but there were still underlying tensions, I think, because it was unresolved. You could still see that as soon as Antillean families and these boys were back in the area there was animosity between them and other people in the area ... As a resident you could see the tension between people living in one area.
>
> What worried me especially was that in the incidents the older guys were doing the fighting, but 3 or 4 years ago their little brothers and sisters

[had] started to throw stones at other kids, because they said: "Oh, you're part of 'this' group." It wasn't very big, but you could feel the tension. On hot summer days you could hear shouting. It never resulted in any real conflict, but it was like: "Stay out of my area!" – that kind of behaviour. I felt like people were a bit scared of each other.

The first episode led to changes in the way the football tournament was organized. "The neighbours talked about it, and they said that they didn't want this to happen again, and so they made some of the Moroccan boys part of the organization." This seemed to work until the close of the tournament the following year:

Then last summer at the end of the tournament there was a football game between The Netherlands and Russia in the European Championship, and the Netherlands were losing. Everyone was watching television together on the big square. The Dutch lost and the Moroccans yelled, "Hurrah," and then some of the Dutch fathers – the adults – became very angry, and then there was a fight.

In a … span of five or six minutes there was a huge fight and a lot of aggression, real fighting and hitting with bottles …

You can imagine: it was a hot night, [and] everyone had [had] a couple of beers too many, and then you're not happy because [your] team loses. There are a couple of stories of what happened, but apparently they started touching each other and there was some [pushing] and in a few minutes there was an enormous fight.

This fight reopened questions about fractured relations in the neighbourhood. What exactly was going on? Kuitenbrouwer continues:

[In] the days after that, you could really feel [that] the whole area was full of tension, and we said: "Oh, my god, what kind of conflict do we have here?" … Nobody really understood what was happening. I actually lived there. I wasn't there that evening, but when I came home I knew immediately, and I could feel the tension of what had happened here.

Doubt, uncertainty, and fear led to pressure to act:

The next day you could see the tension building up. I got e-mails and phone calls. First [of all] it was a fight, and then it [became] a big racial issue.

> The police [collected] a lot of testimonials from people who wanted to make statements. The police got information from [the] two sides blaming each other.

Public officials were divided on whether and how to act, and what, if anything, to do:

> The [neighbourhood director] said: "This doesn't feel right, you have to do something, we don't really know what to do." Other people working at the [borough government] said: "Maybe you should leave it till after the summer." This built up over a few days, and I didn't know exactly what I was supposed to do … My own hunch was to do something immediately, although my colleagues told me to let it [settle] down and discuss it after the summer.
>
> But I was going to be away for summer [holidays,] and I said that I couldn't do that. I thought I needed to do something, thinking back on the previous conflict. I could feel the tension. Then I realized how important it was to deal with it immediately. [I thought:] You cannot resolve this, but you have to address it, otherwise there is going to be this [on-going] feeling again of something that's not resolved.

The residents put pressure on the public officials to define the situation and restore order. To most, this meant enforcing established norms and regulations about what can and cannot be done in the neighbourhood. Someone had a broken nose. Another had a broken rib. The perpetrators should be identified and punished. People voluntarily went to the police to offer testimony on the situation and, implicitly it seems, to define the situation and push for action. Kuitenbrouwer felt this pressure.

> I started to receive lots of e-mails with testimony from both sides. Both sides said that they would go to the police because the other one should be blamed …
>
> I was … talking to the residents … They wanted this guy to be punished and sent away. I could see that they did not understand. The conflict was very bad and they didn't understand why this guy was so angry. This little working guy was so incredibly angry – so angry! …
>
> Each day there were new testimonials reported. The story kept evolving: "I know somebody else who has been hurt, and this guy also hit somebody else, and this guy did this and [so on]."

> You could see that the whole conflict was becoming quite big. The police were brought in … [and] were quite shocked by the amount of aggression. They weren't quite sure what to do, because they kept getting testimony. People went to the police to testify.

The strategy of encouraging community members to air their grievances with the police was incomplete and, so, unstable. Sooner or later irate citizens would press for action and doing nothing would look like evasion. Action required a different way of looking at the conflict – one that would offer a different view of what was needed, what was possible, and what should be done.

> [Seeing that collecting testimonies was not enough] changes the way you perceive a conflict. The conflict is not just about resolving [who was right or wrong]. You can see the delinquent can't be excused, but you have to understand how fundamental the feeling of being an outsider is in order to resolve the conflict.
>
> For example, this meeting that was arranged afterwards [could] only help if the victims and the people who were involved in the conflict all agree that they want to move forward. If that doesn't happen, it's a very bad idea to put them together, because then the tension [would] only get bigger. This I came to understand later, because first you think: "Why can't these guys talk about it?"
>
> [But] it's much more fundamental than that …What I did was start to see if people were prepared to come together – not to resolve anything, but just to look each other in the eye before the summer holidays would start. Because if we did not do that, I wasn't sure about what was going to happen. Otherwise, even though these people were going to be away for the summer – not everyone was going away – you were going away with [only] your own side of the story.

This meant bringing the main actors in the conflict together. There were risks: Were those parties ready to come together? More practically, could they accept the goal of meeting to try to talk with each other about what had happened? Kuitenbrouwer's first step was to test the waters:

> I have someone working here who's very good with handling these kinds of situations, so I asked him to go and see this Moroccan guy, making sure he was there to listen and not to judge, which was very difficult at the beginning, because the Moroccan boy didn't want to talk.

He said: "I didn't do anything." First, he denied everything: "I didn't hurt anybody," even though there were 50 people present seeing that. He kept saying: "I didn't do anything, I wasn't even part of this conflict."

Then my guy had to say: "I'm not going to judge you, but I have to understand what happened." Eventually he came and went to talk to him a few times and then we got him and his friend to talk. He didn't exactly say much.

The other side of the conflict offered a different challenge:

The Dutch guys kept on insisting on punishment ... They said: "He has to be punished! We want to make sure he's punished and if he's not punished, we don't want to talk to him."

The police reassured them that the evidence they had given would remain on file. They were simply delaying taking action on the file. The other residents were also pushing for action. They wanted this guy to be punished and sent away.

Kuitenbrouwer tried to be clear with the parties about what the goal of meeting was, how her own understanding was limited, what participation would mean for each of the participants, and what their realistic alternatives were:

"OK, we'll come together for just half an hour. I want to know that we can, at least, just go on [summer] holidays and find that we can at least agree on the fact that we can resolve this conflict. We don't need to resolve it, but I'm trying to get you to agree, so that we can at least leave it for now, and come back to it in a few weeks' time – so that there would not be more tension over the summer."

We didn't even talk about the whole situation. I just said to them: "I don't want to leave this situation here as it is, and I'm not sure how we're going to find a way out of it, but I don't want to leave it as it is. I don't have a solution yet, I don't even know what happened exactly, but are you willing to come together after the summer and are you willing to talk about the conflict?"

This is what I told them that the purpose of the meeting was: that we couldn't leave the situation as it was. I told the Moroccan boy that he should realize that if he wasn't prepared to at least meet the other side, it would be very bad for his police file, because he'd probably get a police file, which would be huge. I told the other guys: "If you're not prepared to meet the Moroccan boy, I'm afraid that he's going to be so frustrated, that I don't know what's going to happen."

The unfolding events stressed what was at stake: if left alone, the stories about the fight would continue to circulate and would grow as they made the rounds. Early indications suggested such patterns were already beginning to play out, Kuitenbrouwer explained:

> There was a couple [in the neighbourhood] who I knew quite well. The woman of this couple came over one day [and was] quite hysterical. She made the conflict bigger and bigger, and she started to alter the stories. She started to say things like: "Well, you know, I've always known they don't like us, these Moroccans. I've got the feeling they're watching us and that if we're going on holidays, they're going to rob our houses."
>
> She actually called me on a Sunday morning, completely panicking, saying that she had gotten a text message. It said: "I'm going to meet such and such there and there on this address." And the name was quite similar to the name of the Moroccan father. She had gotten completely paranoid and thought that they were conspiring and that this Moroccan man was in some kind of Islamic organization and that they were going to "get us."
>
> She was building story on story, and then it got way out of hand! I went to talk to her, and I said, "It's really not true what you're saying."

Kuitenbrouwer needed to improvise here, but she was in a delicate and vulnerable position. She was working at the margins of the map: there was no clear conventional path forward, and she held the responsibility for what developed:

> This was one of the situations in which I felt very alone. You have to do it all by intuition. You're walking on eggshells. You don't know anything. You only know you have to do something, but you don't know what. You just have to be there and see how it's going to develop at that time. It could have ended up in a huge fight again.

Moving ahead meant thinking about who else could and should be involved. The father of a friend of the young Moroccan man was a leader in the community, and he agreed to attend as support for the young man. Another prominent member of the Moroccan community in Amsterdam – a "self-made youth worker" – was invited as well.

He was, in fact, invited by the "Dutch" man whose nose had been broken and who knew him and said: "Maybe *he* can relate better to these guys so that they feel more comfortable."

Someone from the borough government staff – who "is good at making people really listen to people, in making sure that they've been heard and that everybody is feeling taken seriously" – was invited. One local police officer who knew the parties would attend to keep the police presence small; the broader force would be "on call." A mediator was invited, not to run this initial meeting, but to show that someone was available for follow-up.

The pressure to define the situation did not vanish in the less formal setting. Preserving doubt and resisting the rush to interpretation required an active hand, Kuitenbrouwer explained:

> The mediators we invited felt that we should put forward a plan and that we should present that. They were quite forward in organizing this thing, and I had to sort of lower everybody's expectations because they wanted to say: "We're going to move on and we're going to do this and this."
>
> [But] I said: "Well, I'm not sure if they're willing to. We have to ask them if they're willing to go ahead with it. If you make it too big now they're going to feel like we're pressuring them. Both sides [might feel that way]."

Sustaining doubt amidst the pressure for decisive government action was difficult. It was hard to remain open about what had happened and what might happen. The period between organizing the meeting and the meeting itself was neither easy nor comfortable:

> I felt alone between the Saturday and that Tuesday because I was worried about whether I was doing the right thing. Was this the right way forward? It was very risky to put everyone together. There were people who said: "Oh no, you should just leave it over the summer."
>
> I had this very strong feeling that this was not possible, but maybe they were right! ... I could feel the tension building up; I could see that stories were getting bigger on each side. You could see that if you wouldn't intervene now, the tension was going to be bigger and bigger and after the summer we'd have a whole racial conflict here in this area in which the whole Moroccan and the whole Dutch community were going to do God knows what! That's what I was afraid of.

There were real risks to consider. The plan could make the situation worse – and the government would be culpable.

> I was nervous, because I thought: "Am I going to do the right thing? Is there not going to be so much aggression and tension and shouting that it will make things worse?"
>
> That was my main worry. Because I went there knowing that those people broke each other's noses and that they're going to see each other again – and it [had been] a heavy incident and there [had been] a lot of fighting. So I really thought that it could also end up bad. Maybe they were going to end up fighting, screaming, and maybe it would get worse.

Tuesday evening came. It was a hot evening. The group met in "a sort of community house" in the middle of the neighbourhood where the fight had happened. Kuitenbrouwer recounted:

> There was a round table. We didn't even have the time to do anything else. The main perpetrator and his friend and his father were sitting on one side and there were these Dutch guys on the other side and we were in the middle. There's no other way we could have done that – also because we felt like we had to keep them apart, because we were not sure if they were going to hit each other. My God, I've never been in a room with so much tension – that you could feel you could cut the air with a knife. It was terrible.
>
> I started the evening, because I was going to chair [the meeting]. I said that I was shocked and worried about what had happened and that I didn't quite know what had happened exactly, and that the only thing I could say was that this kind of conflict is not the way we are used to living together. I said, "I feel worried and I look forward to a way out of this situation." …
>
> I asked them if they could agree to listen briefly to each other in order to let everyone tell their part of the story. They would get one minute and I asked them if they could agree not to interrupt each other and not [to] blame each other for the moment. They did. It was difficult, but they did.
>
> I had to intervene quite harshly a couple of times, because of course they started to react: "I didn't do that. I didn't say that!" And: "You did this!" I said, "We're not going to blame each other, because we don't exactly know what happened. It's not about who's right and who's wrong."
>
> For example, first the perpetrator said: "I didn't do anything!" Then the Dutch man said: "What do you mean you didn't do anything? Have you

seen my nose?" I had to intervene when the Dutch man said: "What you did was (strong language deleted)."

The conversation moved back and forth.

The Moroccan father also intervened and … he also chose the side of the Moroccan boys, and I think that was very important. Even though he tried to build a bridge he would say: "My boys felt very threatened, and they felt as if they weren't part of this football tournament, and they felt you made the first step in the conflict."

Then the Dutch man said: "That's not true! You hit me first!"

This sort of thing happened, and they started talking, and I had to intervene all the time by saying things like: "You have to [stop] talking and just listen to him. We're not going to make any judgments today."

As the stories came from both sides and were linked, an unexpected turning point emerged. As ambiguity was acknowledged, certainty broke down:

Then we found out … that the friend of the perpetrator, who was also part of the conflict, had [grabbed] one of the other guys quite strongly during the fight. He had [raised] his hand, and he had a broken bottle in his hand. He was being very dangerous. The reason was that there was somebody else coming, and so he was actually protecting him. This was one of those things that got revealed during that evening. The Dutch man said: "I thought you were going to kill me," and the Moroccan boy said: "I was only trying to protect you!" …

What happened was that for the first time you could see that the three Dutch men got something of a hunch about the other side of the story, just a hunch. They could see that these guys felt threatened. They were very aggressive, they were hurt, they threw bottles, they kicked people's noses. The Dutch men could understand that they felt threatened. The day before, they couldn't understand that.

This led to a next turning point. The Dutch man – who had been pressing for an apology – changed course and apologized himself.

That was very difficult, but he did it and I thought that was very good. The Dutch man apologized to the Moroccan boy who had [grabbed him]. I thought that took a lot of courage, because first it was the other way

> around. First the Dutch men said: "We want an apology from these guys," and then *they* actually apologized.
>
> The response of the Moroccan boys wasn't immediate. The whole evening they were sitting and looking down. When the Dutch man apologized, there wasn't an immediate response ... There was clear physical tension. So they were sitting in a completely tense pose, not making eye contact ... It was an outright apology. And then you could feel the tension disappear, although the Moroccan boys didn't really respond to it, in language, but you could see it really helped. You could see that.

The Moroccan group left almost immediately thereafter. The "self-made youth worker" went after them and talked to them. When he returned, Kuitenbrouwer recounts,

> He explained to the Dutch men how these guys feel. He said: "You have to understand that these guys feel complete outsiders. They look at you and see that you have nice footballs, you have organized these games, you are nice and so on. They feel like they are not part of your party. Of course they shouldn't hit, they shouldn't do that, but that's their fundamental feeling; that they do not feel part of your party."
>
> And then the Dutch men said: "How can they feel that, because they're always invited and we're nice to them! We always have to apologize and that's ridiculous!"
>
> And [the youth worker] said: "You're right in your perspective, but they feel they don't have a way of expressing themselves in the way you do. So they feel left out, and ... when they feel being pushed in a corner, they start to hit. You started to talk, they started to hit."
>
> Then you could see the Dutch men beginning to understand a little bit. They said: "OK. " [The youth worker] said: "It's not acceptable, but you have to understand how it works for boys like these."
>
> That was very good, because [he] could relate to these boys. When the Dutch men asked: "Why doesn't he apologize?" he could explain a little bit about these apologies. He said: "In the Dutch culture apologies are a very normal way of dealing with each other, but in the Moroccan culture, you never apologize. The closest thing to apologizing is ignoring the conflict. It's all about honour, and that makes it very difficult to get an apology."
>
> Then he explained and he said: "... You should not expect an apology, even though that may be very hard for you because that's the way *you* deal with conflicts. Do not expect an apology. The most you

> can expect from this boy is that he agrees to do this conflict resolution and that he probably won't bother you anymore. That's what you can expect."
>
> That was very helpful, because they could lower their expectations.
>
> ... [The youth worker] also told the Moroccan boys, when he was outside, that the fact that this Dutch man apologized was very meaningful. He tried to explain to the others the fact that the Dutch man really meant that, and that the fact that he had apologized was an enormous step. Because the Moroccan guy had become very angry and had said: "You don't mean that! You don't mean that!" And the Dutch man said: "I do mean that." And he had to repeat that several times, and then the Moroccan boy said, "OK." So that was very good.

The group met again after the summer holiday. Kuitenbrouwer did not attend the meeting – "I thought, 'Now it's time for me to step back'" – but she was available. Each side told their version of the story again. There were apologies. The Moroccan "perpetrator" did not actually apologize, but said that he regretted what had happened. "The participants 'signed a form in which they said they were also proud of being able to solve this conflict' and they committed to 'organize a new sort of party, or a get together.'" The latter was proposed by the Moroccan father "in order to sort of cleanse the area, to do a cleansing ritual."

The event was organized around a Moroccan holiday and was held on the same site as the fight. The turnout was not what the organizers hoped for. Kuitenbrouwer said, "From the Dutch side there were a lot of people. From the Moroccan side there weren't many, just a few. That was a bit of a disappointment for everybody." Still, the event was planned, held on schedule, and came off without incident. The reactions of those involved were more dramatic.

> The woman who [had been] hysterical said: "I've learnt so much. I really never understood how it must be for a guy like this to be part of a conflict like that. I really understand how it is to live here and how it is to be part of this community, and I understand the way they look at us." She felt it was terrible, but she said she'd really become a stronger person ... She said that she felt that there were two things that she understood fundamentally better. First, there was the potential for these conflicts. She learnt that you could understand the other side of the story. For example: How come it is so hard for these boys to apologize? This was really an insight.

And second, she was proud that they had done it themselves, that they had the strength of resolving this conflict.

Kuitenbrouwer said that the "self-made youth worker" commented on the role that the government took in the case.

> The fact that you are doing this together and that I was there, being also your resident and being also the leader of the community, is very strong. Because [we did not just have] a … case of professionals who come in and solve this. I've never seen this before; a big conflict like this that needs to be addressed – and then is – by the people themselves. This is very powerful. If you can reach this level of conflict solving, that's enormous. That's really powerful …
>
> He said: "The fact that you're not afraid of addressing the issue and that you go and do that is very important." That helped both sides, because they actually started to feel a little bit proud after this and they said: "OK, we [can] fight but we can also solve the conflict." That was very helpful, and I was very happy that he was present and after half an hour we were all gone and it was finished.

Kuitenbrouwer reached similar conclusions about a subtle, but significant, shift that had developed in the role that the government had taken in the conflict:

> That is something that has been a fundamental learning point for me as well. For me it just felt like this was something that I had to do, but now I would really make more effort to make sure that there are tools put into this community to solve the conflicts themselves – to make sure that they are capable of doing that themselves …
>
> Leadership, I think, is addressing the issues with a proactive attitude: "I feel responsible for the fact that the conflict has happened. I'm not responsible for the conflict, but I feel responsible for the fact that there is tension in the community. I feel responsible for the fact that I should make sure that the community has the capability to resolve this."
>
> So there's a difference between "I'm going to solve this" and "I feel responsible for the fact that there are tools in the community or skills in the community that can help them solve this themselves." I've reflected on that, and there's a difference between taking over and saying: "I'm going to make sure that this is not going to happen again." I can say: "I'm going to make sure that you are capable and that you have the skills to deal with this when this happens again.

This experience was strong, but the shift remains incomplete, and it did not come without resistance.

> There was some resistance at first. Some people in the organization were a little bit resistant ... They were afraid that if we [in government] would step in, [others would] take over and we'd get the blame. And you can see now ... that we were very proactive. We took responsibility and we didn't take over. The civil servants who were hesitant can see now that it works. They can see there's a different way. You can be proactive. You can show your emotions. You can be part of it while not taking over. Because when you start taking over, you also start taking the blame, or you are being the one who has to resolve everything because it's now your fault. If you [in the community] want to try to solve this, go ahead. I think that really has made some sort of shift in the way that [we in government] can understand our own role ...
>
> Now ... to the other people ... who said: "Let's not get too involved, we get blamed for everything" ... I'd point out the difference [between] taking over and addressing a conflict and feeling responsible ... I think that it shows now that you can make it personal. I felt worried, but I never felt really vulnerable. I wasn't sure about what was going to happen, but I didn't feel vulnerable about my own role.
>
> [Now people in] my organization see a shift in role ... [T]he guy who played a central role, although he was hesitant at first, he's become much more enthusiastic about using this method ... He and some others have become really proud in the way this was solved. We can do something. We can intervene as a government. We can intervene in a way that's not taking over. We can give the local people the tools.

Learning from Experience

What can we learn from this case? We draw five points that matter for planning and politics in a multicultural city.

First, the line that divides high politics from the everyday is neither clear nor fixed. Multicultural politics fall along a Möbius strip in which everyday and formal politics move continuously from one to the other. The politics of "street-level bureaucracy" have changed since Lipsky first wrote about them (Laws and Forester, 2015). In addition to the hierarchical, formal dramas of control and resistance that Lipsky described so clearly, we now see basic relations among groups being worked out in the politics of the street (Warren, 2014). The concrete experience of individuals at the street level powerfully shapes their ideas about politics. Lipsky quoted Frances Fox Piven and Richard Cloward writing vividly on this point:

> ... People experience deprivation and oppression within a concrete setting, not as the end product of large and abstract processes, and it is the concrete experience that molds their discontent into specific grievances against specific targets ... People on relief [for example] experience the shabby waiting rooms, the overseer or case-worker, and the dole. They do not experience American social welfare policy ... In other words, it is the daily experience of people that shapes their grievances, establishes the measure of their demands, and points out the targets of their anger. (Piven & Cloward, 1977, quoted in Lipsky, 2010, p. 10)

Second, moments of engagement, like the one that we have described, contain far more than a potential for anger and alienation. The Amsterdam case moved in a different direction. The "concrete experience" of being listened to, of the government being involved without taking over or taking sides, and of being part of a community discovering that it has the resources to address its conflicts may be as transformative as the path to grievances, demands, and anger that Piven and Cloward describe. The citizens in the Amsterdam case directly experienced a kind of democracy in which they played a vital part, in which they formed different relationships with each other and with the government. This experience suggests both what multiculturalism might mean in practice and what possibilities for democracy exist when diverse citizens must share time and space.

Third, this democratizing potential arose in a moment filled with risk. Things went well in Westerpark, but they could have gone badly, with equally, or even more dramatic, lasting consequences for relationships in the neighbourhood and the city. The encounter of mutual accusations giving way to reciprocity of recognition and even apology demanded a level of engagement hardly possible when neighbours merely barbecue or play football together. We saw clearly how alienation developed through the organization and performance of the football tournament, and its impact on identities and relationships became more apparent – and risky – as conflict erupted.

This episode – liminal as it might have been – was also a moment of contingency. Kuitenbrouwer details both the differences in views it elicited and her uncertainty at being neck deep in an improvisation.

The openness and attention that she displayed clearly created opportunities. Yet this alone was not sufficient. Her active resistance to others' facile framing of the situation ("Let the police handle it!") and her ability to sustain doubt about what had happened created a political space for mutual regard and recognition. The distinctive feature of the process that developed – and succeeded – involves Kuitenbrouwer's stubborn determination to suspend the rush to judgment and to commit to listening (cf. Forester, 1999, 2009, 2013). Possibilities exist, she insisted: "I'm not sure how we're going to find a way out of it, but I don't want to leave it as it is. I don't have a solution yet, I don't even know what happened exactly – but are you willing to come together after the summer and are you willing to talk about the conflict?"

Her actions set the stage for the engaged, if difficult, respectful and inquisitive listening that promoted the understanding that parties noted after the case. She helped them begin to grasp how the case was coloured by such taken-for-granted expectations that their influence could only be seen by someone who did not share those presumptions.

Fourth, the government role that emerged adapted to the demands of conflict and moments of contingency. It was active, but not controlling. It was engaged, but not committed to a particular interpretation of the situation. It was forward-looking, but not directive. It was responsible in the sense of holding space, but through Kuitenbrouwer it shared that responsibility with the parties involved in the conflict. It was open to parties' shifting understandings of what happened and to the practical proposals that their encounter produced.

While this clearly does not typify a routine government "best practice," it does highlight the ways such ad hoc episodes can surface intergroup tensions and local leadership might influence how identities become understood and negotiated. For public officials hoping to avoid being the sole responsible party – hoping to invite active contributions from citizens – the case offers a valuable alternative to withdrawing. Kuitenbrouwer was active – too active for many of her colleagues – but she managed to act so that others felt they had the opportunity, the responsibility, and the capacity to resolve the situation. For cases in which changes of relationships and identities are at stake, these are valuable lessons. Perceptive improvisation is likely to be an essential, practical response of governance that enables relationship changes as multicultural relationships stand increasingly near the core of urban politics.

Finally, this case shows that a democratic politics of multicultural deliberation does *not* depend on prior consensus as either a starting point for interaction or a precondition for agreement. Deliberative encounters develop in practical and concrete settings, and the details and demands of those settings matter. The development of perspectives, perspective taking, and emergent self-other relations, while characteristic of deliberation in the abstract, become driving forces that incorporate the concrete details of each case.

We can see, finally, the limits of a communitarian perspective in the Westerpark case. Participants were hardly just group members as they struggled to come to shared interpretations and assessments of their situation. The lessons learned concern not just the different meanings of apology for various participants, but what's required to try to come to an understanding under conditions of difference.

Any "consensus" here is thin and partial, practical and corrigible: not about any harmony but about finding a way forward together to be better able to cope with the next experience of difference. Such working agreements develop from the struggle to make sense together under difficult circumstances that threaten identities and security. They proceed through the telling of grounded stories that, in their details and their differences, create the grist for learning and change.

The view of democracy and democratic deliberation that this case presents is not romantic. It is hardheaded and practical and won with sweat in moments of risk that sensible people might choose to avoid. This may not be what proponents of deliberation anticipated, but it is clearly part of the answer that a deliberative politics of a multicultural democracy must try to find, to embrace, and to enrich.

NOTES

1 Turkish and Moroccan immigrants combined constitute less than 6 per cent of the population in the Netherlands, but almost 18 per cent in Amsterdam. The figures are similar for Rotterdam (16 per cent), The Hague (17 per cent), and Utrecht (14 per cent).
2 Ambrose Evans-Pritchard and Joan Clements, "Fortuyn killed 'to protect Muslims,'" *Telegraph*, 28 March 2003.
3 Source: BBC News in Pictures, "Netherlands Mourns van Gogh." http://news.bbc.co.uk/1/shared/spl/hi/pop_ups/04/europe_netherlands_mourns_van_gogh/html/9.stm

4 The case is based on a practice-focused oral history account provided by Ms Kuitenbrouwer in an interview with the authors (see Laws & Forester, 2015).
5 This is a designation used in the Netherlands, previously also formally in policy and regulations. It comes from the Greek for "self" and "land" or "ground" and is used to distinguish individuals and groups of "original" or "ancestral" inhabitants from more recent immigrant groups. The latter are referred to as "allochtoon" to signify their origins in a foreign land or ground.
6 The full text of the interviews is available in Laws and Forester (2015).

REFERENCES

Forester, J. (1999). *The deliberative practitioner*. Cambridge, MA: MIT Press.

Forester, J. (2009). *Dealing with differences: Dramas of mediating public disputes*. New York: Oxford University Press.

Forester, J. (2013). *Planning in the face of conflict: The surprising possibilities of facilitative leadership*. Chicago: American Planning Association Press.

Gale, R., & Thomas, H. (2015). Multicultural planning in twenty-first-century Britain. In M. Burayidi (Ed.), *Cities and the politics of difference*. Toronto: University of Toronto Press.

Hurewitz, J. (2005). Too tolerant of the intolerant: The Netherlands' multicultural drama posted in *Other* | 19-Jun-05 | Retrieved May 3, 2013 from http://www.worldsecuritynetwork.com/Other/Hurewitz-Jeremy/Too-Tolerant-of-the-Intolerant-The-Netherlands%E2%80%99-Multicultural-Drama.

Laws, David, & Forester, J. (2015). *Conflict, improvisation and governance: Street level practices for urban democracy*. New York: Routledge.

Lipsky, M. (2010). *Street-level bureaucracy: Dilemmas of the individual in public services*. New York: Russel Sage Foundation.

Piven, F.F., & Cloward, R. (1977). *Poor people's movements*. New York: Vintage.

Scheffer, P. (2000). Het multiculturele drama. NRC Handelsblad, January 29.

Warren, M. (2014). Governance-driven democratization. In S. Griggs, A. Norval, & H. Wagenaar (Eds.), *Practices of freedom: Decentred governance, conflict and democratic participation* (pp. 38–59). Cambridge: Cambridge University Press. http://dx.doi.org/10.1017/CBO9781107296954.002

5 Multicultural Planning in Twenty-First-Century Britain

RICHARD GALE AND HUW THOMAS

Introduction

In principle, planning in contemporary multicultural Britain can build upon a body of discussion, interventions of various kinds, and evaluation and research conducted over 30 years. In practice, it is by no means clear that within the planning system – among professionals and other actors – there is a widely shared understanding of what it might mean to help plan for, or bring about, a multicultural city, town, or village. This chapter will draw upon published research, by the authors and others, and more recent involvement on an advisory and professional basis in individual cases, to explain the challenges involved in multicultural planning in twenty-first-century Britain and prospects for the future.

Historically, Britain has been a unitary state, and even after the devolution of legislative and administrative powers (including in planning) to the national-regions of Scotland and Wales and Northern Ireland, it remains the case that local planning – i.e., at the municipal level and below – has to take place within a legal and policy framework established by higher tiers of government. An essential starting point, therefore, is an understanding of the national policy and legislative context in relation to race equality, multiculturalism, and planning and its volatility in recent decades. Devolution of administrative and legislative powers is now well into its second decade, and there are increasingly evident policy and administrative divergences between the constituent nations and regions of the United Kingdom. Consequently, in this chapter we have chosen to focus on England, which is where some 85 per cent of the British population and the overwhelming majority of minority ethnic groups live. This is a pragmatic decision, and does not imply that there is nothing of interest to be said in relation to multiculturalism elsewhere in the UK; on the contrary, multiculturalism and planning are of significance everywhere in Britain, and some issues – for example, relating to linguistic minorities and addressing deep-rooted historic enmities – have been addressed in other parts of Britain with particular sophistication, energy, and care, with many lessons to be shared with others.[1]

Focusing as we will on the national governmental context in England, particularly pertinent at present are the lukewarm attitude of the current Conservative–Liberal Democratic coalition government towards proactive promotion of race equality, the increasing government emphasis in (English) education policy and elsewhere on an essentialist notion of cultural identity, the readiness of press and politicians to slip into moral panic over immigration, including from eastern and southern Europe (cf. the tensions reported from the Netherlands by Laws and Forester in chapter 4), and the emphasis on *localism*, an ambiguous term, but one with overtones at least of empowering local residents, including in planning. Widespread confusion and ambiguity in relation to terminology and political and moral principles, of the kind Harper and Stein decry in chapter 2, also have their effect on the nature of public discussions. Potentially, this could be the framework for an ugly, xenophobic reaction to multiculturalism. But perhaps it offers opportunities for more positive progressive planning, too, as we will discuss.

Developing the progressive planning potential of circumstances, however limited it may sometimes be, depends, as it always has, on a mix of local political circumstances, the attitudes and level of understanding of professional planners, and action and pressure from minority groups. We use short case studies to illustrate some promising developments which are taking place at what might appear to be a generally inauspicious time in British planning.

British Approaches to Multiculturalism

Approaches to multiculturalism in terms of policy and practice, as well as the philosophies on which they rest, have a decidedly chequered history in the British context, exhibiting a much weaker degree of institutionalization than in such avowedly multicultural countries as Canada and Australia (Modood, 2007; Qadeer, chapter 3). Indeed, whereas societies like Canada and Australia, with rather more complex and extensive histories of colonization and immigration, have achieved a degree of settlement with cultural difference of various kinds by defining themselves as multicultural at the level of the state, the British approach to multiculturalism has emerged very much on a pragmatic and incremental basis. In the words of the *Parekh Report* (Runnymede Trust, 2000), itself a controversial plea for a more concerted attempt to reconceive the British "national story" in a way that would allow greater space for an accommodation with ethnic and cultural differences, the British approach to multiculturalism has "evolved as an unplanned, incremental process – a matter of multicultural drift, not of conscious policy" (Runnymede Trust, 2000, p. 14). In this context, it is worth acknowledging a distinction that is often made between *multicultural* as a descriptor of the social and cultural condition of the British population, and *multiculturalism* as a political philosophy that would articulate and express recognition of that condition through various forms of institutional design; the former can be seen as an accomplished fact, whilst the latter is an idea whose appeal has ebbed and flowed with changing political priorities, remaining only loosely embedded in the topsoil of British policy (Hesse, 2000, p. 2).

In large part, the distinctiveness of the British approach to multiculturalism is historically linked to the way in which Britain became a clearly multiethnic society through postcolonial immigration in the aftermath of the Second World War (Spencer, 1997). In other societies in which multiculturalism is more firmly rooted, the need for an explicit commitment to policies recognizing difference at the level of the state has tended to emerge at the nexus of multiple and historically entrenched sources of social and cultural differentiation. Thus, in the "settler" societies of the US, Canada, Australia, and New Zealand, complex histories of immigration dovetailed with the expropriation and resettlement of indigenous peoples, and in the US context, with ongoing concern over the rights of African Americans (Joppke, 1996; Anderson, 2000). In the UK context, by contrast, the periodic nature of debates about multiculturalism provides an index of the dismantling of Empire and the successive development of nationality and immigration laws from the mid-twentieth century onwards, indicating how concerns about immigration and the rights of post-migration minorities have tended to be more tightly coupled in the UK than in many other "multicultural" societies (Castles & Miller, 2009).

Arguably the most important instance of this relationship between migration and multiculturalism can be traced back to the first phase of postwar migration, which ended in the early 1960s with the passage of the Commonwealth Immigrants Act 1962. Prior to this legislation, migration from Britain's former colonies in South Asia and the Caribbean Islands had been primarily economic, with migrants coming to the UK for extended periods to work and remit their earnings, before returning home and sponsoring friends or family members to do the same (Ballard, 1994; Spencer, 1997). With the passage of the Commonwealth Immigrants Act, however, chain migration gave way to permanent settlement, as migrants feared that to leave the UK would result in their being barred from re-entry at a later stage (Ballard, 1994). The implications of this shift in migration practices were profound in terms of creating the conditions for the emergence of multicultural Britain. As Spencer (1997, p. 129) has observed, "the Commonwealth Immigrants Act of 1962, the first legislation supposedly introduced to prevent multiracial Britain from happening, in fact ensured that foundation stones were laid and the Asian and black communities transformed into populations of substantial proportions." The most significant effect of this change in migration patterns for the long term was that migrants already resident in the UK invited their families to join them, setting in train a comprehensive process of social and cultural reconstruction and the formation of substantial ethnic and religious enclaves in the urban industrial centres of London, the Midlands, and the North West (Rose & Deakin, 1969; Dahya, 1974; Anwar, 1979). As we discuss in a later section of the chapter, this process of family and social reconstruction also led to one of the more noticeable effects of the emergence of multicultural Britain on the British planning system, as the reintroduction of religious observances into the life-world of migrant groups resulted in a surge in the demand for premises in which to worship (Peach & Gale, 2003; Gale, 2008).

It is against the background of controlled immigration and the "need" to manage migrant populations that the debate about multiculturalism in the UK has largely taken shape, from the 1960s down to the present. During the 1960s and 1970s, concerns were expressed in a variety of institutional settings, such as public housing and education, as to whether the British state should seek to accommodate or assimilate post-migration minorities, with controversial policies of residential dispersal being applied and subsequently revoked in areas in which the formation of cultural enclaves had been pronounced (see, e.g., Flett,

Henderson, & Brown, 1979, on Birmingham City Council's policy of residential dispersal in the 1960s and 1970s). Whilst the detailed trajectory of the multiculturalism debate in the intervening years may appear complex, there has in fact been a remarkable durability to the ways in which fears over immigration have provided the motivating force behind the question of how far British institutions should go in recognizing social and cultural differences, with concerns over the number, concentration, and citizenship status of asylum seekers and migrants from European Union accession states constituting only the most recent examples (Cohen, 2002; Garapich, 2008).

Current Policy Priorities: The Return of Assimilation?

The current nadir in British enthusiasm for multiculturalism as a political project arguably began with the policy volte-face which numerous commentators have remarked in the way New Labour addressed issues of race and ethnicity following the urban disturbances that occurred in the northern cities and towns of Bradford, Blackburn, Burnley, and Oldham in summer 2001 (Ratcliffe, 2012; Worley, 2005). There was no shortage of early critics who pointed out a lack of coherence in the way New Labour addressed issues of race equality and cultural difference on coming to power in 1997 (Back, Keith, Khan, Shukra, & Solomos, 2002). Nevertheless, prior to summer 2001, the new government made several significant contributions across a range of policy areas, which, when taken together, indicate a genuine, if not exactly radical, commitment to progressive race politics and multiculturalism. Perhaps the most salient of these was the publication of the Stephen Lawrence Report in 1999 (MacPherson, 1999), which concluded an extensive public inquiry, ordered by then Home Secretary Jack Straw, into the failures of the investigation of the Metropolitan Police Force into the racially motivated murder of Black teenager Stephen Lawrence in April 1993. A socially and politically significant document in itself, the report was critical in bringing about changes to the institutional culture of British police forces, whilst also providing a boon to applications of the concept of institutional racism in other settings such as the health and education systems (McKenzie, 1999; Osler & Starkey, 2002). Other significant developments included innovations in the legal field, most notably the incorporation of the European Convention on Human Rights into UK law as the Human Rights Act 1998. Article 14 of the act provides protection against discrimination according to the categories of race, colour, language, and national and social origin; whilst Article 9 enshrines the right of citizens to freedom of thought, conscience, and religion, and has been applied in the context of a high-profile planning case involving the religious minority group the International Society for Krishna Consciousness (Nye, 2001). A further development in the legal field during New Labour's first term in office was the passage of the Race Relations (Amendment) Act 2000. This was effectively an updated version of an earlier piece of powerful legislation, the Race Relations Act 1976, which had also been passed by a previous Labour government, and of which the provisions for preventing "indirect discrimination" had important implications for combating racialized practices on the part of local government and other public bodies (Young & Connelly, 1984).

In the aftermath of the 2001 disturbances, however, a series of inquiries at both local and national scales ushered in what many regard as a paradigm shift in the way New Labour addressed issues of race, ethnicity, and multiculturalism (Back et al., 2002; Worley, 2005; Ratcliffe, 2012). The central concept in the analyses of each of the reports on the factors underlying the disturbances, namely those of Cantle (2001), Denham (2001), and Ouseley (2001), is that of *community cohesion*, a term without precedent in the history of British policy discourses on race and ethnicity (Ratcliffe, 2012, pp. 264–265). Although lacking precise definition, the term is used in a way that identifies the principal source of social conflict and unrest in the social and related residential separation of culturally and ethnically defined sub-populations in the affected areas (assumptions similar to those tested by Agrawal in chapter 13). Accordingly, community cohesion is culturalist in emphasis, suggesting that the solution to social conflict is to encourage different ethnic communities – and notably *minority* ethnic communities – to commit to "core shared values" and hence overcome the tendency to lead "parallel lives" (Bagguley & Hussain, 2006; Phillips, 2006). Not only does this suggest a return to earlier ideas of ethnic and cultural assimilation, as Ratcliffe (2012, p. 265) points out, but also such emphasis significantly downplays the extent to which disparities in the life chances of individuals from different backgrounds are cross-cut by persistent patterns of socioeconomic disadvantage.

New Labour's change of heart with regard to multiculturalism was subsequently compounded by the events in London on July 7, 2005 (often referred to as 7/7), when home-grown radical Islamists engaged in a bombing campaign that killed 52 civilians and left several hundred injured. The public response to the bombings amounted to a comprehensive rethink of British multicultural drift and its putative consequences, with the mood being most graphically captured in a widely publicized speech by Trevor Phillips, then chair of the Commission for Racial Equality and subsequently chair of the Equality and Human Rights Commission. Establishing an unambiguous thematic connection to the community cohesion reports of four years earlier, Phillips located the main source of the problem of home-grown terrorism in what he identified as a drift towards increasing levels of ethnoreligious segregation: Britain, he claimed, was "sleepwalking to segregation" with some districts being "on their way to becoming fully fledged ghettos" (Phillips, 2005). In the aftermath of this speech, a series of detailed analyses of neighbourhood population change emerged, focusing above all on the effects of internal migration on established ethnoreligious residential concentrations. These showed compellingly that concerns about ethnic residential segregation in the UK have been significantly overstated and that the overall trend in the last two decades has been one of movement of ethnic and religious minority populations out of inner-urban concentrations into more culturally mixed areas (Peach, 2006; Phillips, 2006; Simpson & Finney, 2009; Stillwell, 2010; Gale, 2013). Phillips's sentiments appear nevertheless to have had a long half-life, and the terms in which he characterized the link between spatial segregation and social integration, particularly for British Muslim groups, resonate strongly with the way in which the Conservative and Liberal Democrat Coalition, which came to power in May 2010, approach the issue of multiculturalism. The clearest indication of this comes from Prime Minister David Cameron's speech at the Munich Security Conference in February 2011, in which he explicitly linked the putative problem of the "doctrine of state multiculturalism" with Muslim segregation and social separation:

> In the UK, some young men find it hard to identify with the traditional Islam practiced at home by their parents ... But these young men also find it hard to identify with Britain too, because we have allowed the weakening of our collective identity. Under the doctrine of state multiculturalism, we have encouraged different cultures to live separate lives, apart from each other and apart from the mainstream ... We've even tolerated these segregated communities behaving in ways that run completely counter to our values. (Cameron, 2011)

In terms of policy implications, notably, this speech coincided with the Coalition's efforts to adopt and relaunch the controversial Preventing Violent Extremism agenda of New Labour, introduced shortly after 7/7 (O'Toole, Nilsson, DeHanas, & Modood, 2012).

The continuity of the approach to multiculturalism that appears to have straddled the change from New Labour to Coalition governments has been further compounded by a major shift in political priorities for the Coalition, who have embarked on an extensive program of public spending cuts, to reduce the extent of the UK budget deficit (Taylor-Gooby & Stoker, 2011, p. 4). One consequence of this program has been a significant cut in the size of local authority budgets, with many of the services that had historically been at the forefront of efforts to ensure civic participation and redress localized inequalities between people of different ethnic backgrounds – including local authority youth services, and a wide range of government-funded state and civil society partnerships – having been downsized or excised altogether (Lowndes & Pratchett, 2012). The ideological correlate of these fiscal changes finds expression in the Coalition's much-vaunted though ill-defined concept of the Big Society, a communitarian notion that emphasizes the need for an enhanced role to be played by civil society groups in the delivery of core services (Corbett & Walker, 2012). As we discuss later in the chapter, the Big Society idea has specific application to the planning field in terms of the Localism Act, which became operative in England in 2011, and which, *inter alia*, makes provision for community groups to have access to devolved planning powers in the form of neighbourhood planning (Bailey & Pill, 2011). It is as yet unclear, however, how this or other aspects of Localism and Big Society agendas will play out in relation to minority groups, with some commentators fearing that the minority ethnic organizational presence within the wider "third sector" is peculiarly under threat (Craig, 2011).

The British Planning Response to Multiculturalism

Over the last 30 years or so, what has been the UK planning response to greater ethnic and cultural diversity and political struggle around multiculturalism and race equality? There have been governmental initiatives, and in a centralized state like the UK that's not insignificant. However, in England, the questioning of the promotion of multiculturalism that is evident in public policy in general has been extended to planning. If we begin by considering the top tier of the governmental hierarchy, under Labour the government department in England responsible for national planning policy commissioned both research and advice on good practice in relation to planning in a socially diverse society (Office of the Deputy Prime Minister, 2003; Office of the Deputy Prime Minister, 2005). The core of the advice then issued was that planners and planning agencies must avoid any assumptions about homogeneity of ways of life, and aspirations, among the population of the UK, and avoid drawing on stereotypes in their planning. It contained examples of ways in which planners have been sensitive to the socially diverse population they are working with, and have tried to listen to voices too often drowned out or overlooked in discussions of planning matters. This is very positive, and a world away from the governmental silence born of apparent incomprehension in the 1970s and 1980s (Royal Town Planning Institute/Commission for Racial Equality, 1983; Krishnarayan & Thomas, 1993). Moreover, under Labour there was a transformation in the tone of national governmental policy and advice in relation to Gypsies and Travellers.[2] Under the Conservative governments of the 1980s and 1990s, these were viewed as enemies within, a living affront to the values of British society who deserved no tolerance. Under Labour they benefited from an initial concern for respecting cultural difference (a concern which in their case has been immune to the more recent confusion caused by anxieties over the "Islamic threat"). In addition, Gypsies and Travellers benefited indirectly from renewed government interest in issues such as child poverty and health. There was the promise of a more humane approach to planning, one which was prepared to accept that Gypsies and Travellers have aspirations, cares, and concerns which are recognizably human and in general reasonable.

The Conservative–Liberal Democrat coalition government in power since 2010 has drawn back from this approach. Notably, the good practice guidance has been withdrawn and will not be replaced (Lewis, 2013). A variety of reasons are given for this, but some certainly draw upon wider discussions of multiculturalism; for example, the charge that the guidance document

> ...tells councils to translate into foreign language, which undermines integration by discouraging people from learning English, weakens community cohesion and a common British identity, and wastes taxpayers' money. (Lewis, 2013)

On the narrower issue of identifying and granting permission for sites for the caravans of Gypsies and Travellers, the current government's policy in England is to imply that Gypsies and Travellers were receiving unfairly favourable treatment under the previous government and to make it plain that they are to receive the same treatment as every other citizen from now on. What this amounts to in practice is illustrated by a governmental statement making it clear that evidence of unmet need for accommodation is not usually a strong enough reason to grant planning permission for sites for Gypsy and Traveller caravans in the (protected) Green Belt around cities. The practical significance of this is that public hostility, planning policies, and patterns of land ownership typically make it very difficult to find sites for Gypsy and Traveller caravans within the developed envelope of settlements, making urban fringe and rural areas – including the Green Belt – the only options (Thomas, 2000). The wider significance is that it illustrates a broader stance that Gypsies and Travellers, and by extension others deemed to be outside the norms of mainstream British society, however defined, gain benefits only if they engage with society and the state on the terms of the majority (see, for example, Department of Communities and Local Government, 2012).

If the planning policy stance which is explicitly related to multiculturalism suggests a reversion to attitudes of the 1970s, there are other aspects of national planning policy in England which have potential, albeit an ambiguous one, for helping create places which are more reflective and supportive of a diverse citizenry. Put simply, government strategy for planning in England aims to begin to replace what it perceives as the bureaucratic-technocratic, top-down, authoritarian hand of the professionalized state – and especially local government – with opportunities for people to make their own decisions, individually, about how their own property is developed; and collectively, about the planning of their neighbourhoods (Allmendinger & Haughton, 2012). Planning controls over various forms of development – presented as burdens on business and householders – have been loosened, and a legal framework constructed for groups of residents (and other stakeholders such as businesses) to be recognized as legitimate promoters of neighbourhood plans which will be part of the formal planning system. From the perspective of a coalition government, the approach happily marries principles close to the hearts of one or another of the constituent parties: of civic empowerment, with residents being encouraged and enabled to be active citizens; and of allowing already advantaged neighbourhoods to negotiate the terms on which they contribute to meeting pressing national needs for housing and other development land. The history of attempts at neighbourhood governance suggests that the places best positioned to take advantage of an initiative like neighbourhood planning will be those with the kinds of social capital which can support widespread active participation, and who can generate – or hire – an animateur/facilitator to focus and bring to fruition ideas which are generated (Lowndes & Sullivan, 2008). Early examples of neighbourhood planning in fairly homogeneous rural communities suggest that appropriately skilled and experienced professional planners can play the role of animateur (Geoghegan, 2013). The possibility that this system will become largely a vehicle for articulate, relatively wealthy, and (in the UK context) largely White groups to plan for de facto exclusion of people unlike them is obvious but is unlikely to worry a conservative government (Allmendinger & Haughton, 2012). Yet, it is possible that residents of other kinds of areas may take the opportunity to begin to fashion an environment more supportive of their diverse needs. Laws and Forester's account of a diverse population working at ways of fashioning futures in a shared place (chapter 4) illustrates this potential.

In reality, if neighbourhood planning bears any fruit in relation to multiculturalism, it will be more likely in those areas which have a history of initiatives by local government, for as Laws and Forester show, political leadership has a vital role to play in creating conditions for engagement between people who may initially be focused on differences rather than commonalities. Below the national level of government, there have been occasions over the last 30 years where local councils have been sensitive to the importance of race equality. Some places, like Leicester, have a history of reasonably open public debate and some strong political direction in relation to race equality, though ethnographic work suggests that race remains a source of social tension and estrangement here – and if anything is getting worse (Herbert, 2008). In cities like London and Birmingham the local struggle for race equality appears to ebb and flow across various policy terrains, including planning. It is possible to identify good things which have happened, as Leonie Sandercock (2003) has done in relation to Birmingham (Greed (2005) has found isolated examples of progressive practice in relation to gender equality too). But since the abolition of the Greater London Council in 1986, it is difficult to think of many councils which appear to have a reasonably well considered political project within which countering racism is significant and which it is pursuing consistently.

For more than 20 years, when local or national politicians in the UK talk of changing the perceptions of people about themselves and their relations to others, they tend to be thinking about ways of getting people to see "problems" as "opportunities," to see themselves and their locality as genuine competitors in a global economy, to see themselves and their locality as having something to offer. On occasion, what is on offer is ethnic diversity, but there is no evidence that this commodification of ethnocultural diversity connects to any kind of anti-poverty or anti-racism strategy (e.g.,Young, Diep, & Drabble, 2006). Perhaps not surprisingly, when planning or planners have been involved in positive initiatives related to race equality it is because non-governmental organizations are involved, or even ones outside the mainstream of elected bodies – such as Urban Development Corporations or New Deal for Communities projects (Brownill & Thomas, 2001).

Planners still have difficulty in understanding how their work relates to promoting the kind of community which respects diversity and is often labelled in British policy debates as one which supports "race equality" (e.g.,Beebeejaun, 2006). In 2007, the government-sponsored Commission for Racial Equality reported on a formal investigation into physical regeneration in Britain. It involved collecting evidence about a range of agencies in the public, private, and voluntary sectors, and concluded, *inter alia*, that "We were very concerned by the number of officers, at all levels, who said that racial equality and good race relations were irrelevant to the work of regenerating the built environment and that they used a 'colour blind' approach to their work" (CRE, 2007, p. 127).

It is pretty evident that there is a persistent blind spot among many built environment professionals, including planners, when it comes to thinking through the connection between race equality and the development, use, and management of the built environment. Why this might be will be discussed below, a discussion which will be better informed if we first discuss examples of how it is possible for more sensitive and positive planning in a multicultural setting to develop. One example of this is with regard to the provision of places of worship, which, as noted above, has been one of the more conspicuous areas of interaction between ethnic and religious minority groups and the planning system since the era of mass migration in the early 1960s (see Agrawal, chapter 13, for a related discussion of places of worship and planning in Canada).

As noted above, the reunification of migrant workers with their families from the 1960s onwards resulted fairly swiftly in the re-establishment of religious customs, particularly among the South Asian religious populations of Hindus, Sikhs, and Muslims primarily from India and West and East Pakistan (the latter becoming independent Bangladesh after 1971). In built environment terms, this translated into a need for premises in which to pray and to provide children with after-school religious instruction (Ballard, 1994). Albeit with some variation, as Peach and Gale (2003, pp. 479–486) have argued, the emergence and development of this need generally followed a four-stage cycle, with progressive changes in the scale of premises sought, and correlative shifts in the planning response. Stage 1 was defined as "tacit change and planning denial," in which minority religious groups converted the front rooms of houses into prayer facilities, which resulted in the attempts of planners to obstruct what they saw as illegal changes of use. In stage 2, identified as "larger scale conversion and minimalist change," religious groups have established themselves in both size and resources, and sought out larger social, industrial, or other religious buildings for conversion. Occasionally such buildings were adorned with the insignia of the respective group, such as the Sikh *Nishan Sahib* (the symbolic flag of Sikh brotherhood, or *khalsa*), but were otherwise discreet, and from a planning point of view, often advantageously located some distance from residential units.

Stages 3 and 4 of this cycle represent the ultimate flourishing of religious institutional development in the form of the establishment of purpose-built premises, whether Hindu mandirs, Sikh gurdwaras, or Muslim mosques, often on an impressive physical scale, and tending to incorporate distinctive architectural styles. The key differentiating factor between these stages is the planning response. Stage 3, "hiding and displacement," corresponds to the efforts of planners to encourage such developments but also to ensure their location away from residential areas, which often resulted in such buildings being constructed in unprepossessing, hidden-away settings, such as industrial complexes or busy road junctions (see, e.g., the mosques discussed in Gale, 2004). Stage 4, on the other hand, characterized as "embracing and displacement," signifies a relatively new departure in the planning response, occurring approximately from the late 1990s onwards, in which these new buildings have been actively encouraged to be of interesting and distinguishing design, and to be located in prominent positions, often on land sold by local authorities at reduced cost (Gale & Naylor, 2002;

Gale, 2004). Major examples include the Sri Singh Sabha gurdwara in Hounslow and the Dar ul Uloom Islamia mosque in Birmingham (Peach & Gale 2003, pp. 484–485), both of which proudly proclaim their purpose and architectural heritage. The important point with this final stage of the cycle is that it represents one of the more positive ways in which planners have begun to respond to cultural and religious diversity, in a sense making a virtue of necessity as the public presence and demands of particular minority groups have become more confident and assertive.

Nonetheless, there remains the question why, overall, so many British planners find it difficult to make the connection between their work and issues of multiculturalism and race equality or, indeed, promoting equal opportunities more generally (see, e.g., Greed, 2005). We suggest there are a number of factors associated with the context within which they work – understood broadly and narrowly. At a very general level, the racialized and often embittered nature of public debate about multiculturalism in the UK does not encourage any occupation to wade into it. And in planning's case, there are factors which tend to consolidate this distancing rather than countering it. Some key general ones are the role of law in shaping British planning, and the nature of its professionalization.

British planning developed, and continues to develop, as an outcome of the struggle between different class interests. The law around planning has been a key product of, and has helped shape, those struggles (MacAuslan, 1980). It has defined what constitutes property rights, has enshrined particular kinds of safeguards for landowners and others with various interests in property, and has created a complex framework which for very many planners actually defines what planning is. As a consequence, there has been, in practice, a narrowing of planning from a concern with making better places to managing (or, if in the private sector, navigating through) a particular kind of bureaucratic and legal thicket. Notwithstanding attempts in the last decade or so by the profession (Royal Town Planning Institute, 2001), academics (Healey, 2010), and government (see Nadin, 2007) to promote planning as placemaking, the evidence suggests that the core of planning is still heavily procedurally focused. Continued calls by various authoritative voices for culture change among professional planners are a testament to how little has changed (Inch, 2010). Indeed, Allmendinger and Haughton (2012) have suggested that the experiment of the last decade has failed and is dead. This is not to deny the interest of many planners in making the world a better place and their capacity to do this in innovative ways at times, but procedure and process remain central, and the current UK government with its instinctive suspicion of big government will, if anything, hope to limit the role of local government planning even more.

The British planning system has provided a firm (but constraining) base for the development of the British planning profession as that part of local government which administers the statutory planning system (Thomas, 1999). Meanwhile, outside government a group of planners works as advisors to interests wishing to negotiate the planning system. Both kinds of planners focus on planning within its legal framework. This is important because the role of the planner within the legal framework is that of an expert witness, a dispassionate, technically proficient non-political professional.

This invitation to be politically neutered has not been unattractive to the occupation of planning as it has professionalized. Many years ago, Foley (1960) noted how British planning had been presented as a good thing to different, seemingly incompatible, political interests. In seeking broad-based support, professionalism has sought to emphasize its technicality and distance from contentious social issues. A key element of professionalization has also been the defining of a defensible turf for the planner. In general, this has involved narrowing the occupation's concerns, and specifically, narrowing them to a concern with the physical (notably, land and buildings: how they look, how they are distributed and should be distributed, what they are used for). Even when the occupation has sought to colonize new policy areas (economic development and tourism development are examples), it has tended, in the end, to fall back on its core expertise of building-related issues. When especially hard pressed, the core area retreats to simply understanding the (statutory) planning system. Of course, this focus doesn't in itself preclude asking questions relating to social justice: what buildings are used for and where they are sited has all kinds of different implications for the lives of different groups in the population. But the pervasiveness of law and regulation within planning – both in shaping practice and in shaping a dominant notion of professionalism – means that these questions are regarded as inappropriate by most planners. Though the professional institute of British planners, the Royal Town Planning Institute, has consistently been ahead of most of its membership in engaging with emerging issues arising in a multiethnic society, it has tended to emphasize issues of procedural justice rather than visions of what multicultural places might be (e.g., Royal Town Planning Institute/Commission for Racial Equality, 1983; Krishnarayan & Thomas, 1993). Thus the profession as a whole has generally been disinclined to engage with discussions of the desirability or otherwise of residential segregation, for example, but has taken a position in relation to issues of discriminatory practices.

The burgeoning planning literature on substantive social justice (e.g., Fainstein, 2010; Marcuse et al., 2009) can provide guides and reminders to planners of why and how the just city matters in socially diverse cities. Education can continue to sensitize planning students to cultural difference, to the insidious legacies of racist oppression, both within and outside their own countries, and to the place planning can play in addressing these matters (Rios, chapter 14). But all this can be neutered, and undone, unless planners understand how they might work for justice in a sometimes hostile environment. For many years, the work of scholars like John Forester and others has provided exemplars, hints, tips, and encouragement to progressive planners. One important message from this work is the importance of collective action. Yes, planners need to be brave at times, but they are not superheroes; they achieve change by creating networks and coalitions, of more or less durability (see Laws and Forester, chapter 4). This is an important message, but perhaps what has been overlooked in the case of the run of the mill, not too heroic, ordinary planner is the significance of workplace solidarity and support. This is a lesson to be learned from accounts of Cleveland when Krumholz was a planner there (e.g., Krumholz & Forester,1990) but is typically overshadowed by the focus on leadership (also important, of course). In future, planning students and planning professionals need to be reminded of the importance of workplace solidarity in sustaining and promoting progressive agendas. In Britain, this will be especially important given the sometimes bleak conditions under which planners may work in the future.

Prospects for the Future

It is easier to identify the factors which will – in varying degrees – shape the future of planning for multiculturalism than it is to gauge what strength each may have, and how the future will play out. One thing can be said with a degree of confidence: even in a pretty centralized polity, in general, locality is still significant: i.e., the socioeconomic characteristics of areas, their distinctive histories, and their political cultures influence local planning. This was so in the 1980s, under evangelically right-wing Thatcher governments (Brindley, Rydin, & Stoker, 1996), remains the case (e.g., Punter, 2009), and will continue. That said, a key factor in generating local difference is local politics, and in contemporary England there is a significant convergence between political parties around a quasi-assimilationist stance in relation to religious and ethnic difference. This may be resisted in small areas with minorities which are well organized politically, but elsewhere the major concerns often remain expressed in terms of national values, community cohesion, and equal (meaning identical) treatment for all groups. These resonate with the current government's approach to planning (see earlier) and in truth are unlikely to be seriously challenged under any conceivable government in the next five years or so. Equality legislation – some elements of which have been in place for close to 50 years – remains well entrenched, and it will help ensure that planning *procedures* remain sensitive to the social realities of multiculturalism. However, the content of planning policies and initiatives is very likely to be conservative and assimilationist in its thrust. Moreover, a move to the right in national politics cannot be ruled out. The current prime minister is tolerated, rather than loved, by many in his party, as long as he delivers a measure of electoral success. Since he is regarded by many within his own party as far too liberal on social issues, especially those relating to diversity, it is conceivable that, were he to go, then his party would take a harder stance in relation to multiculturalism; this might also be the opportunity for more influence for the many eager to help the Conservative party undertake a dismantling of the post-1947 UK planning system (Pennington, 2005). The result would be a planning system focused on property rights rather than social justice.

NOTES

1 Tewdwr-Jones and Allmendinger (2006) provide an introduction to the implications for planning of devolution. Readers should bear in mind that devolution is a dynamic project, and all three devolved administrations are currently well advanced with plans for reform of the planning system. A flavour of the complexities of religious, cultural, and racialized tensions in Northern Ireland can be gained from Ellis (2001) and Shirlow and Murtagh (2006). Mooney and Scott (2012) place countering racism in a broader Scottish context. Thomas (2002) and Williams, Evans, and O'Leary (2003) assess Wales's reaction to ethnic diversity.

2 It has become conventional in many British policy and research circles to use the term "Gypsies and Travellers" in recognition of the diverse ethnic origins of the groups who have a nomadic way of life or would identify themselves with a tradition of such a way of life. Among these groups are people who might in some contexts refer to themselves as Romany, and might elsewhere in Europe be termed "Roma." Other groups are sometimes called Irish Travellers and Scottish Travellers. Talking in these terms can tend to mislead about the degree of intermingling and intermarriage between these groups. For example, one of the authors has attended a wedding involving a red-haired, fair-skinned young man and an olive-skinned, black-haired young woman, both "Gypsy" or "Traveller" residents of the same city. The term "Gypsies and Travellers" does not include so-called New Age Travellers – i.e. first or second generations of people who are opting out of mainstream society.

REFERENCES

Allmendinger, P., & Haughton, G. (2012). Post-political spatial planning in England: A crisis of consensus? *Transactions of the Institute of British Geographers*, *37*(1), 89–103. http://dx.doi.org/10.1111/j.1475-5661.2011.00468.x

Anderson, K. (2000). Thinking "postnationally": Dialogue across multicultural, indigenous and settler spaces. *Annals of the Association of American Geographers*, *90*(2), 381–391. http://dx.doi.org/10.1111/0004-5608.00201

Anwar, M. (1979). *The myth of return – Pakistanis in Britain*. London: Heinemann.

Back, L., Keith, M., Khan, A., Shukra, K., & Solomos, J. (2002). New Labour's White heart: Politics, multiculturalism and the return of assimilation. *Political Quarterly*, *73*(4), 445–454. http://dx.doi.org/10.1111/1467-923X.00499

Bagguley, P., & Hussain, Y. (2006). Conflict and cohesion: Official constructions of "community" around the 2001 riots in Britain. In S. Herbrechter & M. Higgins (Eds.), *Returning (to) communities: Theory, culture and political practice of the communal* (pp. 347–366). Amsterdam: Editions Rodopi B.V.

Bailey, N., & Pill, M. (2011). The continuing popularity of the neighbourhood concept and neighbourhood governance in the transition from the "big state" to the "big society" paradigm. *Environment and Planning. C, Government & Policy*, *29*(5), 927–942. http://dx.doi.org/10.1068/c1133r

Ballard, R. (1994). Introduction – the emergence of Desh Pardesh. In R. Ballard (Ed.), *Desh Pardesh – the South Asian presence in Britain* (pp. 1–34). London: Hurst and Company.

Beebeejaun, Y. (2006). The participation trap: The limitations of participation for ethnic and racial groups. *International Planning Studies, 11*(1), 3–18. http://dx.doi.org/10.1080/13563470600935008

Brindley, T., Rydin, Y., & Stoker, G. (1996). *Re-making planning* (2nd ed.). London: Routledge.

Brownill, S., & Thomas, H. (2001). Urban policy deracialized? In O. Yiftachel et al. (Eds.), *The power of planning* (pp. 189–203). Dordrecht: Kluwer Academic Publishing.

Cameron, D. (2011). Speech at Munich Security Conference, February 5, 2011. Retrieved May 22, 2015 from https://www.gov.uk/government/speeches/pms-speech-at-munich-security-conference.

Cantle, T. (2001). *Community cohesion: A report of the Independent Review Team.* London: Home Office.

Castles, S., & Miller, M.J. (2009). *The age of migration: International population movements in the modern world.* (4th ed.). Basingstoke: Palgrave Macmillan.

Cohen, S. (2002). The local state of immigration controls. *Critical Social Policy, 22*(3), 518–543. http://dx.doi.org/10.1177/026101830202200308

Corbett, S., & Walker, A. (2012). The big society: Back to the future. *Political Quarterly, 83*(3), 487–493.

Craig, G. (2011). Forward to the past: Can the UK Black and minority ethnic third sector survive? *Voluntary Sector Review*, 2(3), 367–389. http://dx.doi.org/10.1332/204080511X608780

CRE. (2007). *Regeneration and the race equality duty: Report of a formal investigation in England, Scotland and Wales*. London: CRE.

Dahya, B. (1974). The nature of Pakistani ethnicity in industrial cities in Britain. In A. Cohen (Ed.), *Urban ethnicity* (pp. 77–118). London: Tavistock.

Denham, J. (2001). *Building cohesive communities: A report of the independent review team.* London: Home Office.

Department of Communities and Local Government (UK). (2012). *Progress report from the ministerial working group on tackling inequalities experienced by Gypsies and Travellers*. London: DCLG.

Ellis, G. (2001). The difference context makes: Planning and ethnic minorities in Northern Ireland. *European Planning Studies, 9*(3), 339–358. http://dx.doi.org/10.1080/713666480

Fainstein, S. (2010). *The just city*. London: Cornell University Press.

Flett, H., Henderson, J., & Brown, B. (1979). The practice of residential dispersal in Birmingham 1969–1975. *Journal of Social Policy, 8*(03), 289–309. http://dx.doi.org/10.1017/S0047279400009016

Foley, D. (1960). British town planning: One ideology or three? *British Journal of Sociology, 11*(3), 211–231. http://dx.doi.org/10.2307/586747

Gale, R.T. (2004). The multicultural city and the politics of religious architecture: Urban planning, mosques and meaning-making in Birmingham, UK. *Built Environment, 30*(1), 30–44. http://dx.doi.org/10.2148/benv.30.1.30.54320

Gale, R. (2008). Locating religion in urban planning: Beyond "race" and ethnicity?*Planning Practice and Research, 23*(1), 19–39. http://dx.doi.org/10.1080/02697450802076415

Gale, R. (2013). Religious residential segregation and internal migration: The British Muslim case. *Environment & Planning A, 45*(4), 872–891. http://dx.doi.org/10.1068/a4515

Gale, R.T., & Naylor, S. (2002). Religion, planning and the city: The spatial politics of ethnic minority expression in British cities and towns. *Ethnicities, 2*(3), 387–409. http://dx.doi.org/10.1177/146879680200200305601

Garapich, M.P. (2008). The migration industry and civil society: Immigrants in the United Kingdom before and after EU enlargement. *Journal of Ethnic and Migration Studies, 34*(5), 735–752. http://dx.doi.org/10.1080/13691830802105970

Geoghegan, J. (2013). People's choice. *Planning*, March 25, 16–19.

Greed, C. (2005). Overcoming the factors inhibiting the mainstreaming of gender into spatial planning policy in the United Kingdom. *Urban Studies (Edinburgh, Scotland), 42*(4), 719–749. http://dx.doi.org/10.1080/00420980500060269

Healey, P. (2010). *Making better places: The planning project in the twenty-first century*. Basingstoke: Palgrave Macmillan.

Herbert, J. (2008). *Negotiating boundaries in the city*. Aldershot: Ashgate.

Hesse, B. (2000). *Un/settled multiculturalisms – diasporas, entanglements, transruptions*. London: Zed Books.

Inch, A. (2010). Culture change as identity regulation: The micro-politics of producing spatial planners in England. *Planning Theory & Practice, 11*(3), 359–374. http://dx.doi.org/10.1080/14649357.2010.500133

Joppke, C. (1996). Multiculturalism and immigration: A comparison of the United States, Germany and Great Britain. *Theory and Society, 25*(4), 449–500. http://dx.doi.org/10.1007/BF00160674

Krishnarayan, V., & Thomas, H. (1993). *Ethnic minorities and the planning system*. London: RTPI.

Krumholz, N., & Forester, J. (1990). *Making equity planning work.* Philadelphia: Temple University Press.

Lewis, B. (2013).*Planning and travellers.* Written Ministerial Statement delivered on July 1, 2013. Retrieved August 20, 2013 from https://www.gov.uk/government/speeches/planning-and-travellers.

Lowndes, V., & Pratchett, L. (2012). Local governance under the coalition government: Austerity, localism and the "Big Society." *Local Government Studies, 38*(1), 21–40.

Lowndes, V., & Sullivan, H. (2008). How low can you go? Rationales and challenges for neighbourhood governance. *Public Administration, 86*(1), 53–74. http://dx.doi.org/10.1111/j.1467-9299.2007.00696.x

MacAuslan, P. (1980). *The ideologies of planning law.* Oxford: Pergamon.

MacPherson, W. (1999). *The Stephen Lawrence Inquiry – Report of an inquiry by Sir William MacPherson of Cluny.* London: HMSO.

McKenzie, K. (1999, Mar.6) Something borrowed from the blues? We can use Lawrence Inquiry findings to help eradicate racial discrimination in the NHS. *British Medical Journal, 318*(7184), 616–617.

Marcuse, P., Connolly, J., Novy, J., Olivo, J., Potter, C., & Steil, J. (Eds.). (2009) *Searching for the just city.* Abingdon: Routledge.

Modood, T. (2007). *Multiculturalism.* Cambridge: Polity Press.

Mooney, G., and Scott, G. (Eds.). (2012). *Social justice and social policy in Scotland.* Bristol: Policy Press.

Nadin, V. (2007). The emergence of the spatial planning approach in England. *Planning Practice and Research,* 22(1), 43–62. http://dx.doi.org/10.1080/02697450701455934

Nye, M. (2001). *Multiculturalism and minority religions in Britain.* London: Routledge.

Office of the Deputy Prime Minister (UK). (2003). *Planning and diversity: Research into policies and procedures.* London: ODPM

Office of the Deputy Prime Minister (UK). (2005). *Diversity and equality in planning: A good practice guide.* London: ODPM.

Osler, A., & Starkey, H. (2002). Education for citizenship: Mainstreaming the fight against racism? *European Journal of Education, 37*(2), 143–159. http://dx.doi.org/10.1111/1467-3435.00099

O'Toole, T., DeHanas, D.N., & Modood, T. (2012). Balancing tolerance, security and Muslim engagement in the United Kingdom: The impact of the "Prevent" agenda. *Critical Studies on Terrorism, 5*(3), 373–389. http://dx.doi.org/10.1080/17539153.2012.725570

Ouseley, H. (2001). *Community pride not prejudice.* Bradford: Bradford Race Review.

Peach, C. (2006). Islam, ethnicity and South Asian religions in the London 2001 census. *Transactions of the Institute of British Geography,* ns *31*(3), 353–370.
Peach, C., & Gale, R. (2003). Muslims, Hindus and Sikhs in the new religious landscape of England. *Geographical Review, 93*(4), 469–490.
Pennington, M. (2005). Competition in land use planning: An agenda for the twenty-first century. In P. Booth (Ed.), *Towards a Liberal Utopia?* (pp. 101–108). London: Institute of Economic Affairs.
Phillips, D. (2006). Parallel lives? Challenging discourses of British Muslim self-segregation. *Environment and Planning. D, Society & Space, 24*(1), 25–40. http://dx.doi.org/10.1068/d60j
Phillips, T. (2005). After 7/7: Sleepwalking to segregation. Speech to the Manchester Council for Community Relations, September 22. Retrieved August 20, 2013 from www.equalityhumanrights.com.
Punter, J. (2009). *Urban design and the British urban renaissance*. London: Routledge.
Ratcliffe, P. (2012). "Community cohesion": Reflections on a flawed paradigm. *Critical Social Policy, 32*(2), 262–281. http://dx.doi.org/10.1177/0261018311430455
Rose, E.J.B., & Deakin, N. (1969). *Colour and citizenship – A report on British race relations*. Oxford: Oxford University Press.
Royal Town Planning Institute. (2001). *A new vision for planning*. London: RTPI. Retrieved August 20, 2013 from http://www.rtpi.org.uk/media/9321/RTPI-New-Vision-for-Planning.pdf.
Royal Town Planning Institute/Commission for Racial Equality. (1983). *Planning for a multi-racial Britain*. London: Commission for Racial Equality.
Runnymede Trust. (2000). *The future of multi-ethnic Britain*. London: Runnymede Trust.
Sandercock, L. (2003). *Cosmopolis II: Mongrel cities in the 21st century*. London: Continuum.
Shirlow, P., & Murtagh, B. (2006). *Belfast: Segregation, violence and the city*. London: Pluto Press.
Simpson, L., & Finney, N. (2009). Spatial patterns of internal migration: Evidence for ethnic groups in Britain. *Population, Space and Place, 15*, 37–56.
Spencer, I. (1997). *British immigration policy since 1939: The making of multi-racial Britain*. London: Routledge.
Stillwell, J. (2010). Ethnic population concentration and net migration in London. *Environment and Planning A, 42*, 1439–1456.
Taylor-Gooby, P., & Stoker, G. (2011). The Coalition programme: A new vision for Britain or politics as usual? *Political Quarterly, 82*(1), 4–15. http://dx.doi.org/10.1111/j.1467-923X.2011.02169.x

Tewdwr-Jones, M., & Allmendinger, P. (Eds.). (2006). *Territory, identity and spatial planning: Spatial governance in a fragmented nation*. Abingdon: Routledge.
Thomas, H. (1999). Social town planning and the planning profession. In C. Greed (Ed.), *Social town planning*. London: Routledge.
Thomas, H. (2000). *Race and planning: The UK experience*. London: UCL Press.
Thomas, H. (2002). Equality and planning in Wales. *Town and Country Planning* (May), 142–143.
Williams, C., Evans, N., & O'Leary, P. (Eds.). (2003). *A tolerant nation? Exploring ethnic diversity in Wales*. Cardiff: University of Wales Press.
Worley, C. (2005). "It's not about race. It's about the community": New Labour and community cohesion. *Critical Social Policy*, *25*(4), 483–496. http://dx.doi.org/10.1177/0261018305057026
Young, C., Diep, M., & Drabble, S. (2006). Living with difference? The "cosmopolitan city" and urban reimaging in Manchester, UK. *Urban Studies (Edinburgh, Scotland)*, *43*(10), 1687–1714. http://dx.doi.org/10.1080/00420980600888486
Young, K., & Connelly, N. (1984). After the act: Local authority policy reviews under the Race Relations Act 1976. *Local Government Studies*, *10*(1), 13–25. http://dx.doi.org/10.1080/03003938408433124

PART 2

Expanding the Boundaries of the Multicultural Discourse

6 Planning for Sexual and Gender Minorities

PETRA L. DOAN

Being the supreme crossers of cultures, homosexuals have strong bonds with the queer white, Black, Asian, Native American, Latino, and with the queer in Italy, Australia, and the rest of the planet. We come from all colors, all classes, all races, all time periods. Our role is to link people with each other – the Blacks with Jews with Indians with Asians with whites with extraterrestrials. It is to transfer ideas and information from one culture to another. Colored homosexuals have more knowledge of other cultures, have always been at the forefront (although sometimes in the closet) of all liberation struggles in this country and have suffered more injustices and have survived them despite all odds.

Gloria Anzaldua from *Borderlands La Frontera: The New Mestiza*, pp. 106–107

Introduction

In the latter half of the twentieth century, gay and lesbian neighbourhoods developed in many metropolitan areas in Europe and North America, increasing the visibility of many lesbian, gay, bisexual, or transgender urban residents. In the US context, the Stonewall rebellion in 1969 marks a watershed in the gay and lesbian movement (Carter, 2010), stimulating a greater willingness for many LGBT individuals to be open about their sexuality and claim their place in the city. Some scholars use the term *queer spaces* to describe these residential areas, though they are popularly termed gay villages, gayborhoods, lesbivilles, and boystowns. One observer argues that many of these queer spaces are highly commodified habitation zones for White and wealthy gay men (Nast, 2002), and another wonders why gay ghettoes are

White (Nero, 2005). Browne (2006) challenges scholars to embrace a more radical notion of queerness that not only interrogates the racial and class positions of gay and lesbian individuals but also attempts to deconstruct the categories themselves, echoing Peake's (1993) call to challenge the patriarchal structuring of urban areas.

The term *intersectionality* describes the complex network of discriminations, including those experienced by LGBT people excluded from queer spaces by virtue of their race or class (Valentine, 2007). Brown (2012) adds intersections with ability and age as well as gender identity to the increasingly multifaceted web of discriminations. Scholars in the field of Black Queer Studies have added critical new dimensions to this concept by suggesting that "a broadened understanding of queerness must be based on an intersectional analysis that recognizes how numerous systems of oppression interact to regulate and police the lives of most people" (Cohen, 2005, p. 25). In a critical twist, Cohen indicates that "I do not consider myself a 'queer' activist or for that matter a 'queer' anything ... like other gay, lesbian, bisexual, and transgender activists of colour, I find the label 'queer' fraught with unspoken assumptions that inhibit the radical potential of this category" (p. 35). Johnson (2005) goes a step further and argues that queer studies for people of colour should really be called *quare* studies, because most scholarship on LGBT issues completely omits any discussion of race. Finally, Walcott (2007) observes that while Black gay men inhabit the margins of Toronto's queer spaces, they still exert an outsized influence on the gay social scene through influential drag performances and popular house music, and by acting as promoters of the local gay party scene, thereby serving as "imaginative conduits of queerness" (p. 243).

Unfortunately, planning academics and professionals have been slow to adopt a more explicitly multicultural approach that might address minority group concerns in cities (Burayidi, chapter 1). While there is a growing literature on the impact of planning on people of colour in cities (Ritzdorf, 1997, 2000; Thomas, 1994, 1998, 2008; and also Connerly, 2002), there has been virtually no discussion of how planning might or might not be meeting the needs of LGBT people of colour. This chapter cannot really fill this gap (since the author identifies as a White transgendered woman), but it takes a first step by providing notice of this important gap in the literature.

The field of planning has been even slower to recognize that the LGBT population might constitute a marginalized community worthy of attention (Doan, 2011). Frisch's (2002) contention that planning has long been a heterosexist project intended to make the city safe for heterosexuals, often at the expense of LGBT people, still rings true. In support of this argument, Forsyth (2001) provides an extensive treatment of the range of planning issues linked to non-normative (LGBT) populations. Doan and Higgins (2011) provide a specific example of the way that planning can influence redevelopment and gentrification in the Midtown neighbourhood of Atlanta, a long-standing gay village. Their analysis questions the role of planning professionals in a redevelopment process that ignored the presence and needs of the gay community almost completely. Finally, the edited volume titled *Queerying Planning: Challenging Heteronormative Assumptions and Reframing Planning Practice* (Doan, 2011) provides a broader overview of recent LGBT scholarship in planning and geography.

This chapter considers how using a multicultural framework with respect to LGBT people might ameliorate the gaps noted above. The chapter uses case material from Atlanta to explore how planning practice has failed to consider the needs of LGBT-identified individuals, both those people living in queer-identified areas and those living beyond queer spaces, especially those whose identity is shaped by intersectionality. The Atlanta case is an instructive example of how the failure to recognize LGBT populations as key stakeholders in neighbourhood development has led to the gradual transformation of Midtown from a gay neighbourhood to a more upscale environment which has been made safe for heterosexuals and their families, without consideration of LGBT individuals and their families. The chapter then extrapolates lessons for planning practice to incorporate the needs of the entire spectrum of non-normative populations.

Differing Visions of Multicultural Planning

Traditionally, culture is considered linked to family origins or heritage, making language, ethnicity, and religion well recognized as cultural components. However, identity-based groups can also be a vital part of the melting pot of culture that is multiculturalism, although they are often overlooked. Large-scale demographic shifts in the United States (James, 2000) have challenged planners to rethink the utility of one-size-fits-all planning and adopt practices that recognize the diverse publics that any given plan may be trying to serve. Failure to recognize that there is no single undifferentiated community with whom to plan has led to the failures of what Burayidi (2000) calls "monistic planning," which has dominated US planning practice for most of the twentieth century.

The slowness of American planning to recognize and plan for LGBT communities, described above, seems linked to an underlying reluctance to admit that culture is not just a "born this way" characteristic but is linked to identity. The public discourse surrounding sexual orientation and gender identity often reinforces this narrow conception. Many more conservative Americans are reluctant to allow "rights" to be granted to sub-populations that represent a "lifestyle choice" and not an immutable characteristic (Wald, Button, & Rienzo, 1996; Haider-Markel & Meier, 1996). The sharpness of these debates has made some localities unwilling to expand their understanding of culture to include identity groups.

In contrast, multiculturalism in other countries has taken a somewhat broader definition of what constitutes culture. The United Kingdom has a strong commitment to understanding the critical importance of differences. One British planning professor argues that collaborative planning discourse requires the recognition of difference among various stakeholders (Healey, 1997). Another British planner contends that planning should not be restricted "to those with whom we share a common identity and/or encounter through face-to-face contact" (Campbell, 2006, p. 100). In Australia the notion of multiculturalism has been expanded to cultural pluralization, where culture is broader than ethnicity, including the expression of any human social identity (Gleeson & Low, 2000). Sandercock (1998) adds that a "just city" is socially inclusive where difference is given recognition and respect.

In the Canadian context, planners argue that multiculturalism as public policy needs to emphasize "racial and cultural differences in a society" (Qadeer, 1997). Frequently multicultural planning is interpreted to mean that planners should be aware of ethnic and cultural sensitivity in cities (cf. Qadeer, chapter 3). Implicit in these exhortations is the notion that the planning institutions are covertly discriminatory against ethnic and racial minorities. They are guided by the values and preferences of the dominant majority, embedded in the singularity of public interest and incorporated in planning policies and standards (Qadeer, 2009, p. 10).

Fincher and Iveson (2008) provide the most explicit statement of how multicultural planning does in fact work and then suggest three social logics (redistribution, recognition, and encounter) that would better ensure that all city residents can participate in the just city. In their discussion of the second logic, recognition, they argue that recognizing the presence of LGBT residents is a fundamental element in planning for diversity. The discussion of the Atlanta case which follows uses this framework to show the ways that failure to recognize the validity of LGBT people by city government can lead to displacement and discrimination. In addition, when neighbourhood residents encounter sexual and gender minorities, there is often initial turmoil, but also the possibility for subsequent understanding and positive change.

Atlanta Case Study

Over the past 50 years, the city of Atlanta has seen rapid growth in population and employment, transforming itself from a city originally built as a railway hub to a world-class airport hub with three interstate highways enabling the city to become a major centre for logistics and professional services (Ross, Sjoquist, & Wooten, 2009). Atlanta has also been described as a city of contrasts, "two largely separate cities: a mostly white north side of town, where economic activity is vigorous and expanding, and a mostly black south side" (Keating, 2001, p. 8). The African American south side is further segmented into relatively dynamic Black middle-class areas and very poor Black neighbourhoods (Keating, 2001). In recent years, Atlanta has seen substantial growth in other minority populations, especially Latinos and Asians, who have added greater diversity to the population base (Gallagher, 2009), making the city a fine case for examining multicultural planning.

Business interests have dominated development decision making for years, resulting in rapid growth in employment and overall population in the metropolitan area, but also high rates of inequality between these divided cities. Atlanta's history of neighbourhood preservation, especially for minority communities, has been abysmal. Keating (2001) documents the large-scale destruction of smaller African American communities that were inconveniently located in areas the business community wished to see develop. Over the past 50 years, new interstate highways, several stadiums, and the Olympic village were built in the midst of poor Black communities, forcing the dislocation of residents and further dividing the city.

This same attitude of business first and ethnic/cultural community second can be seen in the redevelopment of the Peachtree corridor in Atlanta's Midtown area (see figure 6.1). The Midtown neighbourhood has been the heart of Atlanta's LGBT community since the early 1970s

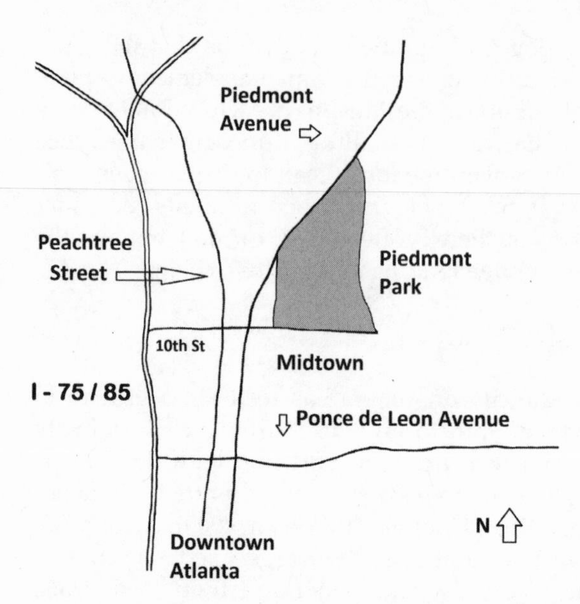

Figure 6.1. Map of Midtown Atlanta showing Piedmont Park, Peachtree Street, Ponce de Leon Avenue, and Piedmont Avenue. This neighbourhood became the centre of gay life in the city in the 1990s, and Atlanta's LGBT community moved into it and started its revitalization.

and more recently has been subject to considerable development pressure (Doan & Higgins, 2011). The western edge of Midtown is defined by the I-75/85 corridor; the eastern edge is formed by Piedmont Park and Monroe Drive; and the southern border is approximately Ponce de Leon Avenue. Peachtree Street is a central north-south corridor from Buckhead through Midtown to downtown, as is Piedmont Avenue. The Metropolitan Atlanta Rapid Transit Authority runs underneath Peachtree for miles, with several stops in the Midtown area.

While there has been no full-length treatment of Atlanta's gay and lesbian history, several contributors to edited books on LGBT issues in the South suggest that Midtown (Howard, 1997) and Candler Park were attractive locations for gay men and lesbians respectively during the 1970s (Sears, 2001; Chesnut & Gable, 1997). In these neighbourhoods, White flight that followed the fear that African Americans might move into nearby neighbourhoods, as well as the desegregation of public transit, parks, and schools (Kruse, 2005), opened up space into which these otherwise marginal communities could move. DeGiovanni (1983) found that during the 1970s Midtown was undergoing intensive gentrification, although he did not discuss the sexuality of those engaged in gentrifying these areas. Anecdotal evidence suggests that gays and lesbians, attracted by the presence of gay bars and other community institutions, began moving into these neighbourhoods and investing in the deteriorated housing stock.

By the 1990s Midtown was widely recognized as the centre of gay life in Atlanta, and nearby Piedmont Park was the location for gay pride events which became a focal point for other LGBT community members across the Deep South. Commercial development oriented to LGBT people soon followed, with a number of gay bars, a bookstore, and other community institutions locating in the neighbourhood.

> For gay Atlantans, the growth in organizations was accompanied by development of their own spaces (originally Midtown, with others added during the 1980s and 1990s) that could be considered gay ghettos or safe havens. These areas were more than cultural or social settings; they also represented important voting blocs. (Fleischmann & Hardman, 2004, p. 418–419).

Redevelopment without Recognition

The expansion of the Atlanta transit system prompted many in the development community to re-examine the Peachtree Street corridor that ran through the Midtown neighbourhood. Increasing demand for commercial and residential properties along this corridor spurred new high-rises, and a wave of intensive gentrification pressure spilled out into the Midtown area. The importance of this corridor to development interest was highlighted by the appointment of the Peachtree Corridor Task Force, which created the Peachtree Vision to guide development. This vision of Peachtree Street as "the Champs-Élysées, the Magnificent Mile, Broadway, and Fifth Avenue with a Southern twist" (City of Atlanta, 2007) purports to recognize the many "faces of diversity," but in reality is concerned with encouraging families from the suburbs to bring their children to the city to see the sights. Such a narrow vision is in direct conflict with the existing use of the area, including the Midtown gayborhood, by LGBT residents and visitors. In Fincher and Iveson's (2007) framework, this conflict in vision constitutes a distinct lack of "recognition" of the existing use of the space by LGBT people.

The lack of recognition of the LGBT nature of the area is exacerbated by the efforts to "clean up" the image of Midtown by ridding the area of gay bars and nightclubs that do not fit the "family-oriented" vision of the Peachtree Corridor Task Force. Consequently, a variety of mechanisms were used to curtail LGBT bars and clubs. Doan and Higgins (2011) suggest that the expansion of special public interest zoning into the Midtown neighbourhood contributed to the closing of gay and lesbian bars in Midtown. Other measures include the strict enforcement of liquor licensing laws (Woods, 2004), shuttering after-hours private clubs (Doan, 2014), and police raids (see for example the raid on the Atlanta Eagle, a gay leather bar cited by Rankin, 2009), as well as the bombing of a lesbian bar called the Otherside Lounge (Gover, 1997).[1]

These efforts to close gay bars were a means of preparing the neighbourhood for more intense gentrification. Specifically, the closures of several large clubs along Peachtree Street (Backstreet, the Armory, and the Metro Video Bar) changed the way gay men in particular used this part of Midtown. One striking feature of Peachtree Street that remains out and proud is a club called Bulldogs. Contrary to Nero's (2005) assertion that most gay ghettoes are generally White, this club caters to gay African Americans, a reflection of the growing political clout of Black gay Atlantans, who in recent years have organized Black Gay Pride Weekend as a separate event from the Atlanta Pride festival (Jarvie, 2006).

In contrast to Bulldogs, the other iconic gay bar along Peachtree Street in Midtown, Backstreet, operated for many years as a private club and served liquor 24 hours a day. The visibility of Backstreet and its all-night party reputation made it ground zero in the efforts to get gay clubs off Peachtree Street. These efforts were orchestrated by an individual who lived across the street from the bar and had a leadership role in the Midtown Neighbourhood Association and also the Midtown Ponce Security Alliance. She made numerous complaints about the bar to demonstrate that it was not a "good neighbour" and encouraged the city to shut it down (interview with Philip Rafshoon).

Other members of the Midtown Ponce Security Alliance were willing to work "to ensure Backstreet remains closed" (Henry, 2004), because removing it enabled the gentrification process to continue so that future condo residents would not have to share the street with noisy gay bars. The alliance was also involved in other efforts to cleanse the neighbourhood and make it safe for middle-class heterosexuals, including efforts to rid the area of the Metro Atlanta Task Force for the Homeless

(Cardinale, 2009). After Backstreet was closed and torn down, a new condominium tower called Viewpoint was erected on the site.

While some commentators question whether gay and lesbian bars can survive in the twenty-first century (Lee, 2007), others wonder whether the gayborhood itself is disappearing, since these prominent bar closures have shifted "gay culture" into more peripheral areas. During the 1990s, as many as 17 establishments were located in Midtown, but by 2011 only four LGBT businesses were still present (Adriaenssens, 2011). Since that study, one of those businesses, Outwrite Bookstore and Coffeeshop, a visible symbol of the queer community for more than 15 years at the corner of Tenth and Piedmont, was closed for financial reasons. The former owner links the closure of the bars with his subsequent decision to close the bookstore:

> As the clubs closed, there were less people out on the street at night, making it, in some ways, feel less safe. At the Outwrite, we could see a sharp decline in LGBT tourists who would come to Atlanta on the weekend. Where the nightlife had been a major attraction to visitors, it now became less appealing, and many of those tourists stopped visiting or shortened their trips. And that was one of the things that cut into our business. We saw our tourist business go from a regular boost from Friday afternoon through Monday morning to a Saturday and Sunday only weekend tourist business. The business just began drying up. (Interview with Philip Rafshoon, March 2013)

The long-term consequences of all these closures on the gay nature of Midtown are not fully clear. Certainly the loss of many bars and clubs as well as Outwrite Books has had a decentring effect on the LGBT community. Planners and developers failed to understand the importance of LGBT social venues for building community in an otherwise heteronormative urban area. A more nuanced understanding of LGBT culture might have enabled a more proactive neighbourhood preservation policy.

Encountering and Demonizing Non-Normative Genders

In addition to the failure of Atlanta officials to recognize the importance of Midtown as an important LGBT neighbourhood, there have also been difficulties linked to the ongoing encounter with sexual and gender differences. Evidently, the heteronormative vision of the Peachtree corridor also problematizes non-normative gender presentations. Some Midtown residents encounter people displaying non-normative genders in public and assume that they would only violate gender norms to engage in sex work or for other undesirable reasons. It is certainly true that many transgendered people in urban areas experience high levels of harassment (Doan, 2007; Doan, 2010), and some of them also have trouble finding employment in a highly gender-normative world.

When transgendered individuals are also people of colour, the intensity of discrimination can be fierce. A recent example of this kind of transgender demonization intersecting with race is the case of Baton Bob, a gay street performer who typically dresses in some kind of costume (often a tutu) and twirls a baton to amuse people passing by. He calls himself the Ambassador of Mirth and has marched in the annual Atlanta Pride parade. However, he is also a lightning rod for disapproval and has received threats of violence (Bagby, 2012b), presumably from those who are offended by his "outness" and his public performance of non-normative gender and sexuality. He has also had several tangles with the law, most recently while performing in Midtown where an off-duty police officer working as a security guard was offended by his performance and, after an exchange of words and threats, had Bob arrested (Bagby, 2013b; Douglas-Brown, 2013). When this kind of encounter results in the arrest of a queer public figure, no matter who initiated the conflict, it contributes to a hetero-normalizing atmosphere.

A more serious example of unfortunate encounters revolves around the presence of transgendered sex workers. Because of the employment discrimination against transgendered individuals, some proportion of the transgendered population may end up engaging in sex work for survival. Furthermore, these sex workers are highly marginalized by other sex workers and are often forced to work in poorly lighted, remote areas that place them at risk (Weinberg, Shaver, & Williams, 1999). As a result, it is not surprising that they would seek out the margins of supposedly safe queer areas. This behaviour has been documented in several cities, including Toronto (Baute, 2008) and San Francisco (Edelman, 2011), where transgendered sex workers frequent the edges of the gayborhood, but are specifically targeted by private security associations in gentrifying areas. This overt discrimination is exacerbated when the transgendered sex workers are also people of colour, since in many cases the residents most offended by the transgendered sex workers are gay White males. It seems that transgendered individuals are perceived as less wholesome than gay men, and when they resort to sex work, they must be driven out of the neighbourhood.

Certain areas of Midtown have been frequented by sex workers for many years, long before the current rush to gentrification. Many of these sex workers are transgendered and attract clients by their visibility on the street. The Midtown Ponce Security Alliance has decided that the presence of what they identify as "cross-dressing prostitutes" constitutes a threat to the neighbourhood and has gone on a virtual crusade that specifically targets transgendered individuals by circulating photos and detailed descriptions of "repeat offenders" on the alliance website and in its neighbourhood circulars. From the website it appears that many of the sex workers who have been targeted are transgendered people of colour. A recent article in the queer press reports that the vice president of the Alliance is an openly gay White man who patrols the neighbourhood in an Alliance pickup truck looking for sex workers, but he characterizes his efforts "not as policing or vigilantism, but as a form of protest" against the presence of cross-dressing prostitutes in the Midtown area (Nouraee, 2008). Unfortunately, according to several transgender rights activists quoted in the same article, he sometimes confuses transgendered individuals who are out at night with sex workers. Several Midtown Ponce Security Alliance circulars described a "gang of transvestite prostitutes" as threatening Midtown residents (the actual reports have been taken down from their website), but the Atlanta police dismiss this claim as hyperbole, reporting that there is neither greater crime in this neighbourhood than in other Atlanta neighbourhoods nor evidence of prostitutes operating in gangs. However, the police spokesperson did suggest that when citizens follow and shine very bright lights on people walking in the street, the response of most people would be fairly strong (Bagby, 2012a).

Decision makers in Atlanta tried to go one step further in February 2013 when the chief of police and the mayor attempted to pass a municipal ordinance, Stay Out of Areas of Prostitution, that was aimed at the problem of repeat prostitution (McWilliams, 2013). This "Banishment Ordinance" would ensure that any individual convicted of prostitution would be fined and required to stay out of areas of known prostitution. A second offence would lead to outright banishment from the city for 120 days. The private individuals pushing this ordinance have not been identified, but the Alliance actively urged members of the Midtown Neighbourhood Association to appear at the February 2013 city council meeting and provide a visible support for the ordinance. A number of other social justice and LGBTQ activists also attended the meeting and raised such an outcry that the ordinance was tabled. These LGBT rights organizations argued that the ordinance was "rooted in homophobia, transphobia, and racism ... By and large, engaging in sex work is an act of survival, not of choice" (Bagby, 2013a).

Other LGBT Midtown residents sent a letter that criticized the Security Alliance for excessive focus on the sex work by transgendered and cross-dressing individuals as reported by *Project Q*:

> We live and work in Midtown, one of the areas most mentioned in this conversation. The news seems obsessed with a few people's claims that there are condoms and syringes littering our sidewalks. This is simply not true ... We understand that some of our neighbors are getting an outsized amount of attention in this debate. As Midtown residents and business owners, we want to make it clear that they do not speak for the entire neighborhood and, in fact, the vast majority of us would much prefer to see the Council lead with solutions to help these women leave the streets for good. Please do not allow this terrible policy to be enacted in the name of Midtown residents and businesses. (Quoted in Hennie, 2013)

Though the Stay Out of Areas of Prostitution ordinance is "dead, and likely to stay dead" (Interview with City Councillor Alex Wan, March 2013), a similar lack of understanding for the powerful impact of intersectional identity is clearly evident. The White upper-middle-class residents of Midtown, including some gay men, who strongly supported this ordinance, appear to have little understanding of the precarious position of transgendered sex workers and in particular sex workers of colour. The fact that key municipal officials agreed to push for such an ordinance suggests a significant failure to understand the real situation of this highly marginalized identity group. A more inclusive multicultural approach might have found a more nuanced middle ground.

Encountering Diversity through Gentrification

Another example of Fincher and Iveson's encounter can be seen in the spread of LGBT residents outside of the traditional gayborhood in Midtown. In this case, encountering LGBT residents at first generated great resistance, but over time this has mellowed somewhat. As the Midtown gayborhood was undergoing considerable development pressure and even experiencing the racist transphobia described above, some LGBT people moved elsewhere for more affordable housing in other parts of the city as well as in nearby communities. Areas like Grant Park, Kirkwood, Edgewood, East Atlanta, Decatur, and East Point have experienced substantial gentrification, much of it by LGBT people unable to afford the high rents in the Midtown area (Doan & Higgins, 2011).

Kirkwood was originally a White working-class community in Atlanta founded around 1910. By 1957 many families had lived on the same block for more than five years, according to a housing survey conducted at that time (cited by Kruse, 2005). However, a few years earlier, Black families had begun moving into adjacent Moreland Heights, where many of the White residents were elderly, setting off a frenzy of consternation. White families in the Kirkwood area rallied around their churches and formed the Kirkwood Churches Committee, later the Kirkwood Community Committee, to try to maintain the all-White nature of the neighbourhood by purchasing the homes of those who wished to sell out and applying community pressure to others who were considering selling (Kruse, 2005). In spite of these efforts, by the mid-1960s the neighbourhood had become almost entirely African American.

It is therefore quite interesting that Kirkwood was once again in the headlines in the late 1990s because of tensions between majority Black residents and LGBT people who began moving into the neighbourhood. Beginning in the 1980s, gay men and lesbians who were attracted to Kirkwood's historic housing stock started to purchase property in the neighbourhood (Chapman, 2001). Fleischmann and Hardman (2004) note the difficulties of "neighborhood transitions when white gays moved into black neighborhoods or when young heterosexuals moved into areas after gay pioneers" (pp. 418–419). By 1998, the conflict reached a tipping point when an African American minister began inciting his parishioners to "put an end to the homosexual and lesbian takeover of our community" (Schrade, 1998). At the same time, the city council member representing Kirkwood jumped on the bandwagon and began criticizing "those homosexuals," and a political firestorm erupted. The tensions lingered until she lost her bid for re-election in 2001 (Bennett, 2001).

In his analysis of gentrification in the neighbourhoods of East Lake, East Atlanta, and Edgewood, which are proximate to Kirkwood, Aka (2010) finds that the Black population decreased between 1990 and 2000 by 20 per cent, 75 per cent, and 15 per cent respectively. In Edgewood (immediately adjacent to Kirkwood) the White population rose from 20 to 252 (an increase of 1160 per cent). In addition to lower prices, many of the LGBT in-movers were actively seeking a more diverse community. For instance, a lesbian interviewee from one neighbourhood just south and east from Kirkwood noted:

> When we moved into our neighbourhood, it was predominantly African American. I had some reservations. Not because I didn't want to live in the neighbourhood that was all African American, but I was a little bit concerned about the way they would look on us. But I was also concerned – and with some reason as it turns out – that the neighbourhood would be changed by the fact that we were there ... The majority of the people living in the neighbourhood were over 60 or 70 ... within a year or less of our moving in, several of the places that, I suspect, had been rented suddenly went on the market, because the owners went, "Oh, we can sell these now because there's White people who are interested." And I kind of went, Oh, shoot, what have we done? It's fairly mixed now. Although the interesting thing to me ... it seems like although White people are moving in, there's some new African Americans moving in as well. So, we're not becoming a totally White neighbourhood, thank God! (White lesbian from south of Decatur)

As this quote makes clear, once places like Kirkwood are seen to be accepting of queer folks, the gentrification process begins to speed up. It is not certain whether neighbourhood preservation programs in places like Kirkwood could have stopped the in-movement of young LGBT people any more than a generation earlier the White residents of Kirkwood could have stopped the White to Black transition. The attraction of this neighbourhood was in part its lower prices compared to more expensive properties in Midtown, Virginia Highlands, and Candler Park, as well as the quality of its housing stock. One African American woman who had returned to Atlanta after a number of years described her dream of buying a house in Kirkwood as follows:

> It's the most beautiful old area I think in the city. It has the biggest Victorian houses. One of them that I want, I got my eye on. It's falling apart though. It's like a haunted house. Nobody lives there, and it's all crappy looking but ... Decatur's right on the other side of that. I love it. It's convenient to I-20. (Straight African American woman)

This respondent subsequently explained that originally her family had lived closer to Decatur but later moved out to the more affluent suburb of Lithonia, which she describes as a land of McMansions, at one time the second most affluent African American community in the country. This movement to the suburbs by African Americans suggests that some of the displacement from gentrification may have been due to African Americans wanting larger houses and better services. However, her interest in returning to in-town living suggests that life in the suburbs is too far from the centre of town.

For gay African Americans this return to the city is even more important. One respondent, a gay African American man, told a similar story. In his interview he explained that he had bought a house in Stone Mountain (a suburb on the east side of Atlanta with a large African American community), but very quickly moved back into the city and rented a house.

> It was just too restrictive out there. I am a city person, so I found a place south of the capital near the stadium. But I am very comfortable at any place in Midtown ... The thing is there are different tastes between the Black and White communities, especially in terms of musical tastes. I like house music, and you just don't find that in most places in Midtown, so I mostly go to these mobile parties that are advertised by a text message the day of the party. (African American gay man)

For many LGBT people, diversity is a kind of code word for those who are more broadly accepting of LGBT issues. Podmore (2001) suggests that in Montreal the lesbian sections of the city are highly diverse places in which there is a broad climate of acceptance of difference. In Atlanta this also appears to be the situation. One lesbian interviewed for this chapter indicated that diversity was the number one factor in the last move she and her partner made. She was happy to live in an integrated neighbourhood, but did not want to be assimilated.

> Well, I think integration means I can live anywhere even though my partner is a woman and still feel part of a gay culture or lesbian culture. I think assimilation is trying to be like straight people, and I don't feel that the people I know necessarily do that. The people that I would hang out with, they're very strong in their identity ... It's just so you're living among, and feel enriched by living in, a diverse community. You know, I feel enriched by my straight friends. I feel enriched by my friends from Iran and Cuba. So I think it's very comfortable to live in a diverse place. And you know, comfortable being queer in a diverse place. (White lesbian from Decatur)

This understanding of diversity is significantly more complicated by intersectionality for people of colour who are also LGBT. Neighbourhoods that already have an established LGBT population are seen to be more welcoming than traditional African American areas.

> I probably always make the assumption that if it is queer-friendly, it's probably going to be African-American-friendly, you know. I've never been proven wrong on that assumption. So, I pretty much am comfortable in any queer neighbourhood. It doesn't have to be queer African American. I would like a neighbourhood that's diverse. But if it was an all-White male queer neighbourhood, I would have no problem with it. My partner would, because she likes diversity. And that would be my preference, but as far as it being all White, I wouldn't have a problem with it. (African American lesbian in a relationship with a White woman)

In contrast, majority African American neighbourhoods pose greater difficulties for LGBT people of colour. Here is where intersectionality really begins to have an impact, and individuals may feel compelled to choose whether to be Black or queer. One African American lesbian said:

> If you're out, as an African American lesbian, living in an African American community, I wouldn't say it's impossible, but it would be difficult, depending on the kind of community you wanted. Maybe if you would be younger, you know by Emory or another university. But if you're talking about your older, traditional African American communities … you're just not going to feel as comfortable as you are, you know, in Decatur. It's just, you have to say, "Do I want to be comfortable as a lesbian, or as an African American?" And lots of people do make that choice, and say, "No, I prefer to be in an African American community." So they would be more closeted. (African American lesbian in a relationship with a White woman)

Atlanta's lack of a multicultural approach to planning has hindered its ability to develop effective neighbourhood preservation programs. This situation is almost certainly a function of its abysmal history of White flight as Blacks began moving into White neighbourhoods in the 1960s. However, it is striking that in the early twenty-first century, the city continues to look the other way and simply allow development to run its course. There is no attempt to recognize or preserve multicultural neighbourhoods. Instead, Doan and Higgins (2011) note that "the politics of property appear to be trumping the more progressive policies that once characterized the Gay Liberation Movement." An example of this property focus is the following statement from gay City Councillor Alex Wan:

> Well, my mandate is to improve the quality of life of my constituents. It is to enhance the community in whatever way I can, and part of that effect is increased property values ... One of the consequences is the folks who can't afford the increases must then find other places to be. And one of the things that gets lost in that are these gayborhoods. (Alex Wan interview, March 2013)

The case of Kirkwood is interesting because in each period there were high levels of hostility expressed by existing residents towards the in-movers, who were seen as "the racial other" in the 1960s and 1970s and as a "cultural or lifestyle other" in the 1990s. Each group failed to recognize any value to that otherness, interpreting the differences as a threat to the status quo. Perhaps a critical role for multicultural planning in situations like these is to create more formalized mechanisms for the encounter and recognition suggested by Fincher and Iveson (2008). A more effective neighbourhood planning process that enabled real dialogue and encounters between different parties might have established an environment where the diversity that many LGBT people crave can be preserved, not gentrified out of existence.

Conclusion

This chapter has used the case of Atlanta to illustrate some of the challenges resulting from the failure to recognize the legitimacy of the LGBT population as a community for planning purposes. It is distressing that despite Atlanta's reputation as the heart of the more progressive "new South," the pervasive history of racial discrimination continues to dominate the urban development scene. As Keating (2001) notes, the delicate alliance forged between the White business community and middle-class African Americans continues to control development decision making. The spread of gentrification is in part a function of this partnership because it prioritizes capital formation over neighbourhood integrity. Because so many poor African American neighbourhoods were dismantled to create new infrastructure (highways and stadiums) and the Olympic Village, there may be a reluctance now to begin taking minority neighbourhoods more seriously. The need for a more progressive planning approach based on a broad awareness of multicultural planning that incorporates the entire range of cultural and identity groups as valid participants in the task of urban governance seems evident, but also unlikely in the present environment.

The consequences of this lack of recognition for the LGBT community include the significant changes to the "heart of the gay community" and the dislocation of LGBT individuals into other less central neighbourhoods in which LGBT individuals, couples, and families can find places where they can establish their homes, feel safe on the streets, and patronize local businesses. There are several important consequences of this displacement. First, it is becoming significantly more difficult to organize LGBT people for protest or for other activist concerns. As one politically active respondent suggested in an interview, it is not clear where and if the LGBT community will gather to celebrate or mourn the next major political or cultural event as Midtown becomes significantly less gay. In fact the LGBT community centre, the Phillip Rush Center, is now located on very busy Dekalb Avenue midway between downtown Atlanta and downtown Decatur in a kind of strip shopping centre with no real possibilities for hosting large community events.

Second, there are a series of spin-off effects on outlying neighbourhoods that are on the receiving end of this cascading gentrification. While encountering LGBT difference created some initial difficulties in Kirkwood, by now these concerns appear to have been reduced dramatically as neighbours came to know each other, or those most resistant to change moved elsewhere. However, as gentrification continues unabated, other areas are likely to feel the same kinds of pressure with no escape valves. If the city of Atlanta had worked to ensure the integrity of the LGBT character of Midtown, as well as discouraged condominium conversion of rental apartments in the area, the pressure to look elsewhere for queer space might have been significantly reduced.

While the Atlanta City Council has declared that it is against discrimination towards LGBT individuals, until recognition of the value of LGBT neighbourhoods is institutionalized as part of a broader multiculturalism, the council's proclamations will ring hollow. A prime example of this hollowness is the vagueness of the diversity description of the Peachtree corridor. Diversity does not just happen; it must be actively encouraged and preserved. The long-standing failures of this governing body to protect other multicultural neighbourhoods does not bode well for the long-term preservation of any community that values its culture over capital.

The lessons for planners elsewhere faced with similar issues hinge on the concepts of recognition and encounter. It is essential that urban planners and city officials expand their understanding of multicultural neighbourhoods to include and specifically to name their LGBT residents as key stakeholders. This recognition should enable and empower residents of these marginalized groups to have a greater voice in planning processes from which they are often excluded. Furthermore, the visibility that is likely to result from such public recognition may in turn promote a process in which neighbourhood residents from diverse backgrounds encounter each other in productive ways. As was the case in the Netherlands, discussed by Forester and Laws in chapter 4, these encounters may initially trigger tensions as in Kirkwood in Atlanta, but they are likely to eventually create greater understanding and neighbourhood cohesion.

NOTE

1 Although the Atlanta Eagle and the Otherside Lounge were not located on Peachtree Street, they were important bars in the vicinity. High-visibility news coverage of the events that occurred at these clubs contributed to an overall sense that the Midtown gayborhood was under threat from outside forces.

REFERENCES

Adriaenssens, Z. (2011). *Mapping the lesbian, gay, bisexual and transgender community in Atlanta*. Atlanta: Unpublished master's paper for the School of City and Regional Planning, Georgia Institute of Technology.

Aka, E.O. (2010). Gentrification and socio-economic impacts of neighborhood integration and diversification in Atlanta. *National Social Science Journal, 35*(1), 1–13. Retrieved May 19, 2015 from http://www.nssa.us/journals/2010-35-1/pdf/2010-35-1.pdf#page=3.

Bagby, D. (2012a, May 17). Police: Despite TV reports, no increased violence among Midtown crossdressing prostitutes. *Georgia Voice*. Retrieved April 4, 2013 from http://thegavoice.com/atlanta-police-no-increased-violence-among-midtown-crossdressing-prostitutes/.

Bagby, D. (2012b, June 14). Atlanta's Baton Bob says he was physically threatened, called anti-gay slur. *Georgia Voice*. Retrieved September 2, 2013 from http://thegavoice.com/atlantas-baton-bob-says-he-was-physically-threatened-called-anti-gay-slur/ .

Bagby, D. (2013a, Mar. 1). LGBT advocates help put Atlanta's "Banishment Ordinance" on hold. *Georgia Voice*. Retrieved March 13, 2013 from http://thegavoice.com/lgbt-advocates-help-put-atlantas-banishment-ordinance-on-hold/.

Bagby, D. (2013b, June 26). Atlanta police arrest gay icon Baton Bob after alleged physical altercation, threats. *Georgia Voice*. Retrieved September 2, 2013 from http://thegavoice.com/atlanta-police-arrest-gay-icon-baton-bob/.

Baute, N. (2008, Sept. 6). Take a walk on someone else's wild side. *Star*. Retrieved May 15, 2013 from http://www.thestar.com/news/2008/09/06/take_a_walk_on_someone_elses_wild_side.html.

Bennett, D. (2001). Dorsey's rise and fall equally monumental. *Atlanta Journal-Constitution* December 2.

Brown, M. (2012). Gender and sexuality I: Intersectional anxieties. *Progress in Human Geography*, *36*(4), 541–550. http://dx.doi.org/10.1177/0309132511420973

Browne, K. (2006). Challenging queer geographies. *Antipode*, *38*(5), 885–893. http://dx.doi.org/10.1111/j.1467-8330.2006.00483.x

Burayidi, M.A. (Ed.). (2000). *Urban planning in a multicultural society*. Westport, CT: Praeger.

Campbell, H. (2006). Just planning: The art of situated ethical judgment. *Journal of Planning Education and Research*, *26*(1), 92–106. http://dx.doi.org/10.1177/0739456X06288090

Cardinale, M. (2009, Nov. 19). Coyle, Wan face off in District 6 Council run-off. *Atlanta Progressive News*. Retrieved April 2, 2013 from http://atlantaprogressivenews.com/2009/11/19/coyle-wan-face-off-in-district-6-council-run-off/.

Carter, D. (2010). *Stonewall: The riots that sparked the gay revolution*. New York: St Martin's Press.

Chapman, D. (2001, Apr. 2). Racial shifts creating harmony and tension – some neighborhoods more successful than others in merging Black and White communities. *Atlanta Journal-Constitution*.

Chesnut, S., & Gable, A. (1997). Women ran it, Charis Books and More and Atlanta's lesbian-feminist community 1971–81. In J. Howard (Ed.), *Carryin' on in the lesbian and gay South* (pp. 241–284). New York: New York University Press.

City of Atlanta. (2007). *Peachtree Corridor Task Force, final report*. Atlanta: Department of Planning, Development, and Neighborhood Conservation.

Cohen, C.J. (2005). Punks, bulldaggers, and welfare queens: The radical potential of queer politics. In E.P. Johnson & M.G. Henderson (Eds.), *Black queer studies: A critical anthology* (pp. 21–51). Durham, NC: Duke University Press. http://dx.doi.org/10.1215/9780822387220-003

Connerly, C. (2002). From racial zoning to community empowerment: The interstate highway system and the African American community in Birmingham, Alabama. *Journal of Planning Education and Research*, 22(2), 99–114. http://dx.doi.org/10.1177/0739456X02238441

DeGiovanni, F. (1983). Patterns of change in housing market activity in revitalizing neighborhoods. *Journal of the American Planning Association*, *49*(1), 22–39. http://dx.doi.org/10.1080/01944368308976193

Doan, P.L. (2007). Queers in the American city: Transgendered perceptions of urban spaces. *Gender, Place and Culture*, *14*(1), 57–74. http://dx.doi.org/10.1080/09663690601122309

Doan, P.L. (2010). The tyranny of gendered spaces: Reflections from beyond the gender dichotomy. *Gender, Place and Culture*, *17*(5), 635–654. http://dx.doi.org/10.1080/0966369X.2010.503121

Doan, P.L. (2014). Regulating adult business to make spaces safe for heterosexual families in Atlanta. In P. Maginn & C. Steinmetz (Eds.), *(Sub) urban sexscapes geographies and regulation of the sex industry*. London: Routledge

Doan, P.L. (Ed.). (2011). *Queerying planning: Challenging heteronormative assumptions and reframing planning practice*. Farnham: Ashgate.

Doan, P.L., & Higgins, H. (2011). The demise of queer space? Resurgent gentrification and the Assimilation of LGBT neighborhoods. *Journal of Planning Education and Research*, *31*(1), 6–25. http://dx.doi.org/10.1177/0739456X10391266

Douglas-Brown, L. (2013, June 28). Baton Bob released from jail, disputes police account. *Georgia Voice*. Retrieved September 2, 2013 from http://thegavoice.com/baton-bob-released-from-jail-disputes-police-account/.

Edelman, E.A. (2011). "This area has been declared a prostitution free zone": Discursive formations of space, the state, and trans "sex worker" bodies. *Journal of Homosexuality*, *58*(6–7), 848–864. http://dx.doi.org/10.1080/00918369.2011.581928

Fincher, R., & Iveson, K. (2008). *Planning and diversity in the city: Redistribution, recognition, and encounter*. Basingstoke: Palgrave-Macmillan.

Fleischmann, A., & Hardman, J. (2004). Hitting below the Bible Belt: The development of the gay rights movement in Atlanta. *Journal of Urban Affairs*, *26*(4), 407–426. http://dx.doi.org/10.1111/j.0735-2166.2004.00208.x

Forsyth, A. (2001). Sexuality and space: Nonconformist populations and planning practice. *Journal of Planning Literature, 15*(3), 339–358. http://dx.doi.org/10.1177/08854120122093069

Frisch, M. (2002). Planning as a heterosexist project. *Journal of Planning Education and Research, 21*(3), 254–266. http://dx.doi.org/10.1177/0739456X0202100303

Gallagher, C.A. (2009). Black, White, and browning: How Latino migration is transforming Atlanta. In D.L. Sjoquist (Ed.), *Past trends and future prospects of the American city: The dynamics of Atlanta* (pp. 205–218). Lanham, MD: Rowman and Littlefield.

Gleeson, B., and Low, N. (2000). *Australian urban planning: New challenges, new agendas*. Sydney: Allen and Unwin.

Gover, T. (1997, Apr. 1). Hate that goes boom in the night. *Advocate*, no. 730.

Haider-Markel, D., & Meier, K.J. (1996). The Politics of gay and lesbian rights: Expanding the scope of the conflict. *Journal of Politics, 58*(02), 332–349. http://dx.doi.org/10.2307/2960229

Healey, P. (1997). *Collaborative planning: Shaping places in fragmented societies*. London: Macmillan.

Hennie, M. (2013, Feb. 27). Midtown's sex workers safe from exile – for now. *Project Q*. Retrieved April 4, 2013 from http://www.projectq.us/atlanta/midtowns_sex_workers_safe_from_banishment_for_now?gid=13069.

Henry, S. (2004, July 29). Dead-end for Backstreet? Notorious party club might close for good. *Creative Loafing*. Retrieved May 19, 2015 from http://clatl.com/atlanta/dead-end-for-backstreet/Content?oid=1248950.

Howard, J. (1997). The library, the park, and the pervert: Public space and the homosexual encounter in post World War II Atlanta. In J. Howard (Ed.), *Carryin' on in the lesbian and gay South* (pp. 107–131). New York: New York University Press.

James, A. (2000). Demographic shifts and the challenge for planners: Insights from a practitioner. In M. Burayidi (Ed.), *Urban planning in a multicultural society* (pp. 15–35). Westport, CT: Praeger.

Jarvie, J. (2006, May 8). Voice of Atlanta's Black gays is emerging. *Los Angeles Times*. http://articles.latimes.com/2006/may/08/nation/na-blackgay8

Johnson, E.P. (2005). "Quare" studies, or (almost) everything I know about queer studies I learned from my grandmother. In E.P. Johnson & M.G. Henderson (Eds.), *Black queer studies: A critical anthology* (pp. 124–160). Durham, NC: Duke University Press. http://dx.doi.org/10.1215/9780822387220-008

Keating, L. (2001). *Atlanta: Race, class, and urban expansion*. Philadelphia: Temple University Press.

Kruse, K. (2005). *White flight: Atlanta and the making of modern conservatism.* Princeton: Princeton University Press.

Lee, R. (2007, Nov. 2). A future without gay bars? Gay clubs put on the extinction list, but will it happen here? *Southern Voice.*

McWilliams, J. (2013, Feb. 11). Atlanta proposes controversial crackdown on prostitution. *Atlanta Journal-Constitution.* Retrieved April 4, 2013 from http://www.ajc.com/news/news/atlanta-proposes-controversial-crackdown-on-prosti/nWLnX/.

Nast, H.J. (2002). Queer patriarchies, queer racisms, international. *Antipode, 34*(5), 874–909. http://dx.doi.org/10.1111/1467-8330.00281

Nero, C. (2005). Why are the gay ghettoes White? In E.P. Johnson and M.G. Henderson (Eds.), *Black queer studies: A critical anthology* (pp. 228–245). Durham, NC: Duke University Press.

Nouraee, A. (2008, Jan. 16). One man's battle against Midtown prostitutes and their johns. *Creative Loafing.* Retrieved May 19, 2015 from http://clatl.com/atlanta/one-mans-battle-against-midtown-prostitutes-and-their-johns/Content?oid=1271636.

Peake, L. (1993). "Race" and sexuality: Challenging the patriarchal structuring of urban social space. *Environment and Planning D: Society and Space, 11,* 415–432.

Podmore, J. (2001). Lesbians in the crowd: Gender, sexuality, and visibility along Montreal's Boulevard St-Laurent. *Gender, Place, and Culture, 8,* 333–355.

Qadeer, M.A. (1997). Pluralistic planning for multicultural cities. *Journal of the American Planning Association, 63*(4), 481–494. http://dx.doi.org/10.1080/01944369708975941

Qadeer, M.A. (2009). What is multiculturalism? *Plan Canada: Special Edition: Welcoming Communities: Planning for Diverse Populations,* 10–13.

Rankin, B. (2009, Sept. 15). Chief defends Eagles bar raid: Undercover cops saw illegal acts, he says. Pennington says patrons' complaints of searches, slurs to be investigated. *Atlanta Journal-Constitution.*

Ritzdorf, M. (1997). Family values, municipal zoning, and African American family life. In J.M. Thomas & M. Ritzdorf (Eds.), *Urban planning and the African American community* (pp. 75–92). Thousand Oaks: Sage.

Ritzdorf, M. (2000). Sex, lies, and urban life: how municipal planning marginalizes African American women and their families. In K. Miranne and A. Young (Eds.), *Gendering the city: Women, boundaries, and visions of urban life* (pp. 169–181). Lanham, MD: Rowman and Littlefield.

Ross, G., Sjoquist, D.L., & Wooten, M. (2009). Tracking the economy of the City of Atlanta: Past trends and future prospects. In D.L. Sjoquist (Ed.),

Past trends and future prospects of the American city: The dynamics of Atlanta (pp. 51–83). Lanham, MD: Rowman and Littlefield.

Sandercock, L. (Ed.). (1998). *Making the invisible visible: A multicultural planning history*. Berkeley: University of California Press.

Schrade, B. (1998, May 20). Flier incites Kirkwood controversy. *Atlanta Journal and Atlanta Constitution*.

Sears, J.T. (2001). *Rebels, rubyfruits, and rhinestones: Queering space in the post Stonewall South*. New Brunswick, NJ: Rutgers University Press.

Thomas, J.M. (1994). Planning history and the Black urban experience linkages and contemporary implications. *Journal of Planning Education and Research, 14*(1), 1–11. http://dx.doi.org/10.1177/0739456X9401400101

Thomas, J.M. (1998). Racial inequality and empowerment: Necessary theoretical constructs for understanding U.S. planning history. In L. Sandercock (Ed.), *Making the invisible visible: A multicultural planning history* (pp. 198–208). Berkeley: University of California Press.

Thomas, J.M. (2008). The minority-race planner in the quest for a just city. *Planning Theory, 7*(3), 227–247. http://dx.doi.org/10.1177/1473095208094822

Valentine, G. (2007). Theorizing and researching intersectionality: A challenge for feminist geography. *Professional Geographer, 59*(1), 10–21. http://dx.doi.org/10.1111/j.1467-9272.2007.00587.x

Walcott, R. (2007). Homopoetics: Queer space and the Black queer diaspora. In K. McKittrick & C. Woods (Eds.), *Black geographies and the politics of place* (pp. 233–245). Cambridge, MA: South End Press.

Wald, K.D., Button, J.W., & Rienzo, B. (1996). The politics of gay rights in American communities: Explaining antidiscrimination ordinances and policies. *American Journal of Political Science, 40*(4), 1152–1178. http://dx.doi.org/10.2307/2111746

Weinberg, M.S., Shaver, F., & Williams, C. (1999). Gendered sex work in the San Francisco Tenderloin. *Archives of Sexual Behavior, 28*(6), 503–521. http://dx.doi.org/10.1023/A:1018765132704

Woods, W. (2004, May 15). Midtown rebirth runs into snags – several lot owners won't part with land. *Atlanta Journal-Constitution*.

7 Planning in Native American Reservation Communities: Sovereignty, Conflict, and Political Pluralism

NICHOLAS C. ZAFERATOS

Planning for difference, this book's overarching theme, concerns the diversity that exists within our communities. This concern has emerged among planning's discourse to encourage a more socially just planning paradigm. But, as so often is the case, rarely is planning truly inclusive of all interests. The inherent plurality of American society has awakened a new vigilance among the profession to address the needs of distinct social groups, particularly those groups that had previously been underrepresented in American planning practice. This chapter looks at the conditions that face one of the least represented and most disadvantaged segments of the American community – the Native American Indian tribal community. Within the diverse field of American planning practice, Indian reservation planning is among the most poorly understood. Until quite recently, the profession has generally ignored the plight of tribes, thereby failing to assist them in their quest for advancement. Native American communities represent a unique segment of American society that has a right to be different. That right was guaranteed in a series of agreements between the United States and the sovereign Indian Nations. Treaties were intended not only to recognize the cultural diversity of Indians in American society but also to guarantee their right to manage their societal affairs, without discrimination or subjugation. American Indian communities are not cultural minority groups; they exist as independent political nations within a larger nation.

During the course of the last 150 years, Indian reservation planning and development have been subjected to a series of federal policies that have disrupted the stability of Indian societies through programs that attempted to both assimilate and terminate tribes. Despite the destabilizing backdrop of those policies, tribal communities continue to exist, and, in light of the current federal policy, which supports Indian self-determination, we are beginning to see unequivocal signs of tribal improvement. However, the path towards tribal self-sufficiency continues to be complicated by the enduring effects of the tribes' historic contact with American society, which has produced reservation conditions that are characterized as both politically and economically underdeveloped.

While Indian tribes have always possessed the powers of self-government, the authority to manage their reservation lands and resources, which was once absolute, has since become limited. The practice of reservation planning continues to be impacted by obstacles stemming from past federal policies, as well as from continuing jurisdictional conflicts in state-tribal relations, and the incorporation processes of the US political economy. An understanding of the contemporary setting that tribes operate in provides a starting point for supporting the work of planning in Indian reservation communities.

As a critical first step towards informing its practice, this chapter exposes many of the complex conditions inherent in tribal planning. The circumstances that tribes face in managing their reservation communities include seemingly insurmountable obstacles that, unless clearly identified, will continue to frustrate tribes as they attempt to improve their reservation conditions. The context in which tribal planning occurs must be made decipherable before effective planning strategies can be selected and employed. Tribal planning should be grounded in a tribe's particular historical experience and should emphasize the development of its political capacity to confront obstacles and reconcile the barriers that stand in the way of its advancement. In this chapter, I examine the nature of the tribal planning context with the aim of shedding light on the difficult task of reservation planning. The approach that I use emphasizes a critical review of the history of federal Indian policy and the doctrine of federal trust responsibility as a way of explaining the causes that constrain tribal sovereignty. This perspective underscores the importance of an understanding of both the breadth and the limitations of tribal governing powers, while concurrently addressing the often adversarial interests of state and local governments that continue to interfere in tribal affairs. The chapter concludes with a discussion of recent dispute resolution experiences that have been successful in overcoming many of the conflicts that exist in tribal and non-tribal government relations and that represent new pathways for fostering greater plurality in regional planning, where outcomes are inclusive of, rather than resistant to, tribal interests.

The Nature of Planning Native American Reservations

Currently, 566 Indian tribes are federally recognized, including more than 200 village groups in Alaska (Bureau of Indian Affairs, 2012). The number of Indian reservations (including federal and state recognized reservations, pueblos, rancheros, and communities) totals 334. While Indian reservations[1] in the United States cover approximately 56 million acres, with reservations varying dramatically in size, many contain a high percentage of land that is owned and occupied by non-Indians. Only about 40 per cent of the reservations have an entirely tribally owned land base where land title is held in federal trust ownership on behalf of Indians. Most reservations contain land parcels that are alienated, or in fee simple title, that were sold to, and occupied by, non-Indians. The major problems that face tribes today can be historically linked to the complex and detrimental federal policies of the past that continue to affect Indian life. While federal laws were intended to benefit Indians, they have paradoxically caused political, social, and economic quandaries that continue to frustrate the efforts of tribes to plan their reservation communities.

Unlike local governments whose authority to plan and manage their territories is clearly established under state enabling laws, in Native American Indian reservation communities, a tribe's authority to regulate its reservation territory is not as clearly evident. The ability to administer tribal planning varies greatly, and is dependent upon the conditions of the particular tribal situation. While a tribe's planning authority is generally established under its own constitutional powers of self-governance, its authority is often subject to challenge by non-tribal governments.

Tribal planning incorporates a variety of strategies to address reservation community needs and the jurisdictional circumstances that are particular to each tribe. What makes tribal planning so complex are the continuous challenges that tribes face regarding the legitimacy of their governing powers. In circumstances where reservation lands are held entirely in federal trust protection, tribes usually face fewer challenges to their right to control reservation lands and resources. In other cases, especially where non-Indian ownership and occupancy of reservation lands exists, where non-Indian governments have usurped tribal authority, or where tribal interests extend beyond reservation boundaries, as in the case of treaty fishing, hunting, and cultural interests, tribal authority is often contested. The causes that led to a diminished tribal governing authority, and which allowed non-tribal interests to enter onto reservations, are the result of past policies which created the complex jurisdictional setting in which tribal planning currently operates. Tribal planning must first clarify a tribe's underlying authority and its capacity to govern before it can effectively plan and manage its reservation territory.

A Brief History of Federal Indian Policy

The ability of a tribe to exercise control over its territory is a fundamental attribute of self-government. Indian tribes derive their governing powers from three basic sources: retained inherent sovereignty, treaty rights, and federally conveyed rights. Prior to the treaties, as sovereign political entities, the tribes exercised absolute autonomy in decision making. The treaties served both to limit sovereignty and to affirm specific rights and powers. The federal history of post-treaty relations, however, has been inconsistent in the treatment of tribes as legitimate governments, and went so far as to actually terminate more than 100 tribal governments during the 1950s and 1960s "termination era." Since 1970, the enactment of the federal self-determination policy and subsequent legislation has reaffirmed many tribal rights and authorities. Past federal policy inconsistencies have shifted between two prevailing and opposing public policy positions. One position regarded the tribes as enduring political entities with protected territories. The other position sought the elimination of tribes, the removal of their territories from federal trust protection, and the assimilation of their land resources and their members into mainstream society. These two contradictory positions have continually emerged to shape legislation, court decisions, and the federal administration of Indian affairs. These policy shifts occurred abruptly at various times, resulting in the tremendous disruption of the social and political cohesiveness of tribes.

Prior to treaty making, tribes remained relatively isolated and enjoyed absolute autonomy. The treaty-making period, from 1787 through 1887, relocated Indians from ceded territories onto exclusive Indian territories known as reservations, as these land areas were self-reserved by the tribes for their own use. Between 1887 and 1934, Congress attempted to reverse, through Indian assimilation, the commitments made in most of its treaties. The General Allotment Act (1887) introduced private land ownership to many reservations by subdividing commonly held lands and distributing those parcels to Indian families. The net effect was disastrous to most tribal communities when those "allotted" lands were later subjected to state taxation, transferred out of federal trust ownership, and eventually sold to non-Indians or taken through auctions. Furthermore, the allotment process declared much of the Indian territories to be "surplus" lands, which were then fully incorporated into the US economy. The immediate effect of this federal "land grab" was the reduction of Indian-held lands from 138 million acres in 1887 to 48 million acres in 1934.

Recognizing the disastrous effects of assimilation, Congress introduced the Indian Reorganization Act (IRA) in 1934, which sought to reconstitute the Indian territory and reaffirm tribal self-governance. In doing so, it encouraged most tribes to adopt a constitutional model that required federal approval for certain governing actions. Approximately half of all tribes adopted an IRA form of constitution, which further limited their rights of self-governance by placing strict federal oversight over tribal legislative actions. The self-governance policy was reversed in 1953 when Congress passed a series of Acts with the intent of terminating tribal governments by disbanding the political authority of certain tribes, foreclosing their tribal territories, and encouraging the full assimilation of those terminated Indians and their resources into the larger US political economy. Other Acts of Congress transferred certain jurisdiction over Indians to designated states. By 1968, Congress and the federal executive branch once more brought about a reversal in its Indian policy by ushering in the era of tribal self-determination and self-governance, a lasting policy that has helped to reconstitute those tribal territories that were lost during termination with a pledge of support for tribal self-governance and reservation community development.

As an evolving political conception, tribal sovereignty has been shaped and reshaped by these past federal policies as well as by legal doctrine. The effects of these policies and court decisions have proven to be detrimental as they have diminished the tribal land base and weakened the powers of tribal self-governance, thereby creating a tumultuous planning setting and a state of jurisdictional uncertainty that still exists. The perplexing jurisdictional ground upon which tribes base their planning is tested whenever a tribe applies its powers over its lands and resources. A tribe's planning authority is particularly contested in the areas of land use regulation, environmental management, and economic development, especially where non-Indian interests are at stake.

The Doctrines of Federal Trust Responsibility and Tribal Sovereignty

The Supreme Court first recognized the existence of a trust relationship in its earliest decisions that interpreted Indian treaties (*Johnson v. McIntosh*, 1823; *Cherokee Nation v. Georgia*, 1831; *Worcester v. Georgia*, 1832) and affirmed the principle of a trust relationship between the US and the Indian people. In almost all of the treaties entered into, Indians ceded their land territories in exchange for promises, including the guarantee of a permanent, self-governed reservation for the tribes and the federal protection of their safety and well-being. The Supreme Court has held that such promises establish a special trust relationship, characterized as being that of ward and guardian and implying the continued promise to create "a duty of protection" for Indians (*US v. Kagama*, 1886). These precedent-setting decisions recognized the tribes as distinct political communities possessing self-governing authority within their territorial boundaries. The Trade and Intercourse Acts prohibited the sale of Indian land without federal consent, thereby establishing tribal territorial rights as trust beneficiaries of the United States. The doctrine of trust over Indian resources protected Indian lands from intrusion by state jurisdiction as well as from non-Indian settlement. Ultimately, the trust relationship created a federal responsibility for the maintenance of Indian lands and natural resources.

Whereas, in the early 1900s, the trust relationship had shifted to a form of transitional protection during the assimilation of Indians into mainstream society, in recent decades the trust relationship has evolved into a permanent doctrine of federal Indian policy that serves to guide the future development of Indian policy and Indian law. The courts have held the federal executive branch to strict fiduciary standards in their management of Indian trust resources (*Seminole Nation v. United States*, 1942; *United States v. Payne*, 1924) and have prevented federal agencies from subordinating the interests of Indians to other public purposes (*United States v. Winnebago Tribe*, 1976). With the proclamation of the federal self-determination policy in 1970, the administration of the trust responsibility has been extended throughout the federal administration to all programs that affect Indians and Indian tribes

(Obama, 2009; Clinton, 1994). Each federal agency has since established its own Indian policy with respect to the federal trust relationship that recognizes tribal sovereignty based on a government-to-government relationship.

Tribal sovereignty refers to the inherent right of tribes to govern themselves, a tribal right that has never been relinquished. In *Johnson v. McIntosh* (1823), Chief Justice Marshall emphasized the affirmative governmental powers of the tribes. In *Cherokee Nation v. Georgia* (1831), the court held that tribes qualified as separate "states" and further characterized tribes as "domestic dependent nations." For 150 years following the Marshall rulings, few further limitations on tribal sovereignty were found to restrict the political status of the tribes. However, in *Oliphant* (1978) the Supreme Court questioned the tribe's ability to exercise criminal jurisdiction over non-Indians living on the reservation. Relying on its inherent sovereignty, the tribe argued that no treaty or Act of Congress had extinguished its criminal authority over non-Indians. But the Supreme Court decided that the exercise of criminal jurisdiction over non-Indians was inconsistent with the domestic status of the tribes. In establishing judicial limitations on tribal sovereignty, the *Oliphant* decision threatened the exclusive nature of tribal authority. Where it cannot be shown that tribal interests are affected, the court in *Montana* (1981) later held that a tribe generally lacked inherent powers to regulate hunting and fishing by non-Indians on non-Indian-owned land within a reservation. In contrast, the courts have affirmed that tribes retain the power to prosecute their own members and to tax non-Indians for activities on the reservation as being consistent with their domestic dependent status (*United States v. Wheeler*, 1978; *Washington v. Confederated Tribes of the Colville Indian Reservation*, 1980).

Furthermore, tribal governments, by virtue of their sovereign status, are not comparable to "local governments," which derive their authority from states. A local government can enact regulations only when the state has conferred such power to the municipality, as the state is the sovereign body possessing those powers. Similar to a state, a tribe also retains its own sources of governing powers. A tribe's power to levy a tax or to grant a land use right through zoning of reservation lands, for example, is not subject to challenge on the grounds that Congress has not conveyed such powers to a tribe. As sovereign political entities, tribes generally require no further authorization from the federal government to exercise such powers.[2] However, the exercise of tribal sovereign powers has certain limitations, which have prompted continuous clarification by the courts.

As sovereigns, tribes are free to act unless some federal intrusion has modified their sovereignty. Tribal sovereignty also prevents the intrusion of state law and jurisdiction within Indian country with certain restrictions. The Supreme Court has consistently prohibited state law from applying to Indians in Indian country. In *Williams v. Lee* (1959), the court ruled that state courts have no jurisdiction over a civil claim by a non-Indian against an Indian for a transaction arising on the Navaho reservation. The court held that state law had application only where "essential tribal relations" were not involved and that "absent governing Acts of Congress, the question has always been whether the state action infringes on the right of reservation Indians to make their own laws and be governed by themselves" (*Williams v. Lee*, 1959). State interference would undermine a tribe's authority over reservation affairs and thereby infringe on the right of Indians to govern themselves. The Supreme Court in *McClanahan* (1973) further held that state law would only be permitted into Indian country if two conditions were met: 1) the intrusion would not interfere with tribal self-government and 2) non-Indians were involved. The ruling in *William v. Lee* that state law may not interfere with tribal self-government serves as a legal test to be applied along with pre-emption analysis.

Tribal Authority for Planning and Regulating the Reservation Territory

The ability of tribal governments to exercise control over their territories became especially clouded when federal assimilation and termination policies created a system of land tenure on many reservations that permitted the subsequent intrusion of state jurisdiction over those lands. The General Allotment Act of 1887 created two irreversible conditions that have continued to affect property interests on many of the reservations established under the treaties. The Act introduced individual Indian ownership and, later, transferred these trust parcels to fee simple ownership. Settlement onto reservations by non-Indians was encouraged under the Act, thereby creating a property rights condition that affected not only the tribes but non-Indians as well. Today, many Indian reservations contain a fragmented, or "alienated," land ownership pattern, with a plurality of property rights interests. Because of this condition, state and local government control exists on many reservations for the purpose of protecting the interests of non-Indian occupants. This land tenure condition represents a pervasive obstacle to tribal planning by challenging a tribe's exclusive control over its reservation. While it has long been recognized that tribes retain inherent rights to manage their reservations, non-Indian property interests and state and local governments continue to contest the exercise of tribal authority over fee simple reservation lands and non-Indian members.

Based upon their reservation land tenure characteristics, Indian reservations experience varying degrees of alienation. Highly alienated reservations contain fee simple lands in private, non-Indian ownership and a large proportion of non-Indian residents.[3] Alienated land conditions occur in reservations that were subjected to the General Allotment Act, where jurisdictional conflicts in land use management are common. In contrast, reservation alienation is virtually non-existent on reservations that were established subsequent to the passage of the Act or in cases where the Act did not apply. A major challenge to tribal planning occurs with regard to alienated reservation lands when states and local governments politically intervene to supplant a tribe's land use authority with their own policies. Given that the control over reservation lands and natural resources is fundamental to the fulfilment of tribal sovereignty and the attainment of tribal development objectives, the overlapping and often subjugating presence of non-tribal interests continues to impede tribal social, political, and economic advancement. A tribe's ability to maintain a cohesive political community is challenged by the erosion of its regulatory authority over its reservation territory.

Recent Supreme Court rulings have further limited a tribe's authority to exercise civil jurisdiction over non-Indians on their reservations. In *Montana v. United States*, the court limited tribal civil jurisdiction over non-Indians by ruling that the Crow Nation did not have the authority to regulate hunting or fishing by non-Indians on non-Indian fee lands within the reservation. A general principle was established in *Montana* that limited tribal authority to what is necessary to protect tribal self-government or to control its internal tribal relations by ruling that tribes have been divested of their sovereignty to regulate relations between Indians and non-Indians by virtue of their dependent status (Goeppele, 1990). However, *Montana* provides two broad exceptions, or tests, where tribal authority may apply to non-Indians: 1) a tribe retains its authority to regulate non-members who have entered into consensual relations with the tribe through commercial dealings, contracts, leases, or other arrangements, and 2) a tribe retains its regulatory authority when that conduct threatens or has a direct effect on the political integrity, the economic security, or the health or welfare of the tribe. The scope of the second exception is similar to the traditional scope of authority found in the police powers.[4] After *Montana*, subsequent lower court decisions granted tribes authority over non-Indians under the second *Montana* exception when the non-Indian activity was found to threaten the integrity of the tribe and its resources. The Tenth Circuit Court in *Knight v. Shoshone and Arapahoe Tribes* (1982) upheld tribal zoning authority over non-Indian fee land under the exception that no competing local or county zoning ordinances existed to restrict land use on non-Indian fee lands on the reservation.

Tribes also retain important powers to regulate activities that threaten to degrade tribal lands, waters, and resources under their proprietary rights, their inherent sovereignty, and federal environmental laws where Congress had conferred environmental protection authority to the tribal governments. The courts had previously affirmed tribal proprietary, aboriginal and reserved water rights by finding that the creation of Indian reservations included the implied reservation of a proprietary water right (*United States v. Winters*, 1908). A corollary to the reserved water right is the right to water of undiminished quality. This right of quality protection is derived from the "equitable apportionment doctrine" that imposes a duty on sister states to protect water quality and prevent the diminishment of quality enjoyed by neighbouring states.

The enactment of comprehensive federal environmental legislation was intended to protect all lands from air and water pollution and to regulate hazardous and solid waste disposal. In the absence of state regulatory jurisdiction in Indian country, Congress instructed the United States Environmental Protection Agency to retain its federal responsibility to protect the reservations' environmental quality. In 1984, the EPA adopted an Indian policy that established a tribe's authority to conduct reservation-wide environmental programs similar to those delegated to the states. Congress later amended most environmental statutes so that the EPA could provide funding to tribes, enabling them to develop the capacity to protect their territories just as it had in the case of states.[5] However, the authority to establish a delegation of program authority to the tribes occurred almost two decades after the initial enactment of the legislation. As a result of the exclusion of tribes in the early period of environmental program development, tribal lands were often ignored and the tribal role and capacity in the implementation of federal programs was poorly defined and largely undeveloped. Since 1984, the EPA Indian policy has been comprehensive, providing extensive guidance for the administration of environmental programs on Indian lands. In particular, the policy directs the EPA to recognize tribal governments as the primary authority for implementing federal environmental programs on tribal lands and to assist the tribes in assuming regulatory responsibility. The policy also encourages cooperation between the tribes and state and local governments in the implementation of federal environmental programs. This last directive is of particular importance as it encourages regional cooperation to comprehensively address common environmental protection problems that are rarely contained within a single jurisdiction's boundaries.

Contentious Histories in State Relations

The history of conflict between states and tribes is largely attributed to questions about the political status that tribes possess. While tribes clearly possess governmental powers, the position of most states, historically, has been that such powers are not equivalent to state powers. Fuelling the conflictive relationship is the fact that non-Indians who reside within the boundaries of reservations are not entitled to participate in the electoral process of tribal governments that enact reservation laws. The situation becomes particularly acute in urbanizing areas where non-tribal residents may constitute a majority of the reservation population. It has been the contention of many states that the lack of direct representation in government violates a fundamental principle of the American democratic tradition: the consent of the governed.[6] The unique status that differentiates tribal residents from non-tribal residents has emerged as a source of conflict and has led to the states' intervention in tribal self-governance. Since passage of federal self-determination policies, the increasing exercise of tribal governmental authority has changed the status quo in the balance of state-tribal power. Congress has contributed to the confusion by vacillating in its Indian policies regarding the extent of tribal sovereignty, ranging from the termination of tribes to the promotion of tribal self-governance. Consequently, states have tended to react to federal policy shifts rather than to focus on the development of a cogent and workable state-tribal policy. Disputes arising between the states and tribal governments have been addressed primarily through three approaches: litigation, legislation, and, more recently, negotiation.

The predominant means for resolving tribal-state conflicts has been through the courts. This approach, however, further aggravates the already tenuous relationship between the parties and often leads to additional litigation. The *Puyallup* "test case" in Washington State distinctly illustrates this problem. Filed in the 1960s to settle the fishing rights controversy, the decision of the lower court was appealed on three separate occasions before the United States Supreme Court (*Puyallup Tribe v. Department of Game*, 1977; *Washington Game Department v. Puyallup Tribe*, 1973; *Puyallup Tribe v. Department of Game*, 1968). Rather than ending the controversy, the litigation evolved into the even more complex *United States v. Washington* (1974) "test case." The emotional and often violent confrontations between the parties' fishers not only contributed to further jurisdictional tensions between the tribes and the state but also resulted in tensions between Indians and non-Indians.

Congress's plenary power provides a second approach for clarifying the balance of tribal and state authority through the delegation and the pre-emption of state laws over tribes. Once Congress enacts laws, legal challenges are balanced against the substantive legislative intent to guide judicial decisions. While this approach can be effective in resolving state-tribal disputes, practical limitations may result, as Congress is often reluctant to limit either state power or tribal rights without the prior agreement of the parties. Further, legislative solutions to problems do not always achieve satisfactory outcomes. The history of tribal-state conflicts suggests that many of the disputes are the result of shifting federal legislative policy. As the recent federal self-determination policies sought to reconstitute tribal political authority and to reaffirm the federal trust responsibility to tribes, this policy shift resulted in reversing earlier legislation that had allowed state jurisdiction in Indian affairs in certain circumstances.

The conflicts between states and tribal governments are multifaceted and are often associated with decades of continuing litigation. As in the case of Washington State, the focus of recent Indian litigation has shifted from treaty rights to current concerns over tribal land claims, natural resource management, land use jurisdiction, environmental management, and the application of tribal civil jurisdiction over non-Indians. While disputes continue in other areas of tribal-state relations, including fisheries litigation and gaming, these current issues pose particular challenges to a tribe's ability to control its future. The recent expansion of tribal authority affects more than the interests of reservation Indians; it also often affects the interests of the state and local governments, as well as the interests of non-Indian reservation residents and property owners.

To resolve disputes that tribes face with adjacent jurisdictions, and as a process for developing improved intergovernmental cooperation, meaningful dialogue is required, based upon a commitment by both the tribes and their regional government counterparts to learn about the differing interests and values that exist in Indian and non-Indian communities. Since 1985, Washington State's approach to resolving conflicts with the tribes has emphasized negotiation and mediation as an alternative to litigation. This trend towards negotiation evolved further, in 1989, when the state's governor proclaimed a new precedent for guiding relations with federally recognized Indian tribes. The precedent sought to reverse a century of hostile relations with tribes by recognizing the legitimacy of tribal sovereignty. The new policy established a government-to-government relationship, creating an innovative avenue for addressing a variety of complex problems. Though these experiences were primarily motivated by the state's need to address Indian treaty rights that were affected by the state's natural resources management authority, the process was also embraced by the tribes as a way to advance tribal interests within and outside the reservations.

While negotiation, under certain circumstances, may be preferable to litigation, other circumstances may preclude the parties from accepting compromised outcomes through the negotiated approach. When disputes arise over the assertion of sovereign control over a resource or territorial area, a compromised solution is often untenable. The state is generally unwilling to negotiate away its sovereign powers or prohibited from doing so. Tribes similarly defend their sovereignty, regarding it as fundamental to their continued existence as autonomous political communities. In these situations, neither the state nor a tribe may willingly concede its control over resource ownership or rights to control land where such a concession may result in a legal precedent leading to the diminishment of its rights. Negotiation is further complicated when uncertainty exists concerning questions about property rights. Such a situation occurred in the Washington fisheries litigation. While negotiation resulted in co-management of the fisheries by the state and the tribes, this outcome was reached through the court's order, which mandated cooperation after first establishing a general allocation standard (*Washington v. Washington State Commercial Passenger Fishing Vessel Ass'n*, 1979). Prior to that decision, the entrenched positions of the parties precluded a negotiated settlement involving the allocation of the resource. In this case, litigation directly provided an opportunity for the subsequent negotiation of the tribal-state dispute by narrowing the context of the negotiation to the means of managing the allocated resource. Negotiation can offer a preferred alternative when the issue is limited in scope and does not involve issues of sovereignty. Litigation may be preferable when negotiation appears unworkable or when resolution requires a definitive determination.[7] Finally, legislation may be preferable when the issue is limited in scope and does not require further clarification through litigation.

The types of conflicts between the states and the tribes can vary greatly, depending upon the circumstances of an individual tribe and the nature of the dispute. One area of state concern is its proprietary interests regarding land ownership. When tribal activities impact these proprietary interests, the state may resist in order to protect the value of the property. Another area of state interest concerns the uniform application of its laws. The state's broad interests include the collection of revenue, the management of natural resources, and the regulation of business and individual activities. In applying its laws universally to all citizens residing within its jurisdiction, the state has resisted the demands of tribal governments for recognition as separate and independent sovereigns. In addition, the state also represents *parens patriae* interests on behalf of its citizens that often require it to choose the interests of one group over others. In some circumstances, the state may not have an explicit interest in how a tribal issue is resolved, only that it be resolved to clarify the uncertainty. The state's interests in defending its sovereignty and property rights are involved in several current disputes in the tribal-state relationship, as summarized in table 7.1, and discussed below.

Land Claims

Several land claim cases were filed during the past three decades involving claims of tribal ownership to the tidelands adjacent to reservations and to aboriginal land title. The states have expressed the opinion that if these claims were upheld, affected land areas previously thought to be free of Indian title would become encumbered, complicating land title and raising questions concerning the states' liability in transferring original title to those lands. Most of the land claims instituted by Washington tribes involve lands underlying "navigable waters," including tidelands. In *Montana v. United States* (1981), the Crow Tribe claimed

Table 7.1. Overview of State-Tribal Conflicts Regarding Reservation Planning

Current Issue	State's Position	Planning Implications/Relevance
Land claims	Clarification of property title; public rights of way and easements	Land tenure uncertainty
Water resources	Universal application of laws; liability associated with granting of rights under common law; reserved right	Absent adjudication of an Indian reserved water right raises questions of validity of junior water rights and may paralyse regional development
Environmental control	Universal application of laws	Multiple regulatory schemes over a common landscape produce conflict in standards and enforcement
Land use and civil regulatory jurisdiction	Control of reservation land use; lack of direct representation in tribal elections	Multiple regulatory schemes for reservation lands based on different ownership types create conflicts and uncertainty

that portions of the Big Horn River located within the Crow Reservation were held in trust for the tribe. The United States Supreme Court rejected the claim, holding that the state owned the riverbed, having acquired it under the "Equal Footing Doctrine" under its admission to the Union in 1889.[8] The court, however, established a general exception to the equal footing doctrine by ruling that when a public need exists at treaty execution, Congress may depart from its general policy of granting ownership of the riverbeds to future states.

In three cases (*Confederated Salish and Kootenai Tribes v. Namen*, 1982; *Muckleshoot Indian Tribe v. Trans-Canada Enterprises, Ltd.*, 1983; *Puyallup Indian Tribe v. Port of Tacoma*, 1983), the Ninth Circuit court held that submerged lands are vested in a tribe rather than the state. In these cases, the court recognized the Indians' historic dependency upon fishing. This dependency proved to be the basis for supporting the tribes' claims, even though the related treaties lacked an expressed intent to convey the submerged lands to the tribes. Several land claims cases are currently pending or have recently been concluded in the federal courts in Washington. In Swinomish, six cases were consolidated, with the United States and the Swinomish Indian Tribal Community as plaintiffs, and the state and various private entities as defendants (*United States v. Cascade Natural Gas Corp*). Swinomish reached settlement through successful negotiations between the parties, resulting in a right-of-way lease to the pipeline companies and railroad, and a purchase option to properties adjoining the claimed tribal tideland properties. In the Swinomish case, the willingness of the parties to seek amicable settlement was motivated largely by the tribe's longer-term interest in acquiring title to the properties in question. In each of these cases, the state had exercised its proprietary interests in protecting the status quo from disruption in property relationships. Should the United States and the tribes win these cases, ownership of tidelands and submerged lands would alter long-established understandings regarding title, public rights of way, and easements, and access rights to marine waters would also be affected. The state's interest in these cases reflects its concern for preserving the status quo in private property ownership rights as those rights were previously granted and protected by the state.

If they remain unresolved, pending Indian land claims can result in land tenure uncertainty, reluctance among title insurance companies to insure land title, and the withholding of financing for land acquisition and development projects by lenders. Resolving land claims is a complex and lengthy process that inevitably requires some form of federal intervention. Land claims have been largely resolved through forms of mitigated compensation, usually in the form of cash settlements rather than the return of lands to the tribes. Most land claims raise complex questions regarding the federal or state government's ultimate liability for the historic transfer of title or granting of access to ancestral or reserved Indian lands.

Water Resources

An ongoing focus in state-tribal discourse involves the state's authority for managing the allocation of water rights. At the time reservations were created, the treaties implied that sufficient water would be reserved to satisfy the primary purposes for which those lands were dedicated. The priority date of this federally reserved water right is the date when the federal reservation was first established. The quantity of the federally reserved water right, however, has not been determined for many reservations. The United States Supreme Court developed a "reserved rights" doctrine in *Winters v. United States* (1908), which held that when the United States reserved lands for special purposes, "it impliedly reserved water in sufficient quantities to satisfy the primary purposes for which the lands were set aside, with priority dating back to the establishment of the federal reservation" (*Cappaert v. United States*, 1976). Further complicating the water rights dialogue in Washington

State is the tribes' demand that proprietary fisheries resources be protected from environmental degradation under phase II of the *Boldt* decision (*United States v. Washington*, 1984). The tribes sought the court's implementation of an environmental impact process that would be triggered whenever a state agency contemplates a state or a private action that may impact the size or quality of a fish run. Since 1984, the environmental right question continues to be unresolved. Should the courts uphold the tribal claim, its water rights that support fisheries habitat would also have priority over water rights granted by the state. Resolving the conflict pertaining to water rights involves two complex issues: the application of the state's codes within the boundaries of the Indian reservations; and the applicability of the state adjudication system to Indian reserved water rights.

Application of state water codes within reservations. Many states have enacted comprehensive codes that cover all phases of water resource management. The State of Washington's surface water and groundwater codes created permit systems for the right to withdraw and use water (Washington RCW 90.03, 90.44). These codes also provide for a general adjudication of existing water rights granted under common law. The state's water code permit system has been applied to waters within Indian reservations since 1917. The first tribal challenge raised objection to the state's allocation of waters flowing through non-Indian owned lands within the Tulalip Indian Reservation. In response to the tribe's assertion that the state had no authority to approve an application for a water right permit submitted by an applicant within the reservation, the attorney general opined that Washington's water right permit system "allowed a non-Indian to divert waters, located on non-Indian lands within an Indian reservation, if those waters exceeded the amounts needed to satisfy prior rights, including the reserved rights of Indians" (Washington Attorneys General, 1985).[9] This analysis of the state's jurisdiction was later upheld in *Tulalip Tribes of Washington v. Walker* (*Snohomish Co. Sup. Ct.*, 1963). Several tribes, along with the United States, initiated litigation in the 1970s against individual water users who were using water from streams or other watercourses that either were entirely within or passed through Indian reservations (*Colville Confederated Tribes v. Walton*, 1981; *United States v. Anderson*, 1984). The state was both a defendant, as a user of water in these actions, and a party, through its regulatory interests. In these cases, the state asserted its jurisdiction to regulate "surplus" or "excess" waters reserved to the tribe under the *Winters* doctrine.

The Ninth Circuit ruled in *Colville* that the state held no jurisdiction because what was involved was a situation "where the watershed was entirely within the boundaries of the Colville Indian Reservation and, therefore, the state's regulatory jurisdiction was pre-empted."

State water rights adjudications. Having waived its immunity to suit, under the *McCarran Amendment* (1982), the United States has been joined as a defendant in several state general adjudication proceedings. These cases have resulted in quantifying claims for water rights of the United States, including claims for reserved rights of Indians. These cases have provided a much clearer understanding of the complexities of water rights by clarifying that state courts are delegated jurisdiction in determining the validity of reserved rights claimed by the United States in a general adjudication for an Indian tribe or an Indian individual (*Arizona v. San Carlos Apache Tribe*, 1983), and that the United States representation, as trustee for a tribe in a general adjudication proceeding, is binding upon Indian tribes (*Nevada v. United States*, 1983; *Arizona v. California*, 1983).

Federal Indian reserved rights are established independent of the state water right system and are generally recognized by the state. These rights have an early date of priority, since many reservations were created in the 1850s. The Washington attorney general has acknowledged an important distinction associated with Indian water rights, since, unlike state-based rights, they are not lost by non-use. Therefore, rights that either have been dormant or have never been exercised remain existent. While several courts have affirmed a *Winters* right for instream uses, including fisheries, the quantification of these rights continues to be unsettled. The question has been advanced in terms of an "environmental treaty right" in the *United States v. Washington* fisheries litigation. The state has maintained that, while it may be possible to reserve a right for fisheries purposes, that right may be limited by a number of factors, including the fact that many fisheries have been destroyed by federal dam construction, and the exercise of a fishery instream right might conflict with the exercise of an irrigation right. Therefore, the state has maintained that an instream fishery right must be determined and quantified in conjunction with the overall scope and purpose of the reservation. Tribal water codes enacted on several reservations generally rely upon the tribe's inherent sovereign authority for the management of their reserved water rights. These codes expressly deny the validity of state-issued permits relating to any excess waters where quantification of the resource has not been adjudicated.

Implications to planning are considerable, particularly in states where the possession of a water right is required as a precondition before development rights may be granted. Absent the adjudication of an Indian reserved water right claim, questions can be raised concerning the validity of junior water rights, including those of municipalities granting land use permits, and may paralyse future urban development where state concurrency laws require evidence of water availability. The issue of water adjudication becomes highly complicated in instances where Indian water rights involve enough water for both future reservation requirements and for conservation of natural resources, as in the case of treaty fishing rights in the Northwest. In those cases, tribes have argued that water resources sufficient to sustain fisheries are primary to other competing uses. In such cases, adjudication appears the only likely remedy for resolving water rights distribution, as attempts to negotiate the competing water demands among regional users have largely proven to be unsuccessful.

Environmental Control

The State of Washington administers comprehensive pollution control programs in the areas of air pollution, water pollution, and hazardous waste disposal. Each relevant federal statute requires a comprehensive federal program unless a state program is developed and approved for delegation by the EPA. Because of the federal trust responsibility, the EPA exempted Indian land from state programs. However, both prior to and following the enactment of these federal statutes, the state administered its state pollution control laws within the boundaries of Indian reservations. At the time of the adoption of the federal-state comprehensive program, the EPA acknowledged such state jurisdiction by approving state water quality standards within a reservation and issuing waste discharge permits to industries within a reservation. It was not until the late 1970s that questions of state authority were challenged. The question of environmental control initially arose when state jurisdiction was challenged on the Muckleshoot and Tulalip Indian reservations. A similar issue was presented to the Ninth Circuit in a jurisdiction challenge by the state against an EPA regulation that denied state jurisdiction over Indian lands. Washington had argued that pollution control programs must be administered by a single agency to avoid a situation of "checkerboard jurisdiction." In 1985, the Ninth Circuit issued its landmark decision (*State of Washington v. Environmental*

Protection Agency, 1985), holding that the state lacked jurisdiction over Indians on Indian lands.

On November 8, 1984, the EPA issued its Indian policy, the first explicit statement by a federal agency supporting its trust responsibilities under the federal self-determination policy. Since 1984, agency programs and several federal statutes have been amended to support tribal self-determination by providing a procedure for the delegation of those programs to the tribes. The core principle of the policy is its commitment to working with federally recognized tribes on a government-to-government basis in order to enhance environmental protection. In carrying out its responsibilities on Indian reservations, the EPA's fundamental objective is to protect human health and the environment while emphasizing a special consideration of tribal interests and ensuring the involvement of tribal governments in managing reservation programs. Following the clarification of the EPA policy and subsequent provisions in federal environmental legislation, tribes have focused on developing tribal environmental protection programs and legislation. Several tribes have since been designated "treatment as a state" status, enabling them to proceed with developing reservation environmental protection programs under federal rules. While the intent of many tribal programs is to operate independent of state program authority, other tribal programs operate in conjunction with state programs.

Notwithstanding recent judicial clarifications regarding the federal responsibility and the authority of tribes to manage the reservation environment, problems persist when multiple layers of government administer concurrent, and sometimes conflicting, environmental management policies within the same geographic area. Effective environmental management requires coordinated efforts as well as consistent policies. While the clarification of jurisdiction is vitally important, it does not solve the problems of inconsistent policies among state, federal, and tribal environmental programs. Cooperative approaches among tribes and federal, state, and local governments in environmental management can have beneficial outcomes when those mutual efforts are inclusive of both the interests of the general public and the particular interests of tribal societies.

Land Use and Civil Regulatory Jurisdiction

The application of tribal civil regulatory authority over non-Indians and fee lands located within a reservation has also resulted in ongoing conflict in the state-tribal relationship. Advancements in tribal governmental capacity, since 1970, have accelerated the exercise of civil jurisdiction over non-Indians through the enactment of zoning, environmental protection, and public health and safety laws, as well as additional restrictive laws that govern the conduct of activities on reservation fee lands. The tribes have maintained that their jurisdiction over non-Indians pre-empts state and local jurisdiction. The state's position has traditionally opposed the assertion of tribal jurisdiction that displaces state regulatory authority.

The United States Supreme Court analyses tribal-state jurisdictional disputes by applying a "pre-emption" test to assess the right of Indians to make and be governed by their own laws. Generally, the court has liberally interpreted the pre-emption test, favouring federal Indian policies that support tribal jurisdiction in a number of opinions (*Rice v. Rehner*, 1983; *White Mountain Apache Tribe v. Bracker*, 1982; *Montana v. United States*, 1981). However, it has also affirmed concurrent state jurisdiction over certain reservation activities by non-Indians unless these activities are expressly pre-empted by federal law. Pre-emption analysis is based on the principle of balancing the competing state, federal, and tribal interests rather than relying solely on Congressional intent to pre-empt state law. In addition, the courts have ruled that federal approval of tribal ordinances that exercise exclusive jurisdiction does not alone confer a delegation of power by Congress (*Washington v. Confederated Tribes of the Colville Indian Reservation*, 1980). If tribal regulation is consistent with prior federal policy and established notions of Indian sovereignty, the courts may favour exclusive application of tribal law. The problem of determining the extent of tribal jurisdiction over non-Indians stems from the defining of tribal interests in exercising its jurisdiction over non-Indians. The subjective nature of the test, itself, obscures the extent of state or tribal regulatory power. The antagonistic position of states, in challenging virtually every attempt by tribal governments to exercise their powers of self-government, continues to perpetuate the distrust between these governments.

The management of land use activities, similar to the coordination of environmental programs, also requires a coordinated approach to foster greater consistency in regional development. The notion of a reservation territory becoming jurisdictionally divided based on whether a particular parcel of land is held in fee or federal trust ownership presents a disastrous outcome for planning. The reservation landscape cannot be adequately managed under a bifurcated and inconsistent land use program that represents the competing interests of Indian and non-Indian communities. The following section examines several attempts by tribal and non-tribal governments to set aside questions of contested land use jurisdiction by emphasizing the attainment of land use policy consistency. As a long-term goal, such collaborative processes seek to build stronger regional communities by reconciling the competing Indian and non-Indian interests that coexist within the reservation community.

Tribal-State Mediating Structures: Lessons from Washington State

Negotiation, as an approach to resolving conflicts, can be successful when all parties are able and willing to negotiate, particularly when the number of parties and the range of issues are limited. Conversely, when diverse interests are involved, negotiations often become more complicated. The divided nature of the parties involved with negotiation presents additional problems. The United States, comprising several agencies, may represent varied interests. The states, likewise, may represent different interests as landowner, sovereign, and representative of individual interests. While litigation often involves great expense and animosity between the parties, it has also led to state-tribal cooperation. The formulation of Washington's Indian Policy emerged after the federal court mandated that the landmark decision in *US v. Washington* be implemented. To reach agreement on how to implement the fisheries decision, the court essentially ordered the state and the tribes to cooperate. New forms of dialogue were encouraged by the court, which led the parties to participate in a new method of dispute resolution to address fisheries co-management.

The dramatic shift in intergovernmental relations between Washington and Indian tribes began in the 1980s with the application of the Comprehensive Cooperative Resource Management (CCRM) principles, and has since greatly expanded to include other areas of governmental cooperation. The principle supporting CCRM is the recognition that natural resources cannot be managed in isolation from other resources management approaches. The approach has resulted in institutional changes affecting the nature of the relationships between the interests in resources management, including the tribes, state and federal governments, industry, the environmental community, and the public. Comprehensive Cooperative Resource Management is based on the view that cooperation is the best way to resolve resources management issues, since it promotes the discovery of common ground upon which to build agreement (Northwest Indian Fisheries Commission, 1991). The process emphasizes a separation between participant group needs and entrenched group positions in order to identify a unified basis for agreement. The CCRM approach, however, needed the support of a non-biased, third-party facilitator to help foster dialogue among parties that were previously adversarial.

The Northwest Renewable Resources Center (NRRC) emerged in 1984 to assist in the resolution of statewide and regional disputes regarding renewable resource management among various interests in Washington State. The NRRC was formed by the tribes, business leaders, and other vested groups to help resolve disputes over renewable resources using the negotiation process. The catalyst leading to the creation of the NRRC was phase II of the *Boldt* decision, where the court concurred that the government has a "duty to prevent the degradation of the habitat" so as not to deprive the tribes of their moderate living needs. Four principal options were identified as political strategies for resolving concerns relating to the effect of a phase II ruling (Gordon, Honeywell, Peterson, & O'Hern, 1981). The first two options considered further judicial action and legislative intervention. It was concluded that both options would prove to be too risky and expensive. The third option, of maintaining the status quo, risked subjecting future resources questions to the court's further test on impacts on fisheries habitats, and likewise was concluded to be too risky. The fourth and preferred option outlined a political strategy containing four stages: 1) the creation of a process that provided for tribal participation in governmental decision making; 2) negotiations to resolve related natural resources policy issues, including timber harvest practices and minimum in-stream flows; 3) the development of fisheries enhancement programs; and 4) the assurance that salmon harvest allocation would meet tribal expectations.

The principal role of the Northwest Renewable Resources Center was defined by the need for a neutral convener. Unlike traditional forms of mediation, the NRRC was formed as a coalition of members from the business, environmental, and tribal communities. By focusing on broad problems with the intent of generating legislative proposals, the NRRC acted as a body for recommending policy alternatives. As a result of the NRRC mediation, the Washington Department of Fisheries and the treaty tribes now successfully co-manage the state's salmon runs. The "Long Live the Kings" project was later introduced to respond to the decline of King, or Chinook, salmon by recommending a series of short-term enhancement projects to rebuild the degraded runs of wild salmon. The approach focused on building a broad-based consensus about the need for wild salmon enhancement with a strategy of initiating demonstration projects.

The successes of these two early projects prompted the NRRC involvement in other environmental policy mediations involving the tribes. Extending the principles of intergovernmental cooperation to the local level, several efforts aimed at establishing cooperative institutional relationships between tribes and local governments were also initiated. The "Tribes and Counties Intergovernmental Cooperation" project assisted tribes and county governments in establishing several problem-solving processes to address complex jurisdictional disputes, ranging from land use and environmental regulation to taxation and service delivery. Table 7.2 summarizes several of the efforts embarked upon in developing intergovernmental cooperation among tribes, local and regional governments, and Washington State agencies.

Rebalancing Interests: Getting to Regional Pluralism

Cooperation between local governments can produce greater efficiencies in the provision of local governmental services while reducing jurisdictional conflicts and helping to attain a unified regional development policy. Many states encourage voluntary agreements by local governments to create joint plans for economic growth, land use, transportation, and other aspects of regional planning.[10] Washington mandated local government cooperation in 1990 when it enacted its growth management law (Washington State, Growth Management Act, 1990). One requirement of the Act calls for local land use plans to be consistent with adjoining local jurisdictions, thereby creating the necessity for intergovernmental cooperation. While cooperation among counties and municipalities has progressed in many states, fewer gains have been made with regard to improving the intergovernmental relationships with tribal governments.

Depending upon the nature of political conflict that tribes experience in their reservation affairs, planning approaches among the tribes can vary greatly. Similar to the CCRM experience in Washington State, planning solutions that are based on regional cooperation can serve as a promising pathway for promoting inclusive planning and for overcoming the jurisdictional stalemates that can impede both tribal and regional development. The resolution of historic conflicts in tribal and

Table 7.2. Cooperative Tribal-Local Government Negotiated Programs in Washington State since 1980 (Northwest Renewable Resources Center, 1992)

Private Project	Tribal-Local Participants	Purpose
West Point Secondary Treatment Upgrade Project	Suquamish and Muckleshoot Indian tribes; Seattle Metro	Agreement to mitigate marine habitat impacts
Columbia River Gorge National Scenic Area Management Plan	Yakima Indian Nation, Nez Percé Tribe, Warm Springs Tribe	Agreement to protect cultural resources and treaty rights
Puyallup Lands Settlement Agreement	Puyallup Indian Tribe; Port of Tacoma, Pierce County, the cities of Tacoma, Fife, and Puyallup Tribe	Agreement resolving land claim and establishing a trust fund for environmental enhancement and tribal economic development
Snohomish County Aquatic Resource Protection Program	Tulalip Tribes, Snohomish County	Agreement to develop wetlands and other protection measures
Seattle Water Department Relations with Muckleshoot Indian Tribe	Muckleshoot Indian Tribe, the City of Seattle, Seattle Water Department	Agreement to foster dialogue supporting tribal fisheries protection objectives
Lake Roosevelt Forum	Colville Confederated Tribes, Spokane Tribe, county, state, and federal agencies	Agreement to achieve cooperative regional planning in the Lake Roosevelt region
Whatcom-Lummi Nation Cooperation in Land Use Planning	Lummi Indian Nation; Whatcom County	Agreement for improved cooperation in land use planning
Skokomish Tribe–Mason County government-to-government discussions	Skokomish Indian Tribe; Mason County	Agreement seeking to minimize jurisdictional and regulatory conflict
Seattle City Light Skagit River Agreement	Upper Skagit, Sauk-Suiattle and Swinomish Indian Tribal Community, Seattle City Light	Agreement for a mitigation plan for fisheries and wildlife protection, recreation, education, and cultural resources
The Swinomish-Skagit Joint Comprehensive Plan	Swinomish Indian Tribal Community, Skagit County	Agreement for joint land use planning, a landmark in cooperative regional planning
Indian Land Tenure and Economic Development Project	Swinomish Indian Tribal Community, Quinault Indian Nation, Skagit County, Clallam County	Agreements for cooperation in land use and timber management
Hood Canal Coordinating Council	Port Gamble S'Klallam and Skokomish Indian Tribes, Kitsap, Mason, and Jefferson counties	Policy coordination for protection and restoration of Hood Canal Waterways

regional government relations must begin with a meaningful dialogue intended to reconcile the differences between tribal and non-tribal interests. The negotiated approach to conflict resolution as manifested by the CCRM experience demonstrates how new public policy responses can emerge to overcome conflicts and provide opportunities for meaningful tribal participation in regional issues.

The experiences of tribes and states illustrate a process of tribal political development that began with the tribes' assertions of inherent sovereignty and treaty rights and evolved into cooperative relations and co-management agreements for resolving long-standing disputes in the management of natural resources. While the experiences in Washington originally focused on the protection of off-reservation tribal treaty rights, the process was later expanded to resolve on-reservation conflicts involving comprehensive planning and the delivery of general public services. The case of the Swinomish tribe and Skagit County regional governments in Washington demonstrates the success of the negotiated approach in resolving land use inconsistencies through mutual planning approaches. Since 1984, more than a dozen cooperative agreements between the tribe and state, county, municipal, and regional governments have resolved former conflicts in land use planning and regulation, building code administration, water and sanitary sewer utility services, environmental protection, habitat restoration, transportation planning, forest practices, mutual aid, and other essential governmental services. These approaches are particularly useful in reservations where high degrees of alienation exist and where the jurisdictional reach of state and local governments has been prevalent. The radical transformation that has begun to take place over the past few decades through improved relations between tribal and non-tribal governments offers a promising avenue for overcoming the obstacles that tribes face in their pursuit of self-determination. The application of cooperative or "concurrency" agreements – in which tribal and non-tribal interests are simultaneously considered – is an effective approach for regional planning, one that not only transcends regional conflicts with tribes but also, perhaps more importantly, helps bridge cultural divides by creating a region where a plurality of interests may flourish.

Final Thoughts

The control of a tribe's reservation territory and its resources is fundamental to its long-term survival. Since the 1970s, the implementation of the federal self-determination policy has resulted in the restoration of many powers that had either previously been removed or remained dormant. The progressive development of the tribal political community continues to be shaped by court decisions that delineate the delicate balance of competing interests that operate within the tribal planning situation. In an attempt to make tribal planning more effective, tribes continue to rely on the federal courts for guidance. While the courts have clarified that tribal sovereignty has been narrowed by virtue of their dependent status, they have also affirmed that tribes clearly possess sufficient governing powers over their affairs, and, in some cases, have considerable influence over the management of off-reservation resources when tribal interests are at stake. Notwithstanding the tribes' powers to manage their reservation affairs, conflicts with states and local governments will undoubtedly continue to persist. As a way of mediating the differences between tribal and non-tribal communities, a commitment to purposeful dialogue signifies a critical initial step towards building a more tolerant and inclusive regional community.

The negotiated approach to conflict resolution has led to meaningful tribal participation in regional decision making. Washington's experiences illustrate the evolution of cooperative relations that arose from long-standing natural resources management disputes. While these initial experiences were motivated by the tribes' desire to protect their treaty rights, the process expanded to agreements for the resolution of many other types of conflicts. These experiences represent a progression in state-tribal public policy development that reached an institutional pinnacle when the tribes and the state signed the Centennial Accord to formally bring closure to a 100-year period of animosity and to finally acknowledge the political legitimacy of tribal governments. The emerging precedent favouring negotiated solutions through inclusive dialogue and co-management approaches offers an important pathway for the reconciliation of the historic exclusion of Native Americans and their interests. It is a crucial step forward as we, in the planning profession, seek to break down the historic barriers that Indians and their governments have faced, and as we work towards building culturally tolerant and politically pluralistic regional communities.

Before inclusive planning can take hold as a universal paradigm in American planning practice, meaningful insight is required to understand the basis of differences that exist among communities. Planning for diversity requires more than a simple acknowledgment of difference – it requires, as well, a respect for and acceptance of difference, without which planning risks remaining a largely discriminatory and exclusionary practice. Planning with Native American communities requires the recognition of the legitimate political right of tribes to plan their lands and resources. Planning with Native Nations means building a planning capacity that can accept their political self-determination and where the goal of regional planning is to facilitate the integration of Native visions within the broader regional vision and to overcome the often conflictive nature of diverse planning policies. As planning encounters the presence of tribal interests, a capacity for cultural tolerance is not nearly enough. Planning also requires a deeper understanding of the differences that exist among communities and the eventual acceptance of both cultural and political plurality. Without this, planning risks serving as a vehicle for the continued subjugation of Native American communities.

NOTES

1 "Indian reservation" defines the area of land set aside by the federal government for the use, possession, and benefit of an Indian tribe or group of Indians. Most reservations were created through Congressional Act, federal treaty, or presidential order. Indian lands are also referred to as "Indian Country."

2 Notwithstanding the provisions of the Indian Reorganization Act of 1934 requiring certain tribes to obtain Secretarial approval in certain governmental actions.

3 For example, 95 per cent of the Puyallup Indian reservation in Washington State comprises alienated lands and a non-Indian population of 96.8 per cent.

4 Police powers are defined as that which "enables the people to prohibit all things inimical to comfort, safety, health, and the welfare of society" (*Drysdale v. Prudden*, 195 N.C. 722, 143 S.E. 530, 536 (1928)).

5 Congress reaffirmed the EPA's policy of working on a government-to-government basis with tribes when it amended the provisions of the Safe Drinking Water Act in 1986, the Clean Water Act in 1987, and the Comprehensive Environmental Response, Compensation, and Liability Act in 1986, to recognize the EPA's obligation to treat tribes as states for the purpose of delegating responsibility for implementing environmental programs and regulating the reservation environment.

6 While ballot box participation in general elections is limited to the membership of a tribe, participation in other areas of tribal governance does not preclude the appointment of non-members to governmental committees and commissions to provide for resident participation in the affairs of the reservation.
7 Another example demonstrating the inability of tribes and states to find resolution, without the intervention by the courts, for the protection of treaty fishing rights concerns the tribes' ongoing lawsuit in *United States v. Washington* (1984) contending that the state's system of road culverts prevents fish from reaching spawning habitat areas. On March 29, 2013, a US District Court judge ruled in favour of the tribes and ordered the state to provide fish passage in 817 road culverts within a 17-year period. The repair work is estimated to cost $1.9 billion.
8 The Equal Footing doctrine applies to those states entering the Union on an "equal footing," with the same rights and powers as the original thirteen states. Since the original thirteen states owned the lands beneath navigable waters, subsequent states likewise owned those lands.
9 In 1928, the Office of Attorneys General of Washington issued an informal opinion concluding that state water right laws could not be applied within the Colville Indian reservation because of the pre-emptive impact of 25 U.S.C. sec 381 (1982). Letter of February 28, 1928. This opinion, however, has not been followed.
10 Wisconsin's intergovernmental cooperation law requires local governments within metropolitan areas to sign compact agreements with neighbouring municipalities or counties for the provision of joint public services (Wisconsin Statutes, Chapter 66, Subchapter III: Intergovernmental Cooperation. Sections 66.0301–66.0315).

REFERENCES

Arizona v. California, 460 U.S. 605 (1983).
Arizona v. San Carlos Apache Tribe, 463 U.S. 545 (1983).
Bureau of Indian Affairs. (2012). Indian entities recognized and eligible to receive services from the Bureau of Indian Affairs. Federal Register, Volume 77, Number 155, FR 47868, FR Doc. 2012-19588.
Cappaert v. United States, 426 U.S. 128, 138–142 (1976).
Cherokee Nation v. Georgia, 30 U.S. 1 (1831)
Clinton, W. (1994). Memorandum on government-to-government relations with Native American tribal governments. *Weekly Compilation of Presidential Documents*, *30*(17), 936–937.

Colville Confederated Tribes v. Walton, 647 F.2d 42 (9th Cir. 1981), aff'g in part and rev'g in part 460 F. Supp. 1320 (E.D. Wash. 1979), cert. denied, 454 U.S. 1092.
Confederated Salish and Kootenai Tribes v. Namen, 665 F.2d 951 (9th Cir. 1982)
General Allotment Act (or Dawes Act), 24 Stat. 388, ch. 119, 25 USCA 331, 49th Cong. Sess II, sec. 331–334, 339. 341, 348, 354, 381 (1887).
Goeppele, C. (1990). Solutions for uneasy neighbors: Regulating the reservation environment after *Brendale v. Confederated Tribes and Bands of Yakima Indian Nation*, 109 S. Ct. 2994, 1989 [Seattle, WA]. *Washington Law Review (Seattle, Wash.)*, *65*(2), 417–36.
Gordon, T., & Honeywell, M. Peterson, and O'Hern (1981). U.S. v. Washington, Phase II: Analysis and Recommendations. Seattle, WA.
Johnson v. McIntosh, 21 U.S. (8 Wheat) 543 (1823).
Knight v. Shoshone and Arapahoe Tribes, 670 F.2d. 900 (10th Cir. 1982.).
McCarren Amendment, 43 U.S.C. sec 666 (1982).
McClanahan v. Arizona State Tax Commission, 441 U.S. 164 (1973).
Montana v. United States, 450 U.S. 544; 564–65 (1981).
Muckleshoot Indian Tribe v. Trans-Canada Enterprises, Ltd., 713 F.2d 455 (9th Cir. 1983).
Nevada v. United States, 103 S.Ct. 2906 (1983).
Northwest Indian Fisheries Commission. (1991). *Federally recognized tribes of Washington State alternative approach to water quality management: A cooperative intergovernmental watershed strategy*. Olympia, WA.
Northwest Renewable Resources Center (1992). *NRRC tribes and counties: Intergovernmental cooperation project*. Seattle, WA.
Obama, B. (2009). Memorandum for the heads of executive departments and agencies: Tribal consultation. Presidential documents. Federal Register, Volume 74, Number 215, FR Doc. E9–27142
Oliphant v. Suquamish Indian Tribes, 435 U.S. 191 (1978).
Puyallup Indian Tribe v. Department of Game, 391 U.S. 392 (1968).
Puyallup Indian Tribe v. Department of Game, 433 U.S. 165 (1977).
Puyallup Indian Tribe v. Port of Tacoma, 717 F.2d 1251 (9th Cir. 1983).
Rice v. Rehner, 103 S.Ct. 3291 (1983).
Snohomish Co. Sup. Ct. No. 71421 (1963).
United States v. Anderson, 738 F.2d 1358 (9th Cir. 1984).
United States v. Cascade Natural Gas Corp. No. C76–550V W.D. Wash.
United States v. Kagama, 118 U.S. 375, 384 (1886).
United States v. Washington, 384 F. Supp. 312 (W.D. Wash. 1974), aff'd, 520 F.2d 676 (9th Cir. 1975), cert denied, 423 U.S. 1086 (1976).
United States v. Washington, 506 F. Supp. 187 (W.D. Wash. 1980), en banc appeal dismissed (9th Cir. No. 91–3111, 1984).

United States. v. Wheeler, 435 U.S. 313, 327 (1978).

United States v. Winnebago Tribe, 542 F.2d 1002 (8th Cir. 1976).

United States v. Winters, 207 U.S. 563 (1908).

Washington Attorneys General. (1985). *The State of Washington and Indian tribes*. Office of the Attorney General, Indian Litigation Coordinating Committee. Olympia, WA.

Washington Game Department v. Puyallup Tribe, 414 U.S. 44 (1973).

Washington State, Ground Water Code, 90.44 RCW.

Washington State, Growth Management Act, RCW 36.70a (1990).

Washington State, Surface Water Code, Chapters 90.03 RCW.

Washington v. Confederated Tribes of the Colville Indian Reservation, 447, U.S. 134, 152–56 (1980)

Washington v. Environmental Protection Agency, 752 F.2d 1465 (9th Cir. 1985).

Washington v. Washington State Commercial Passenger Fishing Vessel Ass'n, 443 U.S. 658 (1979).

White Mountain Apache Tribe v. Bracker, 448 U.S. 136 (1982).

Williams v. Lee, 358 U.S. at 220, 223 (1959).

Worcester v. Georgia, 31 U.S. 515 (1832).

PART 3

Planning the Inclusive City

8 Majority-Minority Cities: What Can They Teach Us about the Future of Planning Practice?

MICHAEL A. BURAYIDI AND ABBY WILES

Introduction

Big cities have always been the vanguard of social change. Smaller communities look to see how they succeed, or fail, in a myriad of issues. Statistics Canada projects that by 2031 one in three Canadians will be a visible minority with Toronto and Vancouver becoming majority-minority. As cities in North America change from a mostly White nation to one populated by more and more minorities, cities like New York, Houston, and Phoenix will grapple first with what these changing demographics mean for planning. "Large metropolitan areas will be the laboratories for change," noted demographer William Frey of the Brookings Institution told the *Washington Post*. "The measures they take to help minorities assimilate and become part of the labor force will be studied by other parts of the country that are whiter and haven't been touched as much by the change" (Mellnik & Morello, 2011).

Collectively, minority groups – Hispanics, Blacks, and Asian and Pacific Islanders, in particular – are now the majority in more than half of America's largest cities, providing plenty of lessons for the nation's smaller communities to study (Frey, 2011). The term "majority-minority," used to describe these communities where Whites are outnumbered by non-Whites, already describes the 10 largest American cities. These planning laboratories are New York, Los Angeles, Chicago, Houston, Philadelphia, Phoenix, San Antonio, San Diego, Dallas, and San Jose.

The change has happened quickly in many places: six of the 50 most populated US cities transitioned from a White majority to majority-minority in the decade between 2000 and 2010. These quick-change cities include Austin, Charlotte, Las Vegas, Albuquerque, Tucson, and Arlington.

This growth in minority population is expected to continue and will outpace growth in White communities. The US Census Bureau projects that by 2043, minorities will make up 57 per cent of the total US population, and more cities will become majority-minority. Potential impacts will be felt in politics, education, housing, governance, land use, and urban planning. When the Metropolitan Washington Council of Governments provided information to homeowners and renters facing foreclosure, the brochures were printed not just in the expected English and Spanish languages but also in Mandarin Chinese, Vietnamese, and even Amharic.

As most new immigrants to the US now emanate from non-European countries, the growth of the immigrant population will require cities to consider unfamiliar cultures, religions, and living arrangements. For most of the US, the future has not yet arrived. However, for those cities that now have a majority-minority population, the future is now. These cities are the testing grounds of policies and programs that can be models for cities in transition. Thus, other cities can utilize successful strategies and programs implemented by the majority-minority cities as they make the transition themselves. In this chapter, we examine the planning practices of the largest majority-minority cities in the United States to find planning strategies, policies, and programs that effectively address the diverse needs of a multicultural society. We review comprehensive plans, zoning regulations, development codes, citywide policies, governance committees, economic development policies, historic preservation laws, cultural plans, and other relevant planning documents to shed light on their planning practices. We conclude with lessons and recommendations for addressing the needs of multiple publics.

Selection of Cities

The cities evaluated in this chapter were selected based on size and collective share of minority populations. Data provided by the US Census Bureau and analysed by the Brookings Institution provided the basis for selection. The top 25 largest majority-minority cities were selected for the study. Fresno, California, the 25th largest majority-minority city, was removed from this study in favour of Honolulu. The inclusion of Honolulu provides us with a unique opportunity to examine a multicultural city with a high percentage of Asian and Pacific Islander citizens. For the purpose of this chapter, only cities are evaluated, not metropolitan statistical areas or regions. The cities evaluated in this chapter are shown in table 8.1. Hispanics have the largest share of the

Table 8.1. Largest Majority-Minority Cities in the United States.

City	State	2010 Pop. (1000s)	Hispanic	Black	Asian	White
			Share of Population, 2010			
New York	NY	8,175	29	23	13	33
Los Angeles	CA	3,793	48	9	11	29
Chicago	IL	2,696	29	32	5	32
Houston	TX	2,099	44	23	6	26
Philadelphia	PA	1,526	12	42	6	37
Phoenix	AZ	1,446	41	6	3	47
San Antonio	TZ	1,327	63	6	2	27
San Diego	CA	1,307	29	6	16	45
Dallas	TX	1,198	42	25	3	29
San Jose	CA	946	33	3	32	29
San Francisco	CA	805	15	6	33	42
Austin	TX	790	35	8	6	49
Fort Worth	TX	741	34	18	4	42
Charlotte	NC	731	13	34	5	45
Detroit	MI	714	7	82	1	8
El Paso	TX	649	81	3	1	14
Memphis	TN	647	6	63	2	27
Baltimore	MD	621	4	63	2	28
Boston	MA	618	17	22	9	47
Washington	DC	602	9	50	3	35
Milwaukee	WI	595	17	39	3	37
Las Vegas	NV	584	31	11	6	48
Albuquerque	NM	546	47	3	3	42
Tucson	AZ	520	42	4	3	47
Honolulu*	HI	339	n/a	1	57	18

Source: (Frey, 2011); Retrieved March 16, 2013, http://www.brookings.edu/~/media/research/files/papers/2011/5/04%20census%20ethnicity%20frey/0504_census_ethnicity_frey.
*Data for Honolulu accessed by searching "Urban Honolulu CDP, Hawaii" US Census Bureau – 2010 American Community Survey 1-Year Estimates

population in seven of the 25 cities, and African Americans are the largest demographic group in six.

Policies and Governance

Governance provides a yardstick for measuring the role of a group in a community and for understanding the politics of difference. As Bird, Saalfeld, and Wust (2010) observed:

> If there is a growing segment of the population that neither participates electorally nor has access to elected office, there is an increasing risk that the interests of such persons are not part of the electoral preference aggregation mechanisms of modern liberal democracies. If this is the case, the process of democratic representation is incomplete and the legitimacy of public policy is in doubt. (p. 4)

Taebel (1978) found that small city councils and at-large elections adversely impacted minority group representation in government. In her study of immigrant integration and participation in civic activities in Canada and the United States, Bloemraad (2006) concluded that where governments provide material and symbolic support for multicultural groups there is an increased level of participation of these groups in government and in electoral representation. Interestingly, of the 25 majority-minority cities in this study, 20 (80 per cent) have an aldermanic form of city council representation, and four have a combination of aldermanic and at-large election. Detroit is the only city in the sample where council members are elected entirely at large.

All the 25 majority-minority cities studied have strong minority representation in city government. This includes elected officials, such as the mayor and city council members, as well as municipal boards and commissions who have a direct influence on planning and development. Eight of the 25 cities, 32 per cent, have a minority mayor (see table 8.2). Of these eight mayors, five are Black, two are Hispanic, and one is Asian. For the most part, the minority mayors are representative of the diverse groups within each city.

Table 8.2. Minority Mayors (January 2013)

City	State	Name	Ethnicity
Los Angeles	CA	Antonio Villaraigosa	Hispanic
San Antonio	TX	Julian Castro	Hispanic
San Francisco	CA	Edwin Lee	Asian
Charlotte	NC	Anthony Foxx	Black
Detroit	MI	Dave Bing	Black
Memphis	TN	A.C. Wharton, Jr	Black
Baltimore	MD	Stephanie Rawlings-Blake	Black
Washington	DC	Vincent Gray	Black

City council representation is equally diverse. All 25 cities have minority representation on city council. Some of the cities with a high percentage of minority council members include San Antonio, San Francisco, Detroit, Baltimore, and Honolulu. All nine of Detroit's city council members are Black. Eight of the nine city council members in Honolulu are minority, namely Asian and Pacific Islander.

Equally well represented is membership on the boards and commissions that make major planning and zoning decisions for the cities. Los Angeles, for example, has several Hispanic members and one Armenian member on its plan commission. Mayor Villaraigosa's appointment of George Vahe Hovaguimian to the city's plan commission is significant because of Los Angeles' sizeable Armenian population. Andrew Kizirian, chair of the Board of Directors of the Armenian National Committee of America – Western Region, was appreciative of the gesture:

> We applaud Mayor Villaraigosa's decision to include George in his administration. The Mayor has been a leader on Armenian-American issues and George's appointment is yet another example of the Mayor's vision of a Los Angeles that is built on diversity, strengths and knowledge of all Angelenos. (Asbarez.com, 2011)

The majority-minority cities have policies and programs specifically tailored to making immigrants welcome. The city of New York has the Mayor's Office of Immigrant Affairs, whose mission is stated thus:

> Building on its Charter mandate, the Mayor's Office of Immigrant Affairs promotes the well-being of immigrant communities by recommending policies and programs that facilitate successful integration of immigrant New Yorkers into the civic, economic, and cultural life of the city. (City of New York, n.d.)

One goal of the Dallas Cultural Affairs Commission is to ensure that "people of all ages enjoy opportunities for creative expression and the celebration of our community's multicultural heritage." San Francisco boasts an Immigrant Rights Committee, and Fort Worth has a Commission on Immigrant Affairs, both charged with advising the city's leadership on issues and policies related to immigrants. Similarly, Charlotte's International Cabinet promotes the city as a welcoming place for all and works as a "consultant to elected officials and city staff on local activities of international scope and is a reflection of our community's changing face."

Honolulu's citywide diversity statement is inclusive on many levels. It reads:

> One of the greatest assets of the City and County of Honolulu is the ethnic, cultural, and social diversity of its population. The City and County of Honolulu takes great pride in this diversity, and values and respects all of its residents and welcomes all of its visitors, regardless of race, color, sex, marital status, religion, national origin, ancestry, age, disability, gender identification, or sexual orientation. (City of Honolulu, 2013)

Several mayors and city managers of the majority-minority cities have made formal proclamations on the importance of diversity in their cities. In Las Vegas, for example, City Manager Elizabeth N. Fretwell stated that the city is committed to upholding the diversity of its population by developing "programs that represent the needs and interest of the residents" and by reaching out to the diversity population to "forge ties and create open lines of communication" (Fretwell, 2013).

Most of the majority-minority cities have a formal statement, guiding principles, or policies to promote equal opportunity in their hiring practices. Mayor Michael A. Nutter of Philadelphia has a formal statement of policy for equal employment opportunity which stresses the "fair and equitable treatment" of all current and prospective employees (Nutter, 2011). The city's human resources department in Dallas upholds a vision for "a diverse, vibrant, progressive, and engaged workforce." The city of Phoenix's Office of Equal Opportunity Programs states its mission is to "ensure equal opportunity, diversity, and the elimination of discrimination within the city of Phoenix through education, enforcement, and community involvement" (City of Phoenix, 2013).

Many cities have a formalized statement on the importance of inclusivity not only in the hiring process but also for businesses that contract with the city. Philadelphia's Business Services Department's Statement of Economic Inclusion encourages awarding city contracts to "minority, women, and disabled owned business enterprises," in an effort to advance economic opportunities for these businesses. Houston's Office of Business Opportunity seeks to create "a competitive and diverse business environment in the City of Houston by promoting the growth and success of local small businesses, with special emphasis in historically underutilized groups by ensuring their meaningful participation in the government procurement process."

Despite the rising numbers of the minority population and exemplary city policies, there is a palpable "political lag" well behind the observable demographic transition. As Myers (2008) pointed out, despite the fact that California became a majority-minority state in 1999, "in terms of political power, little has changed in subsequent years" and "where it counts – among the voters – non-Hispanic whites still hold a dominant majority, in some recent elections as high as 71 percent. Despite the demographic forces of change, political power sharing is not proceeding at the pace that some might assume" (p. 123). This political lag is significant in planning because it shapes power relations and inhibits the freedom of planners to act. Managing this disparity in the planning process will continue to challenge planning practice for decades to come.

Physical Development and Land Use

The changing demographic landscape in the United States will impact land use and the physical development of communities. The majority-minority cities are already seeing this effect as they adapt their zoning and land use regulations to accommodate the cultural and demographic changes. Myers (2013, p. 12) observed that "diversity is the rightful focus of many programs, because it leads us to better define our program clientele and to keep our eye on the goal of achieving greater equity." He goes on to note that in the next two decades "as a mostly white population ages, requiring more services and paying fewer taxes, a younger and more ethnically diverse group (including recent immigrants) must be welcomed and encouraged as wage earners and tax payees" (p. 12). Here we discuss how the majority-minority cities are accommodating their multicultural populations in their planning and land use regulations.

i) Accommodating Multigenerational Living

The growing population of Asians and other new immigrants is boosting demand for multigenerational living arrangements. Multigenerational households are defined by the US Census Bureau as "family households consisting of three or more generations" (Lofquist, 2012). Most immigrants to the US are from cultures where multigenerational living is the norm. In Asia, South America, and Africa, adult children usually live with their parents and grandparents in the same housing unit. It is therefore not unusual for these immigrants to want a similar living arrangement when they arrive in the US. According to census data, multigenerational households are "more likely to reside in areas where new immigrants live with their relatives" (Lofquist, 2012). In addition, multigenerational households are more likely to be minority – namely Asian, Hispanic, Black, and American Indian. About a quarter of Hispanics and 28 per cent of Asians live in multigenerational households compared to 14 per cent of non-Hispanic Whites (Lofquist, 2012).

Spivak (2012) noted the appeal of Lambert Ranch in suburban Los Angeles to Asian immigrants because "ethnic families are more apt to combine relatives and generations in the same residence" (p. 11). He estimates that one in every six Americans now lives in a multigenerational household arrangement. The National Association of Home Builders considers multigenerational living one of the fastest-growing living trends in the country. Notably, the number of multigenerational households increased by 14 per cent between 2007 and 2010, and the number of people living in such households doubled since 1980. This trend stands in sharp contrast to planning regulations, which make it illegal to have accessory units in single-family residential neighbourhoods. However, cities with majority-minority populations appear more willing to modify planning regulations to permit such uses.

In *Making Room for Mom and Dad*, Spivak describes the growing popularity of housing that is designed to accommodate multigenerational households in Los Angeles. He highlights the Chos, an Asian American multigenerational family in the Los Angeles suburb of Irvine. The Chos were in the process of building a 3,000-square-foot house in Lambert Ranch. The New Home Company is the developer of the Chos' home and will feature a 1,000-square-foot guest house with a separate bedroom, bathroom, and kitchen for the family's grandparents (Spivak, 2012).

Lambert Ranch is a "master-planned community of 169 luxury residences featuring compound estates, guest houses, and private quarters in each of the three neighborhoods." The company offers "life space options," specifically designed for multigenerational households. Nine floor plans in three distinct neighbourhoods are available. According to the company's website, life space homes are "designed to provide room for everyone on the family tree, making Lambert Ranch one of the most forward-looking communities in the country" (Reuters, 2012).

In majority-minority San Jose, the Duane Court townhouses by Taylor Morrison were "designed to cater to a fast growing niche of Asian buyers" who need extra space for multigenerational households (Archers Homes, 2015. According to the company's website, Duane Court offers "townhouse floor plans with a bedroom and bathroom on the first floor, the kitchen and main living area on the second floor, and bedrooms on the third floor to allow multiple generations to live together with sufficient independence" (Archers Homes, 2015).

ii) Code Modifications to Accommodate Accessory Dwelling Units

The growth of multigenerational households in the United States is driving changes to building and zoning codes across the country. A new housing development that continues to gain popularity and is conducive to multigenerational households is the accessory dwelling unit, or ADU. Accessory dwelling units, often referred to as "granny flats," are secondary dwelling units located on a single-family lot and independent of the primary unit. ADUs have traditionally been zoned out of single-family residential neighbourhoods. This is, however, changing, especially in majority-minority cities. Seattle, for example, allowed backyard cottages in Southeast Seattle on an experimental basis in 2006, with the condition that the owner must live in the main house or the accessory unit. The city defines a backyard cottage as "a small residential structure sharing the same lot as a house, but self-contained and physically separate from the primary house" (City of Seattle, 2010). In 2009, the city legalized the development of backyard cottages throughout the city.

Between April 2009 and March 2011, a total of 151 permits were either filed or issued for accessory dwelling units and backyard cottages in Seattle. The city explained the rationale for allowing backyard cottages this way: "An increase of independent, affordable housing within a community can add to the diversity of people living in an area and provide attractive role models for others" (City of Seattle, 2011, p. 5). Interestingly, and contrary to concerns of planners about the potential for overcrowding and traffic congestion resulting from such housing, the city found no such problems in its evaluation of the program, noting: "As the overall trends observed are positive, no changes to standards are recommended to address siting or design concerns" (City of Seattle, 2011, p. 6).

In San Antonio, a city whose Hispanic population increased from 58.7 per cent to 63.2 per cent between 2000 and 2010, zoning regulations were modified to allow granny flats in medium density residential zones in the South Central San Antonio Plan to accommodate changing demographics and culture:

> Medium density residential mainly includes single-family houses on individual lots, however, zero-lot line configurations, duplexes, triplexes, fourplexes, and townhomes may be found within this classification. Detached and attached accessory dwelling units such as granny flats and garage apartments are allowed when located on the same lot as the principal residence. (City of San Antonio, 2005, p. 6)

Charlotte also permits accessory dwelling units on single-family lots. In 2012, its city council approved Ordinance 4931, allowing accessory dwelling units in "both new and existing neighborhoods throughout the city" (Real Estate and Building Industry Council, 2012). Section 12.407 of the city code requires that ADUs should "clearly be subordinate to the single-family structure" in terms of height, floor area, etc. Only one ADU is permitted per lot, and both the principal dwelling unit and the accessory unit must be owned by the same person.

Two other majority-minority cities, Philadelphia and Memphis, are exploring the use of accessory dwelling units to provide additional housing opportunities for residents. The Philadelphia Corporation for Aging published *Proposed Zoning Code & Older Philadelphians* to encourage age-friendly zoning codes. The publication pointed out that 27 per cent of Philadelphia households with at least one older adult are multigenerational households and called for zoning codes that permit accessory dwelling units (Philadelphia Corporation for Aging, 2011). The Shelby County Board of Commissioners, Tennessee, in 2010 approved the provision of accessory dwelling units in the county's Unified Development Code (Shelby County Board of Commissioners, 2010).

Not all cities welcome these regulatory changes. In the 2004 session of the California state legislature, the city of Long Beach and more than 100 other California cities opposed AB 2702, authored by Assemblyman Darrell Steinberg (D, Sacramento) and passed by the state legislature. The bill would have limited a city's ability to zone out such uses as granny flats and second homes on single-family residential lots. Jerry Miller, Long Beach city manager, provided the following rationale for opposing the bill in a letter to Assemblyman Steinberg:

> Though we acknowledge your courage to fight for affordable housing, we do not believe AB 2702 is the way to come about this change. Rather than encourage local, balanced, planned patterns of development that respect local land use priorities outlined in the General Plan, AB 2702 simply imposes a one-size-fits-all approach to second unit development on every community in the state.
>
> AB 2702 mandates standards that have proven unacceptable to the residents of single-family housing. AB 2702 will limit local ability to prohibit absentee-landlord duplexes to be created in single-family neighborhoods, create a substantial problem with parking for the residents, dictate irrational minimum lot and unit sizes regardless of lot dimensions, and lead to a decrease in property value. For these reasons, we oppose AB 2702 and strongly urge you to reconsider this bill. (LBReport.com, 2004)

The bill was also opposed by the League of California Cities, the American Planning Association's California chapter, and the California Association of Counties because of its "one size fits all" requirement. Although the bill was passed by the state legislature, it was vetoed by Governor Arnold Schwarzenegger. The "problems" Jerry Miller notes about accessory uses are, of course, not universally considered to be problems. McCrary reported on the health and psychological gains of an elderly woman in Bucks County Township in Philadelphia after she moved into a granny flat. The apartment in which she lived was added by the owners of the house without a permit and is therefore considered illegal. Katrinka Sloan, a housing specialist with the American Association of Retired Persons in Washington, is quoted as saying: "The reason these apartments are illegal is because zoning doesn't permit them, but they are created because people need the incomes, and a very large portion of them are lived in by elderly people interested in staying in their own communities" (McCrary, 1986). McCrary estimates as many as 2.5 million illegal accessory apartments exist throughout the country. An unofficial survey by a member of the board of the Philadelphia Shared Housing Resource Center revealed as many as 40 per cent of single-family homes have such illegal dwellings.

iii) Cohousing Development

Cohousing is another form of housing that is gaining ground because of the growth in the multicultural population. Cohousing is defined as "a type of collaborative housing in which residents actively participate in the design and operation of their own neighborhoods." According to the Cohousing Association of the United States, "The cohousing idea originated in Denmark and was promoted by U.S. architects Kathryn McCamant and Charles Durrett in the early 1980s." Residents in cohousing have individual apartments or houses, but often share communal outdoor or courtyard space, as well as a common house for group gatherings. Also unique to cohousing developments is the community participatory process. Cohousing residents are involved in "planning, operating, maintaining, and governing the neighborhood" (Schacher, 2006).

The cohousing model has a long history of being attractive to immigrants. In Brooklyn, New York, the cooperative movement started in the 1920s for immigrant garment workers from Russia and other Eastern European countries to obtain housing that was socially equitable and just. Cohousing can have many benefits for immigrants acclimating to life in the United States.

Cohousing developments have a focus on diversity. For example, Sonora Cohousing seeks a "diversity of backgrounds, ages, and opinions" (Sonora Cohousing, n.d.). The Los Angeles Eco-Village features a diverse mix of households, with 15 ethnic groups represented and incomes ranging from very low to middle income (Los Angeles Eco-Village, 2013).

Several of the majority-minority cities studied have established cohousing developments. According to the Cohousing Association of the United States, 14 of the 25 majority-minority cities have or have had a cohousing group in the city (see table 8.3). Many of the majority-minority cities that do not have cohousing groups are currently in the development or planning stages. Cohousing is one tool that majority-minority cities are using to develop neighbourhoods that are sensitive to the diverse needs of multicultural populations.

iv) Accommodating Places of Worship

Handling the religious diversity that comes with a multicultural population is another area of challenge for majority-minority cities. As W. Cole Durham, a law professor at Brigham Young University who specializes in religious freedom, observed: "Planning boards are so used to looking at things through the filter of what they're used to regulating that when religious value comes into play, it's given no weight, or not the heightened weight that it really deserves." One study found

Table 8.3. Majority-Minority Cities with Cohousing Developments

City	Development	Status
Los Angeles	Los Angeles Eco-Village	Formed 1999; established 1993
Chicago	Prairie Onion Cohousing	Formed 2003; re-forming
Phoenix	Vesta Community Concepts	Formed 2011
	Phoenix Commons	In development (2013)
San Antonio	San Antonio Cohousing	Formed 2005
San Diego	Enchanted Garden Intentional Community	Formed 1980; established 1989
Dallas	Dallas Cohousing	Formed 2010
	White Rock Crossing	Formed 2008
	DFW Urban Eco-Village	Formed 2011
San Jose	Silicon Valley Cohousing	Formed 2005
San Francisco	San Francisco Backyard	Formed 2009
	Neighborhood and Learning Center	Formed 2011
	San Francisco Cohousing Project	
Charlotte	Champlain Valley Cohousing	Formed 2000; established 2006
Detroit	Detroit Cohousing	Formed 2002; established 2002
Boston	Jamaica Plains Cohousing	Formed 1999; established 2005
Washington, DC	Takoma Village Cohousing	Formed 1998; established 2000
Albuquerque	Albuquerque Cohousing Group	Formed 2008
Tucson	Milagro Cohousing	Formed 1994; established 2002
	Sonora Cohousing	Formed 1993; established 2000
	Stone Curves Cohousing	Formed 2001; established 2004

minority faith groups to be vastly overrepresented in zoning discriminatory cases. More than 40 per cent of zoning-related complaints came from minority religions, a group that represents only 9 per cent of the churches studied (Lampman, 2000).

Los Angeles learned this the hard way. As Luo (1999) narrated, for more than 30 years Etz Chaim, a group of Orthodox Jews, gathered in a residence in Hancock Park, Los Angeles, for prayer. The religious practices of Orthodox Jews require that members pray together and that they walk to the place of worship on the Sabbath and other holy days. For the sake of the elderly, the disabled, and children, then, housing must be located near to a place of worship. Etz Chaim members held their prayer service in a home at the intersection of Third Street and Highland Avenue, one of the busiest and noisiest intersections in LA. The city ordered the group to stop holding services there because religious uses are not permitted in residential neighbourhoods.

The Etz Chaim argued that the city's request violated their religious freedom and filed a federal law suit. Subsequently, Congress passed the Religious Land Use and Institutionalized Persons Act, 42 U.S.C. § 2000cc. Fearing this new law would favour the religious group, the city settled the case and gave the Congregation a conditional use permit to operate the synagogue.

Rather than zoning on the basis of use, a multicultural population may force cities to recast zoning laws to control impacts rather than the mere use of land. There is also a need to broaden the language of zoning. For example, some zoning codes define places of worship narrowly to include only churches and synagogues. As Smith (2000, p. 27) observed, "one effect of increased diversity is a need for a more inclusive vocabulary." Thus in some cities municipal codes that previously used terms such as "church" or "synagogue" are now using a more inclusive term such as "place of worship," a phrase that also includes mosques and temples.

Use of Public Space

It is increasingly evident in majority-minority cities that different ethnic groups use public space differently. The lifestyles of Hispanics, the largest minority group in the United States, for example, will impact how cities are designed and built in what has been called Latino new urbanism. For one thing, Latinos are comfortable living in mixed-use neighbourhoods where shops, homes, and other uses coexist. This has necessitated the reconsideration of ordinances governing the use of such space. James Rojas, pioneer of Latino new urbanism, says Latinos enjoy an "open air culture" in which social life takes place predominantly in public spaces such as sidewalks, parks, plazas, and parking lots. Latinos use public space for communal gatherings such as family reunions, birthday celebrations, and recreation. And, unlike non-Hispanic Whites and Blacks, who use public space for passive and solitary recreational activities such as walking and jogging, Latinos are more likely to use public space for social and congregational activities.

As Stephens (2008) noted:

> Even if past generations of planners did not deliberately dismiss minority communities, it's safe to say that most American cities were built in the era when American society did not value diversity as much as it does

> today. Planning strategies geared towards auto-oriented cities, detached houses, and scarce public space has nonetheless given rise to a sometimes awkward and sometimes elegant relationship between Latinos and American cities, in which streetcorner entrepreneurship is but one example of Latinos' efforts to make a home in someone else's environment. (Stephens, 2008)

One is more likely to find fluidity in land use and more mixed uses in Latino neighbourhoods than is the norm. Such uses include informal activities such as seasonal food stands, mobile truck restaurants, and retail from garages of residential homes. The Associated Press recounted the story of Narcisa Marcelino, a 34-year-old immigrant and single mother from Mexico who lives with her two young daughters in Martinsburg, West Virginia. Although it is against the city's zoning codes, Marcelino sells food from her home to make ends meet and hopes to open a restaurant in the future (Yen, 2012). In Los Angeles, a court commissioner threw out a city ordinance that prevented taco trucks from being parked in a residential neighbourhood for more than half an hour at a time.

Such differences in land use have prompted some developers to take the special needs of Hispanics into consideration in the design of neighbourhoods. American CityVista is a San Antonio–based real estate development firm chaired by Henry Cisneros, former secretary of Housing and Urban Development under President Clinton. The firm designs and builds houses that cater to Hispanics, "From big kitchens with gas stoves for grilling tortillas to courtyards for social gatherings, multiple bedrooms for large and extended families, and driveways that accommodate numerous cars" (El Nasser, 2005). The goal is to build neighbourhoods that are denser and facilitate social interaction.

Arts and Culture

Diverse and culturally rich cities recognize and celebrate their diversity. A number of the majority-minority cities have designated departments, organizations, or alliances dedicated to promoting diversity of the community. Many use their diversity and cultural assets as tools to promote arts, cultural and heritage tourism, and economic development. These are, in turn, used as means to strengthen residents' quality of life and provide visitors with a truly unique experience.

Leveraging Diversity to Promote Arts and Culture

Most of the majority-minority cities have cultural plans that explicitly address the importance of including all residents. Others stress the role diversity plays in developing cultural tourism. Chicago capitalizes on its diversity and rich cultural assets to strengthen the city's arts and culture. The Cultural Plan 2012 recognizes that diversity is an asset in developing the arts sector. One of five main tenets of the planning process of the Cultural Plan 2012 is that "the process mirrors Chicago's vitality," so the "planning process focused on the city's diversity and breadth of residents" (City of Chicago, 2012).

Another community capitalizing on its diversity to promote arts and culture is San Jose, where the cultural plan for 2011–20 is "founded on San Jose's distinct cultural identity and its abundant cultural resources" (City of San Jose, 2011). Furthermore, the plan states that diversity was the "most frequently mentioned element of San Jose's culture, both as a demographic fact and as a long-held civic commitment" (City of San Jose, 2011).

San Antonio's Cultural Collaborative: A Plan for San Antonio's Creative Economy puts a strong emphasis on the inclusion of diversity in arts and culture and in economic development as a means to attract and keep the creative class in the city. The document is a plan for the entire spectrum of the city's creative economy, including non-profit arts and cultural organizations, creative businesses, and creative individuals. Cultural equity is core to the city's planning effort. The executive committee of the San Antonio Cultural Collaborative defines cultural equity as "an ongoing goal that includes the shared value of mutual respect for diverse cultures and the fair distribution of resources among cultural communities" (City of San Antonio, 2005). In addition to the concept of cultural equity, the plan reinforces the importance of leveraging the city's diversity to develop arts and cultural tourism. Participants in the collaborative's planning process "expressed great pride in San Antonio's authentic identity" and "appreciated the majority Mexican/Mexican American culture, bilingual atmosphere, international relationships, and extensive cultural diversity" (City of San Antonio, 2005).

Public Art, Festivals, and Cultural Events

Public art is one way that majority-minority cities recognize, engage, and celebrate the diversity of their citizens. San Diego's Public Art Master Plan stresses the importance of ethnic and diverse representation in public artworks and recognizes that San Diego is "a truly bilingual and bi-national city" (City of San Diego, 2004). According to the plan, 21.9 per cent of the population self-identifies as Latino and 21.5 per cent are foreign-born (City of San Diego, 2004). The representation of citizens from diverse backgrounds is at the core of the city's public art plan.

The vision statement of the city's public art plan reads in part: "We envision a city with artwork that celebrates the extraordinary diversity and history of our community." Thus, the goal of the plan has been to "create artworks that reflect San Diego's cultural diversity and work with artists of diverse cultures" (City of San Diego, 2004).

In Albuquerque, the Art in Municipal Places Ordinance was adopted in 1978 by the city council, and a Public Art Strategic Plan was put in place in 2013 (City of Albuquerque, 2013). The mission statement for the public art program emphasized the importance of diversity: "Enhancing our community through exceptional public art by embracing a transparent process that reflects the diversity and interest of Albuquerque" (City of Albuquerque, 2013). The plan defines diversity as "broad representation of the cultural, historic, and ethnic make-up of the community" (City of Albuquerque, 2013). The city's public art collection features Native American and Hispanic artists, among others.

In Honolulu, the Commission on Culture and the Arts in City Buildings Program aims to "encourage and provide equal opportunity for the development of cultural and artistic talents of the people of Honolulu" (City and County of Honolulu, 2014). Equal opportunity in the program ensures that all populations within the city are fairly represented.

Multicultural cities also use festivals and other cultural events to showcase and celebrate the diversity of their residents. For the last 30 years, Baltimore has hosted an African American music and culture festival (Visit Baltimore, 2013). Chicago hosts a variety of ethnic and cultural events, festivals, and parades which recognize and celebrate Mexican, Jewish, Ghanaian, Celtic, Norwegian, and Polish heritage, among others (Centerstage Chicago, 2013). Memphis hosts ethnic, cultural, and other festivals which highlight its diverse populations. These include Africa in April, a cultural awareness festival, an annual Israel Festival, and an India Fest, among several others (Simpson, 2013).

Conclusion

Based on the analysis of majority-minority cities, several lessons can be gleaned for other cities contemplating a growth of their minority population. Cities must pay special attention to the diverse needs of their residents and seek ways to meet these needs. They cannot assume that the same processes that were used in the past will continue to prevail in the face of changing demographics. Many of the communities we reviewed changed their governance practices and city policies to reflect their commitment to diversity and fair, equitable treatment and opportunities of all persons. Some of the policies are material, while others may be symbolic. These policies are reflected in the cities' mission and vision statements, diversity and affirmative action statements, hiring practices, and, to a certain extent, economic development practices.

With changing demographics, cities also need to adapt their physical development and land use practices to accommodate the diverse needs of multiple publics. In particular it is important to revisit zoning laws to ensure that they facilitate the accommodation of different ways of living. Accessory dwelling units, cohousing developments, and other housing forms that are favoured by multigenerational households need to be allowed in city zoning codes. We've seen how some of the majority-minority cities have changed zoning regulations or granted variances for religious and cultural reasons.

Lastly, cosmopolitan cities must recognize and celebrate the diversity of residents in the arts and cultural sector. The cultural assets and diverse populations can be used as a basis for enriching the city and for promoting economic development. Cities must also take proactive measures to ensure fair representation of all populations in the art and culture that is displayed in public places. Given their experiences, majority-minority cities provide a guiding light for other cities across the US as they begin to grapple with a multicultural population and the politics of difference that ensues.

REFERENCES

Archers Homes. (2015). Taylor Morrison: Homes inspired by you. Retrieved May 19, 2015 from http://www.archershomes.com/2011/11/taylor-morrison-townhouses-at-duane-court-in-sunnyvale/.

Asbarez.com. (2011). Villaraigosa appoints George Hovaguimian to City Planning Commission. Retrieved May 2, 2013 from http://asbarez.com/94042/villaraigosa-appoints-george-hovaguimian-to-city-planning-commission/.

Bird, K., Saalfeld, T., & Wust, A.M. (2010). *The political representation of immigrants and minorities: Voters, parties and parliaments in liberal democracies.* New York: Routledge.

Bloemraad, I. (2006). *Becoming a citizen: Incorporating immigrants and refugees in the United States and Canada*. Berkeley: University of California Press.

Centerstage Chicago. (2013). Chicago ethnic/cultural – festivals & fairs. *Centerstagechicago.com*. Retrieved April 30, 2013 from http://www.cityofchicago.org/city/en/depts/dca/provdrs/chicago_festivals.html.

City and County of Honolulu (2014). The Art in City Buildings Program. Retrieved May 19, 2015 from http://www.honolulu.gov/moca/moca-artincitybuildings.html.

City of Albuquerque. (2013). *City of Albuquerque: 2013–2014 Public Art Strategic Plan*. Retrieved April 23, 2013 from http://www.cabq.gov/culturalservices/public-art/documents/Public%20Art%20Strategic%20Plan%202013%20-%20Final.pdf.

City of Chicago. (2012). *City of Chicago: Cultural Plan 2012*. Retrieved April 23, 2013 from http://www.cityofchicago.org/content/dam/city/depts/dca/Cultural%20Plan/ChicagoCulturalPlan2012.pdf.

City of Honolulu. (2013). Diversity Statement, City of Honolulu. Retrieved May 2, 2013 from https://www.honolulu.gov/footer-links/59-diversity.html.

City of New York. (n.d.). About New York City Mayor's Office of Immigrant Affairs. Retrieved May 2, 2013 from https://goodpitch.org/orgs/new-york-city-mayor-s-office-of-immigrant-affairs.

City of Phoenix. (2013). Equal Opportunity Department, Phoenix, AZ. Retrieved May 2, 2013, from https://www.phoenix.gov/eod.

City of San Antonio. (2005). South Central Community San Antonio Community Plan Update, South Central Community and City of San Antonio Planning Department, San Antonio, TX. Retrieved May 2, 2013 from http://www.sanantonio.gov/planning/pdf/SouthCentralSanAntonioPlanUpdate.pdf.

City of San Diego. (2004). *Public art master plan*. Retrieved April 23, 2013 from http://www.sandiego.gov/arts-culture/pdf/pubartmasterplan.pdf.

City of San Jose. (2011). *City of San Jose's cultural plan for 2011–2020*. Retrieved April 23, 2013 from http://sanjoseculture.org.p11.hostingprod.com/downloads/CulturalConnectionFinal_FullPlan.pdf.

City of Seattle. (2010). *A guide to building a backyard cottage*. Seattle Planning Commission, Department of Planning and Development, City of Seattle, Seattle, WA. Retrieved May 19, 2015 from http://www.seattle.gov/Documents/Departments/SeattlePlanningCommission/BackyardCottages/BackyardCottagesGuide-final.pdf.

City of Seattle. (2011, Apr.). Backyard cottages annual report. Seattle Department of Planning and Development. Retrieved April 30, 2013 from http://clerk.seattle.gov/~public/meetingrecords/2011/cobe20110512_1.pdf.

Nasser, E. (2005). "New Urbanism" embraces Latinos. *USA Today*. Retrieved May 3, 2013 from http://usatoday30.usatoday.com/news/nation/2005-02-15-latinos-usat_x.htm.

Fretwell, E. (2013). Commitment to diversity. City of Las Vegas. Retrieved March 7, 2013 from http://old.lasvegasnevada.gov//information/5260.htm.

Frey, W. (2011). Melting pot cities and suburbs: Racial and ethnic change in Metro America in the 2000s. *Brookings Institution.* Retrieved March 7, 2013 from http://www.brookings.edu/research/papers/2011/05/04-census-ethnicity-frey.

Lampman, J. (2000, Sept. 23). Religion in America – a test of faith. *Desert News* (Salt Lake City, Utah), A1.

LBReport.com. (2004). Bill to make it harder for cities to stop second units on single residential lots breaks out of committee and passes state senate (Karnette voting no). Retrieved May 2, 2013 from http://www.lbreport.com/news/aug04/ab2702d.htm.

Lofquist, D.A. (2012). Multigenerational households: 2009–2011. *American Community Survey Briefs*. United States Census Bureau. Retrieved May 19, 2015 from https://www.census.gov/prod/2012pubs/acsbr11-03.pdf.

Los Angeles Eco-Village. (2013). *Reinventing how we live in the city.* Retrieved April 9, 2013 from http://laecovillage.org/.

Luo, M. (1999, Aug. 29). When regulations and religious freedom collide: Special report: Battles over zoning codes that limit where houses of worship can locate and how they can operate are on the rise, showing what happens. *Los Angeles Times*. Retrieved April 30, 2013 from http://articles.latimes.com/1999/aug/29/local/me-4845.

McCrary, L. (1986). Over 65, and looking for a home: Cheap housing options are few, and sometimes illegal. *Philly.com.* Retrieved April 30, 2013 from http://articles.philly.com/1986-03-28/news/26083875_1_apartments-affordable-housing-housing-problem.

Mellnik, T., & Morello, C. (2011, Aug. 31). Minorities become a majority in Washington region. *Washington Post.* Retrieved February 22, 2013 from http://www.washingtonpost.com/local/minorities-become-a-majority-in-washington-region/2011/08/30/gIQADobxqJ_story.html.

Myers, D. (2008). *Immigrants and boomers: Forging a new social contract for the future of America.* New York: Russell Sage Foundation.

Myers, D. (2013, Mar.). Diversity and aging in America. *Magazine of the American Planning Association*, 11–15.

Nutter, M. (2011). Mayor's statement of policy. *City of Philadelphia.* Retrieved March 7, 2013 from http://www.phila.gov/personnel/eeo/mayormem2.htm.
Philadelphia Corporation for Aging. (2011). *Proposed zoning code & older Philadelphians: An age-friendly Philadelphia report.* Retrieved May 19, 2015, from http://www.pcacares.org/files/Report_on_Zoning_Code_and_Aging_June_2011_FINAL.pdf.
Real Estate and Building Industry Council. (2012). City Council approves accessory dwelling unit ordinance. Retrieved May 19, 2015 from http://baltimore.org/member-blog/african-american-festival-returns-sponsored-visit-baltimore.
Reuters. (2012). The New Home Company acquires 16 acres from DMB Pacific Ventures in San Jose's Berryessa district. Retrieved May 2015 from http://www.reuters.com/article/2012/06/07/idUS205305+07-Jun-2012+BW20120607.
Schacher, C. (2006). Overview. Cohousing Association of the United States. Retrieved March 8, 2013 from http://ala-apa.org/newsletter/2006/10/17/the-good-and-the-bad-of-cohousing/.
Shelby County Board of Commissioners. (2010). The Memphis and Shelby County Unified Development Code. Retrieved May 19, 2015 from http://shelbycountytn.gov/DocumentCenter/View/19789.
Simpson, T.R. (2013). Cultural festivals in Memphis. *About.com.* Retrieved April 30, 2013 from http://memphis.about.com/od/thingstoseeanddo/tp/culturalfestivals.htm.
Smith, N. (2000). Diversity: The challenge to land use planning. *Plan Canada, 40*(4), 27–28.
Sonora Cohousing. (n.d.). *About us.* Retrieved April 9, 2013 from http://sonoracohousing.com/coho/.
Spivak, J. (2012). Making room for Mom and Dad: Multigenerational families are seeking new housing types. *Planning,* 9–13.
Stephens, J. (2008). Out of the enclave: Latinos adapt, and adapt to, the American city. *Planetizen.* Retrieved May 19, 2015 from http://www.planetizen.com/node/35091.
Taebel, D. (1978). Minority representation on city councils: The impact of structure on Blacks and Hispanics. *Social Science Quarterly, 59*(1), 142–152.
Visit Baltimore. (2013). African American festival. Retrieved April 23, 2013 from http://baltimore.org/member-blog/african-american-festival-returns-sponsored-visit-baltimore.
Yen, H. (2012, May 17). Minority birth rate tops whites for first time in the U.S. *Associated Press.* Retrieved April 6, 2014 from http://www.thestar.com/news/world/2012/05/17/minority_birth_rate_tops_whites_for_first_time_in_the_us.html

9 Community Planning for Immigrant Integration

MAI THI NGUYEN, HANNAH GILL,
AND ANISHA STEEPHEN

Introduction

Traditional models of immigrant "assimilation," posited by scholars in the early twentieth century, were unidirectional, focusing on how immigrants adapt to the receiving society (Park & Burgess, 1925). These studies indicated that, over time and after several generations, through improved language skills and upward socioeconomic mobility, immigrants take on characteristics of their host communities (Alba & Nee, 2003 Frisbie & Kasarda, 1988; Joppke & Morawska, 2003; Ireland, 2004).

More contemporary scholarship suggests that integration is a two-way process whereby the region, type of neighbourhood, and level of receptivity of the receiving community all matter in shaping integration trajectories (Portes & Borocz, 1989; Portes & Rumbaut, 1996; Jones-Correa, 2005). Beyond understanding the trajectory of integration patterns, scholars have also examined different dimensions of immigrant integration, including differences across regions (Pastor, Ortiz, Carter, Scoggins, & Perez, 2012), using different measures (Myers & Pitkin, 2010; Pastor et al., 2012); in ethnic enclaves (Wilson & Portes, 1980; Portes, Fernández-Kelly, & Haller, 2005); and across a variety of racial and ethnic groups (Kasinitz, Mollenkopf, Waters, & Holdaway, 2008). Less attention has been paid to the role that local actors play in accelerating or decelerating immigrant integration. However, an emerging literature is providing insights into how local governments are becoming more actively involved in immigrant integration by adopting myriad policies and practices that affect the life trajectories of immigrants (Varsanyi, 2010; Coleman, 2012). For example, in 2011, then-mayor Michael Bloomberg created the Mayor's Office of

Immigrant Affairs in New York City, which "promotes the well-being of immigrant communities by recommending policies and programs that facilitate successful integration of immigrant New Yorkers into the civic, economic, and cultural life of the City" (New York City Mayor's Office of Immigrant Affairs, 2011). As noted by Burayidi and Wiles in chapter 8 and by Harwood and Lee in chapter 10, Chicago, Houston, Seattle, and many other cities have similar directives from their elected officials to facilitate immigrant integration. Street-level bureaucrats, who serve immigrants in their daily duties, are also at the forefront of altering policies and procedures to accommodate growing immigrant populations (Jones-Correa, 2008a; Jones-Correa, 2008b; Marrow, 2009).

While immigrant integration strategies abound, there has been little discussion about the role of urban planning in facilitating immigrant integration. Furthermore, most integration strategies are piecemeal, involving one organization or having narrow policy goals, such as improving access to health care. Rarely do local governments spearhead immigrant integration planning efforts that are more comprehensive, involving multiple agencies and sectors. For the purposes of our study, we borrow from the Grantmakers Concerned with Immigrants and Refugees' definition of immigrant integration: a two-way process that involves immigrants and the host society working together to enable economic mobility and civic engagement for immigrants in order to build a vibrant and cohesive society.

This chapter has two purposes. The first is to assess local immigrant integration policies and practices around the United States, developing a better understanding of how local governments and administrators are facilitating or discouraging immigrant integration. The second is to evaluate how community planning processes in a "new destination" state, North Carolina, can work to build capacity, strengthen social and organizational networks, and provide concrete and sustainable action strategies for immigrant integration. Collaborating with local stakeholders, the researchers facilitated community planning processes in three local jurisdictions throughout North Carolina. Their participant observation in this process offers insights into the benefits and challenges of community planning for immigrant integration.

Expanding Immigration Authority of State and Local Governments

Political debates over immigration policy and immigration reform have occurred for hundreds of years in the US. Central to these debates are questions about which groups are allowed to migrate and settle in the host country, immigrant rights and privileges, and, most recently, the role of local and state government in immigration legislation and enforcement. Historically, the federal government held primary authority. This was reinforced by the 1976 ruling in *DeCanas v. Bica*, in which Supreme Court Justice William Brennan wrote that the "[power] to regulate immigration is unquestionably exclusively a federal power" (*DeCanas v. Bica*, 1976). In the last few years, state governments in Arizona and Alabama have adopted state-level legislation, S.B. 1070 and H.B. 56, respectively, that expands the authority of state and local law enforcement officers to regulate immigration. Local governments have also attempted to pass punitive anti-immigration policies, such as the Illegal Immigration Relief Act, first adopted in 2006 in Hazleton, Pennsylvania (City of Hazleton, 2006a; City of Hazleton, 2006b; *Lozano v. City of Hazleton*, 2007). Furthermore, local public officials and administrators are using traditional planning tools, such as zoning and housing ordinances, to discourage large immigrant families from locating in jurisdictions (Varsanyi, 2008; Varsanyi, 2010). The vast majority of the state and local policies mentioned here have been challenged in the courts, and for the most part deemed unconstitutional because they breach the federal government's authority of immigration legislation and enforcement. Unlike the policies and programs mentioned above, partnerships between the US Department of Homeland Security and local law enforcement agencies, such as Immigration, Customs and Enforcement (ICE) ACCESS 287(g) and Secure Communities Programs, have not been challenged in court (Gill, 2010; Nguyen & Gill, 2010). The 287(g) and Secure Communities programs are different and have not been challenged because they are still federal programs that expand the authority of local and state government agencies, rather than usurping federal authority.

This recent era of subfederal immigration legislation is radically different from previous eras because of the sheer volume of new legislation proposed after 2006. In 2005, just 300 state-level immigration measures were introduced and 45 adopted, but in 2011, 1,607 measures were introduced and 318 adopted (Immigration Policy Projects, 2011). Furthermore, between 2006 and 2011, there were an estimated 8,244 state-level immigration-related bills proposed, and 1,325 passed (Immigration Policy Projects, 2011).

This most recent wave of state and local government involvement in immigration legislation was fuelled primarily by the economic crisis sparked by the Great Recession. But there are two other notable historical events that markedly affected public opinion about immigrants. The first event, the terrorist attacks of September 11, 2001, created a heightened awareness of homeland security threats and raised concerns about the large numbers of undocumented immigrants residing in the country. The second was the failure of the US Congress to pass H.R. 4437 in November 2006. H.R. 4437 proposed additional enforcement measures to curb illegal immigration and offered some financial relief to local jurisdictions impacted by rising rates of undocumented immigration (US Congress (109th), 2005). Since H.R. 4437 did not pass, there has been little progress towards passing comprehensive federal immigration reform to address problems, perceived and real, associated with the roughly 12 million undocumented immigrants living in the US. Consequently, state and local decision makers have attempted to wrestle with the challenges associated with immigrant settlement on their own.

While jurisdictions adopting anti-immigration policies have received much of the attention, local responses to immigrants vary considerably. In most jurisdictions, agencies and organizations adopt policies and procedures to serve new residents, without fanfare. Schools, hospitals, and social service agencies are at the forefront of integrating recent immigrants into the community. This form of "bureaucratic incorporation" of immigrants, often led by street-level bureaucrats, varies significantly across organizations and institutions because of their missions and ethics and across contexts because of migration trends, socioeconomic characteristics, and the history of the receiving community. It also varies depending on how bureaucrats respond to pressures by local elected officials or whether they adhere to state and federal laws (Jones-Correa, 2008a; Jones-Correa, 2008b; Marrow, 2009).

Local Elected Officials, Public Agencies, and Immigrant Integration

A growing number of local elected officials have publicly committed to immigrant integration by helping both the immigrants *and* the receiving communities adjust to the changing sociocultural dynamics of their cities. In our attempt to understand the universe of immigrant integration policies or programs, we conducted a web search from January to May 2013 to identify cities and public agencies that addressed issues related to immigrants, new Americans, or major ethnic groups, such as

Hispanics and Asian Americans. We also conducted a literature review of reports, newspaper articles, and academic papers to identify cities with immigrant integration initiatives. We then reviewed the immigrant integration practices in the central cities of the 193 largest metropolitan areas in the United States.

Based on our research, we found that 55 out of 193 cities have pro-integrative immigration policies. These policies are often specific to a policy area, such as health, education, or housing, and are usually contained within one public agency. There were occasions in which agencies collaborated with one another and worked on inter-agency strategies or integrated policy areas. As a result of this variation, the categories of immigrant integration policies and programs we developed are not mutually exclusive. In table 9.1, we list the most common pro-integrative immigrant policies and practices found within local government agencies. We classify the integration strategies by policy area and provide examples of programs and policies focused on assisting immigrants to integrate as well as aiding the receiving community to adapt its policies and practices to promote greater immigrant integration.

The two most common pro-integrative practices among local governments are providing greater access to public services and representing the needs of immigrants in local governance practices and decisions. Providing greater access to public services to immigrants involves providing better information, conducting outreach, and offering services in the immigrants' native language. Although local governments are supposed to serve the public, there are many sociocultural barriers that immigrants face in accessing these services. Recognizing this, many cities across the US have developed a specific office within city government that focuses on immigrant affairs, multicultural affairs, or a specific ethnic group, such as Latinos. Other policy areas that local governments are engaged in include developing greater cross-cultural understanding between immigrants and the receiving community, promoting greater civic participation among immigrants, reducing language barriers, providing immigrants with alternative forms of identification accepted within the municipality, offering banking access and services, providing housing programs catering to immigrants, engaging in workforce development and protecting workers' rights, and developing law enforcement programs targeted towards immigrants. It should be noted that while the local public education and health systems typically are active in promoting immigrant integration, our

Table 9.1. Immigrant Integration Policies and Practices for Immigrants and Receiving Communities

Priority Area	Program or Policy Areas	Immigrants	Receiving Community	Primary Integration Domain(s)
Access to city services	Housing, transportation, solid waste disposal, and other public resources/ services	Establish information centres for immigrants to learn about city services.Provide multilingual telephone service to access city services	Hire bilingual/bicultural employees	Social, economic, spatial
		Provide multilingual telephone service to access city services		
Representation of immigrants' needs and views	Public policies and decision making, local governance	Mayor's Office of Immigrant Affairs Mayor's Office of New Americans Multicultural Affairs Office	Immigrant advisory board	Civic
Cross-cultural Understanding	Social and cultural life	Citizen or resident academy One-stop information centre	Lunchtime meet and greet: government staff and immigrants	Social, civic
Civic participation	Voting and participating in civic life (e.g., volunteering, attending public meetings)	Leadership training and development	Outreach to immigrant leaders and immigrants to increase public participation	Civic
		Citizenship courses	Holding public meetings and events where immigrants frequent	
Language Barriers	All	Translate documents, websites, signs		
		Education and outreach to immigrants about language assistance services	Determine rates of illiteracy among immigrants	All

(Continued)

Priority Area	Program or Policy Areas	Immigrants	Receiving Community	Primary Integration Domain(s)
Identification	Economic, criminal justice	Courses on how to obtain government-issued photo identification or a tax identification number	Municipal ID cards issued and accepted by municipal staff and law enforcement officers	Economic, social
Banking access and services	Financial literacy	Financial literacy education	Alternative forms of ID accepted by banks	Economic
Housing programs	Housing	Home ownership workshops	Enforcing fair housing laws in concentrated immigrant areas	Economic, spatial
		Housing discrimination education		
Workforce development and workers' rights	Employment	Support worker centres for day labourers	Citywide wage floors	
		Minority business development	Protecting workers' rights regardless of immigrant status	Economic
		Support day labourer or worker centres		
Law enforcement programs	Crime and public safety	Translate public safety signs into language spoken by the local immigrant community	Appoint an immigrant affairs liaison officer for the police and fire departments	Social
		Prepare and distribute multilingual brochures and presentations to disseminate information regarding public safety and police policy	Community policing programs	

research showed that local government entities are not the primary authority over these program areas.

Our research also found that 21 cities adopted anti-integrative policies. The most common involved partnerships with the US Department of Customs Enforcement to expand the authority of local law enforcers to detect, detain, and deport undocumented immigrants. Sometimes, anti-integrative policies were symbolic, such as mandating local staff to enforce existing policies that prohibit undocumented immigrants from receiving public benefits (most public benefits or services require legal immigration status anyway), yet were intended to send a message that undocumented immigrants are unwanted. There were also more punitive measures, such as forcing businesses to use the E-verify system to check the immigration status of new employees and severely penalizing businesses if they did not comply.

Some metropolitan areas have both pro- and anti-integrative policies. Some have an ICE partnership with the county sheriff's department while the local police department has a pro-integrative policy that explicitly states that local police will not ask for immigration documentation and will not pursue civil immigration violations. Other times, state immigration policy contrasts with metropolitan policies. Thus, our research shows that the local immigration policy landscape across metropolitan regions in the US is complex.

Community Planning for Immigrant Integration

While the vast majority of immigrant integration strategies are narrowly focused, a few comprehensive immigrant integration plans have been developed around the country. Two notable plans were created in Boise, Idaho, and Dayton, Ohio. Boise's "Refugee Resource Community Plan" was developed in 2010 to better enable refugee resettlement and integration and as a response to service providers' concerns that refugees were disproportionately negatively affected by the Great Recession. The plan focuses on six areas: education (for children, adults, and families), employment, health care, housing, transportation, and social integration of refugees and the receiving community (Idaho Office for Refugees, 2010). The plan, while developed by the Idaho Office for Refugees, a private sector organization, involves partnerships with organizations across different sectors, including local, state, and federal government agencies, non-profits, and faith-based organizations.

While the central focus of the Boise plan was the integration of the refugee population, the Dayton plan emphasized the community-wide benefits of immigrant integration. Spearheaded by the Human Relations Council, the 2011 "Welcome Dayton: Immigrant Friendly City" plan had as one of its primary goals to engage community members in a conversation about how they could actively participate in integrating new immigrant residents. While Dayton's plan focused on linguistic, social, and economic integration, it also emphasized the potential role that integrated immigrants can play in neighbourhood revitalization and economic development (City of Dayton, 2011). The plan was developed in eight months, and subcommittees were created thereafter to implement the plan. A total of 133 participants from a range of sectors – representing businesses, local government, the criminal justice system, social services, arts, and education – were involved in developing the plan. The plan received the support of the Dayton City Commission, which declared the resolution to adopt it "an emergency measure" and said the plan "shall take effect immediately upon its passage" (City of Dayton, 2011).

While the ultimate goal of these two plans – immigrant integration – is the same, the motivation and rationale behind the plans were different. The Boise plan was developed by a non-governmental organization and was motivated by the realization that the economic recession was hitting refugees the hardest, resulting in high unemployment rates. Advocates providing services to refugees believed that a more strategic, community-wide effort was needed to address the needs of this vulnerable population and avoid problems associated with intergenerational poverty (McLeod, 2012). Although the plan received widespread support and participation from public officials, it did not necessarily need political support given that it was not a government-funded activity. Architects of the Dayton plan argued that immigrant integration would have community-wide benefits in terms of neighbourhood revitalization and economic development. They argued that if immigrants felt welcomed, supported, and fully integrated into the civic, social, and economic life, they could contribute to the larger society through home ownership and home renovations, small business start-ups, and consumer spending (Wahlrab, 2012). This would, in turn, stimulate the local economy. A major factor in garnering political support for the immigrant integration planning processes in Dayton is that the city is shrinking in population and employment and thus has a struggling economy. Attracting immigrants to promote economic development is a strategy that other cities have employed. For a detailed discussion about immigrant-friendly initiatives that promote economic development, see chapter 10 by Harwood and Lee.

Methods and Case Sites

The contexts of our study sites are quite different from those found in chapters 8 and 10 in this volume. In chapter 8, Burayidi and Wiles focus on major US cities with majority-minority populations. These cities have long histories of immigration and extensive experience adapting to and integrating immigrants. In chapter 10, Harwood and Lee examine immigrant integration initiatives in cities promoting economic development and community revitalization. Two of these cities, Dayton and Detroit, are shrinking in population and jobs. Our study examines three immigrant integration community planning processes in a "new destination" state that has experienced rapid rates of immigrant settlement, mostly Hispanic, in the last few decades. The climate in many local jurisdictions in North Carolina has been hostile towards immigrants, thereby making those jurisdictions that are willing to publicly work on immigrant integration strategies stand out as exceptions. Considerable anti-immigrant backlash is evident among state lawmakers, as the North Carolina state legislature has adopted a number of punitive policies directed at immigrants, particularly undocumented immigrants. Therefore, developing pro-integrative policies and plans goes against the current policy tides within the state.

The reason for selecting North Carolina as a study site is also based on convenience and necessity – the researchers live and work in the state, and funding for the project comes from a philanthropic organization dedicated to addressing issues of social justice in the state. Beyond these reasons, North Carolina is also an interesting state in which to study local immigration policies and practices: the US Department of Homeland Security has taken a keen interest in the state and has partnered with a large number of local jurisdictions to enforce immigration violations, and in the 1990s, North Carolina had the fastest-growing Hispanic population of any state in the country (Nguyen & Gill, 2010).

This study complements chapters 8 and 10 because it provides insights into local immigrant integration planning processes and the role that planners can play in facilitating such a process. Over the course of two years, from 2010 to 2012, the researchers, two planners and an anthropologist, partnered with the North Carolina cities of High Point and Greenville and with Orange County. These communities were chosen to participate through an RFP (request for proposal) application process based in part on the willingness and commitment of elected officials, particularly the mayors, to engage in a long-term planning process that would result in actionable strategies. Communities were selected based on their prior experience working with immigrants and on their organizational and institutional capacity to complete the one- to two-year planning process. In each community, staff members from the human relations departments worked closely with the researchers to identify stakeholders, which included elected officials, local government staff, police, fire, and emergency medical staff, immigrant leaders, immigrants, and community members.

The research team facilitated a series of preparatory meetings with community partners to coordinate outreach efforts and identify stakeholders for future planning meetings. Once stakeholders were invited, a series of three planning meetings was held over the course of a year or more, depending on each community's needs. Each meeting was facilitated by a professional, who was not an urban planner but was skilled in working with groups to build consensus around contentious issues. In preparation for these meetings, our research team identified promising immigrant integration practices across the nation that could be tailored to the specific context and needs of each local jurisdiction. In between the planning meetings, our researchers provided technical and organizational expertise through bi-weekly conference calls and emails to our community contacts in each site. Each plan that was developed was context specific, sensitive to institutional capacities, tailored to the specific immigrant populations identified in each jurisdiction, and geared towards sustainable practices. The second year of the project involved implementation of the community action plans.

Benefits of Community Planning for Immigrant Integration

Even though conducting a community planning process and developing multifaceted plans was time consuming and required careful relationship building, the planning process itself offered a variety of benefits. Collaborations between sectors within a community enabled actors to identify overlapping or duplicative services, leverage existing resources, save costs, share expertise, and establish communication and social networks. In Greenville, after the first planning meeting, several connections were made to enable actors to further immigrant integration practices. In High Point, the stakeholders quickly agreed to form an interfaith council that could address existing and new conflicts regarding religion and culture. High Point is a multicultural, interfaith community where stakeholders are learning to address their differences in a more inclusive and understanding manner. Having an interfaith council in place has allowed for much faster resolution of conflicts and misunderstandings that arise between different racial, ethnic, and religious groups.

Another benefit of employing a long-term community planning process with diverse participants is that the resulting plans offer more holistic integration strategies that address multiple domains of integration: social, economic, and civic. In the High Point plan, for example, the action strategies included (but were not limited to) providing immigrants with access to city services and information to build wealth, creating an international advisory board to the city council, developing leadership among immigrants, and addressing cultural competency among immigrants and city service providers. These action strategies tackle multiple domains of integration and, therefore, may synergistically work to more rapidly integrate immigrants.

The planning process also allowed stakeholders to develop a shared understanding of the needs and assets in the immigrant community. We found that knowledge about the immigrant community was uneven among stakeholders participating in the planning process; therefore, there was little common understanding of the sociodemographics, geographic settlement patterns, needs, and concerns of the immigrant population. Some participants did have substantial knowledge about specific immigrant populations, such as Hispanic immigrants, but very limited knowledge about other groups. A common theme that arose from all of our sites was the sense that the stakeholders needed a better common understanding of their immigrant population. Thus, in some of the communities, the stakeholders in the planning process decided to learn more about the immigrant communities by initiating focus groups and surveys of immigrant groups.

This process of developing a shared understanding about the needs, assets, and concerns of the immigrant community helped the stakeholders in 1) building capacity for data collection; 2) developing relationships with other stakeholders from various sectors (e.g., nonprofit, for-profit, faith-based, public) that are interested in immigrant integration; 3) building confidence in making decisions with adequate knowledge about the immigrant community; and 4) developing more realistic expectations and goals about immigrant integration strategies.

A major goal of the planning process was to develop immigrant integration strategies that would affect institutional change and be sustainable even with staff turnover or a change in political leadership. To achieve this goal, we invited stakeholders who represented a broad constituency, so our coalition of immigrant advocates came from a diverse sector of the community. In addition, we invited local officials to our planning meetings. Even when local elected leaders did not attend, we made sure that they were kept abreast of actions and developments from our meetings. We wanted to garner support for our plan from local elected officials and also ensure that the action strategies in our plan were not dependent on their approval or funding. This allowed us to continue our planning efforts even when there was a change in leadership. Midway through our planning process, the mayor in Greenville was unseated, which could have destabilized our project had we structured it differently. Instead, the new mayor was supportive of our immigrant integration planning process and was well informed about our progress. Although she could not attend our meetings, she made sure to send staff members to show support for our efforts. Switching leadership midway through the planning process could have slowed or averted our progress, but because we maintained frequent communication with local elected leaders, this did not happen.

Challenges of Community Planning for Immigrant Integration

While there are many benefits to conducting a community planning process for immigrant integration, there were also many challenges and lessons to be learned, particularly because we were working in communities located in a new destination state where the immigrant community is relatively new and the network of immigration advocates is weak. In the following sections, we discuss how to gauge community readiness, balance an efficient planning process with immigrant participation, overcome the difficulty of maintaining an inclusive planning process, and, finally, temper expectations and outcomes.

Gauging Community Readiness

Of the three selected sites, the city of Greenville progressed most rapidly in its planning process. There were some key indicators that the city was the most prepared site to start the process of developing an immigrant integration plan. From its application materials, Greenville demonstrated a commitment to inclusivity by already having a Human Relations Council with a diverse multiethnic representation of immigrants. In addition, the city had partnered with local advocacy groups to examine the challenges experienced by underserved populations. Greenville submitted application letters of commitment from local community institutions and the mayor, which indicated wide-ranging community and political support that could provide strong social and political capital. Once the stakeholder meetings began, there was also evidence that the Greenville city government and the Human Relations Council had taken significant steps to conduct outreach because staff members in the city departments that provide services to immigrants were in attendance along with immigrant community leaders.

Alternatively, in Orange County, the elected officials agreed to participate in the planning process and directed county staff members to become involved in it. While the county staff followed orders from their supervisors, they did not spearhead the effort and thus appeared to be less committed to the process. Communication was much slower in Orange County, and organizing meetings with stakeholders was difficult. Participation at the meetings was low, particularly among immigrants, suggesting either that staff members did not conduct outreach or that their outreach techniques were unsuccessful. It became evident that Orange County did not yet have the organizational capacity and commitment to immigrant integration planning that the other communities had. While we still attempted to complete the community planning process and held three planning meetings, a plan was not created within the first year because of low participation and a lack of broad-based support from both local government and the community.

Balancing Stakeholder Engagement with Immigrant Participation

During the first stakeholder meeting in two of our sites, the number of people in the room exceeded expectations. In High Point, for example, the first meeting involved more than 100 individuals from city service agencies, prominent community institutions, local businesses, and the immigrant community. However, in subsequent meetings that involved the challenging task of coming together to craft priorities and objectives for an action plan, the number of stakeholders declined, which proved to have both negative and positive consequences. On one hand, large community meetings can be hard to manage and can also make it difficult to get through all pertinent material in the allotted time. On the other hand, the loss of key stakeholders skewed representation towards individuals who felt more comfortable contributing in such a format (e.g., public staff members), and away from some valuable voices, such as recent immigrants.

The drop-off in participation may be due to connections made between stakeholders that allowed them to accomplish their goals in the first meeting, making attendance at subsequent meetings unnecessary. For example, an emergency medical services employee wanted a forum to introduce herself and her staff to immigrants in order to build trust. Undocumented immigrants who fear law enforcement or immigration officers sometimes mistake other uniformed staff members, such as EMS, for law enforcement and therefore do not understand that EMS staff have no policing or immigration authority. After the first stakeholder meeting, the EMS staff member connected with the community college English as a Second Language director and discovered that this person could introduce the EMS staff and describe emergency services to ESL students. The EMS staff member did not participate in the next two planning meetings because her needs were met during the first meeting. While these informal connections are useful towards the goal of immigrant integration, it would be more beneficial to have stakeholders stay engaged throughout the entire process for purposes of consistency. One way to do this is to be transparent about expectations for stakeholders – in particular, the length of the planning process, the number of meetings, the dates and location of meetings, and the amount of individual effort involved. To encourage participation and engagement, explaining what the ultimate outcome will be and how participants may benefit could also offer incentives for participation and engagement.

The benefits of having immigrants participate in decision-making processes are invaluable. There should be representative participation by all immigrant groups. However, this poses several challenges. First, identifying and reaching out to different immigrant groups to encourage participation is often challenging. Second, language barriers must be overcome. The more groups there are, the greater the variation of languages, and the more difficult it is to facilitate an inclusive meeting. Third, if there is broad representation across immigrant groups, considering how to account for and prioritize the needs and concerns of different groups will be necessary. Questions such as "Will larger groups have a greater voice?" and "Should strategies be implemented only if they benefit all groups?" need to be addressed. Finally, a more inclusive planning process involving large groups will be slower and take longer. Keeping participants engaged in such a long process may be challenging.

These considerations raise questions about the balance between representing immigrants and having a manageable community planning process that results in actionable decisions. In High Point, the first stakeholder meeting was attended by a wide range of immigrant groups and immigrant advocates. With more than 100 participants, the diversity in the room was impressive and inspirational. It was clear that immigrant integration was a salient topic for the stakeholders. However, with only a five-hour window for the first meeting, the large number of participants, and the need for language interpretation, the process was slow and not engaging. For example, having to conduct translation forced us to cut out half of what we presented in other places. Fortunately, the High Point human relations staff had conducted other smaller focus groups with immigrants earlier in the process, which provided them with a better understanding of immigrants' needs throughout the planning process.

In contrast, during the first meeting in Greenville, there was significant representation from public sector leaders and leaders from the immigrant community; however, only a few immigrants were present. While these meetings were more predictable and moved along as planned, the local leaders expressed discomfort about being the voice for the immigrant community. Thereafter, they decided to hold focus groups of immigrants in ESL classes at the community college to get a reality check on their ideas.

The planning processes in High Point and Greenville, although different, showed that one way to balance immigrant representation and have a manageable community planning process is for stakeholders to conduct focus groups with a wide range of immigrant groups to assess the needs and challenges of immigrant integration, develop relationships, and establish lines of communication between stakeholders and immigrants. For the strategic planning meetings, a select number of immigrants who are committed to the entire planning process should be encouraged to attend. This method offers the opportunity for both immigrant representation and manageability in planning meetings.

Creating an Inclusive Planning Process

The sensitive topics and, at times, contentious issues that arose during planning meetings highlighted the necessity of professional facilitators adept at guiding discussion, encouraging participation from all attendees, and steering large group discussion forward. The professional facilitators devised engaging meetings that employed a variety of techniques to break down power imbalances and encourage knowledge sharing between participants. These techniques included paired interviews, small group discussions, larger group discussions, and allowing individuals to write down responses for the facilitator to read anonymously. While such a facilitator may be a planner, in our study we hired professionals who were not planners. In one community meeting, a disgruntled Hispanic business owner attempted to coopt it by disregarding the topic under discussion in order to vent his frustration about not being formally invited to the meeting and his dissatisfaction with prior attempts by the city government to develop immigrant-friendly practices. The facilitator skillfully reminded him of the goals of the session and adeptly brought the discussion back on topic without becoming flustered.

The planners who were present during the meetings provided both technical and substantive expertise without steering the stakeholders in a predetermined direction. The technical expertise came in the form of map making, quantitative and qualitative research analyses, and plan writing. The planners also provided substantive expertise on local immigrant integration strategies, immigration trends, plan making, and plan implementation.

The challenge for the planners in this situation was to know when to assert expertise and when to allow the stakeholders to work through the process. Because the planners did not work or live in these communities and were not involved in implementation, it was important for them to remind stakeholders of the goals of the planning process and provide necessary expertise, but not to suggest what decisions should be made or the types of integration strategies that were best. Instead, the planners stressed that the plan should 1) rely on existing institutional and organizational capacities, 2) build on available resources and community assets, 3) be feasible, and 4) be sustainable even with staff turnover or changes in local elected officials. The planners allowed the stakeholders to determine which immigrant groups were served and prioritize the types of integration strategies in the plan.

One key lesson learned from this process is that this type of planning process is not linear. New stakeholders would arrive at each meeting, thereby raising questions that had already been addressed and moving the direction of the conversation sometimes backwards or sideways instead of forward. In each community, the stakeholders requested examples of best practices and ideas for immigrant integration. When the planners presented these at each session, this often piqued curiosity and inspired new integration strategies, which slowed down the planning process. We would caution against presenting too many ideas and best practices from other contexts because they may not be applicable, and they risk veering the discussion in directions that may not be fruitful.

We learned from this process that we needed to summarize the accomplishments of each previous meeting at the beginning of each new meeting, to outline next steps at the end of each meeting, and to direct and redirect stakeholders so that progress was made. And we had to do all this while not marginalizing new stakeholders, who were often of different immigrant groups from those who had been participating previously. Being sensitive to newcomers to the process, who may also represent different immigrant groups, while trying to move the planning process along requires patience and practice at inclusive facilitation techniques.

Tempering Expectations and Outcomes

In undertaking an immigrant integration planning process, it is important to manage the expectations and outcomes of stakeholders. Some stakeholders may have unrealistic expectations or goals that are infeasible owing to political, financial, or institutional barriers. These expectations and goals may be listed in the plan as long-term goals, but if so there should also be a clear outline of the steps involved in realizing them. If this is not done, stakeholders may become discouraged because they feel their ideas have not been taken seriously. For example, in each of our sites, the idea of developing a multicultural community centre was raised. This would require significant capital funding and take many years to develop. Such an ambitious project, if included in the plan, should outline who would be responsible for leading the project, where funding might come from (e.g., through grant writing or the local government), and how long it might take to accomplish. Once these issues were discussed and written in the plan, the stakeholders realized that such a large-scale project should be a long-term goal and be given lower priority than the more pressing issues that needed to be addressed.

Our research team encouraged stakeholders to think about policies and practices that could affect institutional change and be sustainable for the long term. It was difficult to build consensus around these types of strategies because there were factions that had their own interests. For example, in High Point, a group of stakeholders expressed a desire for a community garden and a recreational site for soccer. While these may be important goals for some stakeholders, they do not address institutional change. It was important for our facilitator to steer them towards prioritizing other strategies, such as developing immigrant leaders and allowing immigrants to participate in city governance, that could address multiple domains of immigrant integration. The stakeholders were successful in identifying strategies that could affect institutional change in both High Point and Greenville.

Conclusion

In the last decade, a greater number of local governments are realizing the importance of integrating institutions across various sectors to help immigrants settle and become productive members of their new communities. And an increasing number are using a community planning process that involves the public, private, non-profit, and faith-based sectors. These local governments realize that they need to alter their policies and practices in order to make their services accessible to immigrants and to educate the larger community about creating an inclusive environment for all residents. Thus, immigrant integration is a two-way process that alters the fabric of immigrant communities and the host community that they reside in. Planners can play an essential role in developing immigrant integration plans that engage multiple sectors and are multifaceted.

Our study found that there are many benefits to engaging in a community planning process for immigrant integration, but that there are a number of challenges that actors engaging in this process should be aware of. First, not all communities are ready to participate in such an endeavour. Communities should have experience working with (and thus, already have working relationships with) multiethnic or socially vulnerable groups and have strong leadership that is committed to carrying out the planning process. Second, an immigrant integration planning process must utilize creative participation techniques to engage multiple publics, including immigrants from different sociocultural backgrounds. It is also important when engaging such a diverse group of stakeholders to develop an inclusive process that is able to break down power imbalances. Finally, managing expectations by developing integration strategies that build on existing assets, that alter existing institutional structures, and that are sustainable is important for identifying strategies that have potential to be successful at promoting immigrant integration.

Our assessment of local immigration policies across 193 of the largest US metropolitan areas also revealed a complex policy landscape, both across and within metropolitan areas. Even within a metropolitan area, institutions and agencies may have different perspectives and goals regarding immigrant integration, thus sending mixed messages to the immigrant community. Given this increasing complexity, the lessons from this study can assist planners, policy makers, elected officials, and other local actors who want to develop more welcoming, integrative policies and practices to build a stronger socioeconomic fabric in their communities.

REFERENCES

Alba, R.D., & Nee, V.G. (2003). *Remaking the American mainstream: Assimilation and contemporary immigration*. Cambridge, MA: Harvard University Press.

City of Dayton. (2011). Welcome Dayton plan: Immigrant friendly city. Human Relations Council. Dayton, Ohio.

City of Hazleton. (2006a, Sept. 8). Ordinance 2006–18, Illegal Immigration Relief Act Ordinance. Retrieved from http://www.hazletoncity.org.

City of Hazleton. (2006b, Dec. 28). Ordinance 2006–40, Illegal Immigration Relief Act Implementation Amendment. Retrieved from http://www.hazletoncity.org.

Coleman, M. (2012). The "local" migration state: The site-specific devolution of immigration enforcement in the US South. *Law & Policy*, *34*(2), 159–190. http://dx.doi.org/10.1111/j.1467-9930.2011.00358.x

DeCanas v. Bica. (1976). 424 U.S. 351.

Gill, H. (2010). *The Latino migration experience in North Carolina: New roots in the old North State*. Chapel Hill: University of North Carolina Press. http://dx.doi.org/10.5149/9780807899380_gill

Frisbie, P.W., & Kasarda, J.D. (1988). Spatial processes. In Neil J. Smelser (Ed.), *The Handbook of Modern Sociology* (pp. 629–666). Beverly Hills, CA: Sage.

Idaho Office for Refugees. (2010, July). Refugee Resource Strategic Community Plan. Retrieved from http://www.idahorefugees.org.

Immigration Policy Projects. (2011, Jan. 5). National Conference of State Legislatures. Retrieved from http://www.ncsl.org.

Ireland, P.R. (2004). *Becoming Europe: Immigration, integration, and the welfare state*. Pittsburgh: University of Pittsburgh Press.

Jones-Correa, M. (2005). Bringing outsiders in: Questions of immigrant incorporation. In R.E. Hero & C. Wolbrecht (Eds.), *The politics of democratic inclusion* (pp. 75–101). Philadelphia: Temple University Press.

Jones-Correa, M. (2008a). Immigrant incorporation in suburbia: Differential pathways, arenas, and intermediaries. In L.M. Hanley, B.A. Ruble, and A.M. Gardland (Eds.), *Immigration and integration in urban communities: Renegotiating the city* (pp. 19–47). Baltimore: Johns Hopkins University Press and Woodrow Wilson Center Press.

Jones-Correa, M. (2008b). Immigrant incorporation in suburbia: The role of bureaucratic norms in education. In D. Massey (Ed.), *New faces in new places* (pp. 308–340). New York: Russell Sage Foundation.

Joppke, C., & Morawska, E.T. (2003). Toward assimilation and citizenship [electronic resource]: Immigrants in liberal nation-states. Basingstoke: Palgrave Macmillan.

Kasinitz, P., Mollenkopf, J., Waters, M., & Holdaway, J. (2008). *Inheriting the city: The second generation comes of age*. New York: Russell Sage Foundation.

Lozano v. City of Hazleton. (2007). 496 F. Supp.2d 477 (M.D. Pa. 2007)

Marrow, H.B. (2009). Immigrant bureaucratic incorporation: The dual roles of professional missions and government policies. *American Sociological Review*, 74(5), 756–776. http://dx.doi.org/10.1177/000312240907400504

McLeod, T. (2012, Sept. 23). National Immigrant Integration Conference, Baltimore, Maryland. Panel on engaging state and local leadership in new gateways. Presentation on refugee resource community plan, Boise, Idaho.

Myers, D., & Pitkin, J. (2010). *Assimilation today: New evidence shows the latest immigrants to America are following in our history's footsteps*. Washington, DC: Center for American Progress.

New York City Mayor's Office of Immigrant Affairs. (2011). http://www1.nyc.gov/.

Nguyen, M.T., & Gill, H. (2010). The cost and consequences of local immigration enforcement in North Carolina. Center for Global Initiatives, UNC–Chapel Hill.

Park, R.E., & Burgess, E. (1925). *The city*. Chicago: University of Chicago Press.

Pastor, M., Ortiz, R., Carter, V., Scoggins, J., & Perez, A. (2012). *California immigrant integration scorecard technical report*. Center for the Study of Immigrant Integration, University of Southern California.

Portes, A., & Borocz, J. (1989). Contemporary immigration: Theoretical perspectives on its determinants and modes of incorporation. *International Migration Review*, 23(3), 606–630. http://dx.doi.org/10.2307/2546431

Portes, A., Fernández-Kelly, P., & Haller, W. (2005). Segmented assimilation on the ground: The new second generation in early adulthood. *Ethnic and Racial Studies*, *28*(6), 1000–1040. http://dx.doi.org/10.1080/01419870500224117

Portes, A., & Rumbaut, R.G. (1996). *Immigrant America: A portrait* (rev. ed.). Berkeley: University of California Press.

US Congress (109th). (2005, Dec. 6). H.R. 4437: Border Protection, Antiterrorism, and Illegal Immigration Control Act of 2005. Retrieved from https://www.govtrack.us.

Varsanyi, M. (2008). Immigration policing through the backdoor: City ordinances, the "right to the city" and the exclusion of undocumented day laborers. *Urban Geography*, *29*(1), 29–52. http://dx.doi.org/10.2747/0272-3638.29.1.29

Varsanyi, M. (2010). City ordinances as "immigration policing by proxy": Local governments and the regulation of undocumented day laborers. In M. Varsanyi (Ed.), *Taking local control: Immigration policy activism in U.S. cities and states* (pp. 135–156). Stanford: Stanford University Press.

Wahlrab, T. (2012, Sept. 23). National immigrant integration conference, Baltimore, Maryland. Panel on engaging state and local leadership in new gateways. Presentation on Welcome Dayton, City of Dayton, Ohio.

Wilson, K., & Portes, A. (1980). Immigrant enclaves: An analysis of the immigrant labor market experience of Cubans in Miami. *American Journal of Sociology*, *86*(2), 295–319. http://dx.doi.org/10.1086/227240

10 Immigrant-Friendly Community Plans: Rustbelt Efforts to Attract and Retain Immigrants

STACY ANNE HARWOOD AND SANG S. LEE

In the face of widespread anti-immigrant agitation, some localities are taking a different tack: encouraging immigrants to settle and start businesses, through supportive services and other incentives. Immigrants have always played an important role in the growth and development of American cities (Muller, 1993), and awareness of that history encourages cities to see immigrants as stimulating growth and revitalizing neighbourhoods. Immigrant-friendly policies are quickly becoming a new strategy for economic development in rustbelt cities that have lost population. This chapter examines the content and public discourse of the plans to create immigrant-friendly environments in three Midwestern cities: Dayton, Ohio; Detroit, Michigan; and Chicago, Illinois.

Background

The movement of people, including immigrants and migrants, both shapes and is shaped by flows of capital, trade, and technology (Sassen, 1988; Sassen, 2002). Such major US cities as New York, Chicago, Los Angeles, and Miami have rich histories of immigration surges going hand in hand with economic expansion. Newer immigrant gateways such as Denver, Atlanta, and Minneapolis that have emerged since the 1990s have attracted immigrants to their lower-density and more suburban layouts (Singer, Hardwick, & Brettell, 2008). Pushed out of gentrified city centres, immigrants have been drawn to suburbs both by the affordable housing and by the new economic opportunities opened by global restructuring (Hardwick, 2008). In 2010, when the US foreign-born population reached 40 million, approximately half were living in the suburbs of large metropolitan areas (Wilson & Singer, 2011).

Thus new migration and immigration patterns are changing the landscapes of cities.

In the national economy, immigrant workers fill key gaps in agriculture, construction, food service, health care, high-tech manufacturing, information technology, and life sciences (Singer, 2012) but mostly in low-wage jobs. That is particularly the case in agriculture, construction, and service jobs, where immigrants with limited education are overrepresented. However, immigrants also fill essential jobs in information technology and the life sciences; in these high-skill sectors their education levels match those of the native-born (Singer, 2012). Such demographic and economic change stirs controversy among "local and national governments, and service providers [and] unleash[es] new debates about multiculturalism, integration, and exclusion" (Price & Benton-Short, 2007, p. 8).

Cities and towns with rapid population change have always been sites of political controversy. Particularly, "during periods of economic distress and social conservatism, policymakers tend to enact more restrictive immigration legislation and to undertake more repressive anti-immigrant actions" (Massey & Sánchez, 2010, p. 72). For example, in the 1920s the public, influenced by "unchallenged acceptance of theories about racial inferiority," demanded tighter immigration controls (Muller, 1993, p. 5). Such nativism re-emerged in the 1990s: California passed Proposition 187, the Save Our State Initiative, which restricted public funds for undocumented immigrants and ultimately denied their children access to state-funded health care and public schools (Jacobson, 2008; Lennon, 1998). Even though Proposition 187 was challenged in court and significantly scaled back, the message was that voters wanted immigrants, especially the undocumented, out.

More recently, anti-immigrant sentiment has led to the enactment of hundreds of local ordinances that target undocumented immigrants. In 2006, Hazleton, Pennsylvania, for example, passed the Illegal Immigration Relief Act, which prohibited businesses from hiring and landlords from renting to undocumented immigrants (Steil & Ridgley, 2012). Other communities have followed Hazleton's lead (Kirchhoff, 2007; Price, 2006). In 2010, Arizona reaped infamy for State Bill 1070, which required all immigrants to register and to carry such documentation at all times – an egregiously discriminatory policing policy (Preston, 2012; Provine & Sanchez, 2011). As Hazleton's ordinances had inspired others, SB 1070 inspired similar state bills to constrain immigrants and immigration (Gomez, Wolf, Cauchon, & Raasch, 2012; Crary, 2012).

Both these and other measures to regulate and control immigrants were contested by a variety of organizations and went to trial over their constitutionality. Ultimately, the question of city and state rights in regulating immigration, a federal purview, was under fire. In Hazleton, the Illegal Immigration Relief Act was immediately struck down by a federal district court in 2007, and the judgment was upheld again in 2010 and 2013 despite appeals (Preston, 2007; Preston, 2010; Morgan-Besecker, 2013). In Arizona, the US District Court filed a temporary injunction against specific SB 1070 provisions. The provisions targeted were the ones that gave the police power to stop, arrest, or detain people based on suspicion of illegal status and to make warrantless arrests; required immigrants to carry documentation of their legal status at all times; and prevented undocumented immigrants from seeking and applying for employment. In 2011 the Ninth Circuit Court of Appeals upheld the injunction, then the Arizona Supreme Court officially overturned most of the bill in 2012 (Magaña, 2013). However, the remaining provision of SB 1070 still allowed "police officers to make a reasonable attempt to determine the immigration status of a person stopped, detained, or arrested. The Supreme Court also maintained that there must be reasonable suspicion that the person is in the country illegally" (Magaña, 2013, p. 159). The unconstitutionality of ordinances and policies like the Illegal Immigration Relief Act and SB 1070 prevented the enforcement of these acts, but the impact was to create an unwelcoming environment for immigrants living in those cities and states.

Although voters in some states had supported anti-immigrant ordinances and policies, their economic repercussions have led communities to reconsider (Esses, Brochu, & Dickson, 2012). For example, organizers of some conventions boycotted Arizona, leading to the loss of an estimated $90 million just for the city of Phoenix (Archibold, 2010). The economic costs of anti-immigrant measures have also been measured in losses of both workforce and consumers. After passing a house bill similar to Arizona's, Georgia lost immigrant agricultural workers who fled to other states with less fearsome environments. The result was hundreds of millions in crop loss due to the shortage of vegetable harvesters (Redmond, 2011). A New Jersey town decided to rescind its anti-immigrant ordinance after immigrants' departure hurt the revenue of local businesses (Belson & Capuzzo, 2007). Cities' and states' attempts to restrict immigrants and immigration have had unintended consequences.

Yet, while some communities have evinced hostility to immigrants, elsewhere efforts have arisen to promote integration and to challenge anti-immigrant legislation. An early example in the civic domain is the Sanctuary Movement of the 1980s. In the context of human rights violations and US military intervention in Central America, the Sanctuary Movement helped undocumented immigrants to seek asylum and refugee status in the US under the protection of churches (Perla & Coutin, 2009). Churches have again engaged in protecting immigrants in the more recent New Sanctuary Movement (Barron, 2007; Freeland, 2010), and some cities have adopted local sanctuary resolutions to protect undocumented immigrants from deportation (Resolution would make Cook County "sanctuary," 2006; McKinley, 2009; Sutterman, 2010).

Immigrants themselves have been active in efforts to gain and protect their rights. Examples include labour-organizing campaigns such as Justice for Janitors and garment worker campaigns (Cranford, 2005; Sullivan, 2010). Immigrants have also been activists in the push to reform immigration policy and have organized and participated in protests against anti-immigrant legislation (Paulson, 2006; Benjamin-Alvarado, DeSipio, & Montoya, 2009). A recent effort by young undocumented immigrants is advocacy for the Development, Relief, and Education for Alien Minors. This group, the DREAMers – which is led by young people who entered the US undocumented at an early age – triumphed when President Barack Obama secured passage of Deferred Action for Childhood Arrivals in 2012. Now immigrants have organized to push for passage of comprehensive immigration reform, currently stalled in Congress.

Diversity as an Asset

With immigration policy bitterly contested across America, some local governments and corporations have adopted policies of building diversity from the conviction that diversity is an asset. Perhaps one of that stance's most famous cheerleaders is Richard Florida (2002, 2005), who expounds how diversity, most strikingly in the arts, improves cities. Florida describes how San Francisco attracts creative people because a diverse array of people feel comfortable living in the city. In a tolerant and welcoming environment, moreover, newcomers can contribute new ideas and ways of doing things. Immigrants not only are a vibrant part of a creative segment but are acknowledged to be risk-taking and innovative drivers of economic growth in technology and science as well (Florida, 2002; Florida, 2005).

Of course, Florida's critics are quick to note his emphasis on a uniquely talented group of people as valued drivers of urban growth. A common criticism excoriates his assumption that this creative milieu can be supported by the city's lower economic sectors – an inherent structural and social inequality in the "creative class" (Peck, 2005; Peck 2010, pp. 192–230; Leslie and Catungal, 2012). Whatever economic advantages a city derives from welcoming the immigrants in a creative milieu must be recognized as underpinned by the energy that immigrants contribute to the other essential sectors of an urban economy. Although Florida values the presence of immigrants in a fashionable arts scene, he seems uninterested in how, elsewhere in the city's economy, they encounter social and economic injustice. In contrast, Hackler and Mayer (2008), examining how effective the "creative class" model may be in generating entrepreneurship among women, Blacks, and Latinos, urge a focus on creating structures of opportunity rather than on the attempt to foster a creative milieu. They argue that "Developing a creative milieu might be more difficult to achieve for urban planners and policymakers than developing programs and policies that facilitate market access, provide capital resources, or contribute to education and skill development" (Hackler & Mayer, 2008, p. 285).

However, a supposed creative milieu isn't the only advantage cities see in growing their diversity. Population diversity is frequently hailed for making localities and regions more competitive through new resources in the form of workers and thinkers, although in reality the results may be mixed (Syrett & Sepulveda, 2011). This "diversity advantage" is posited as a way for cities to gain the upper hand in addressing local and global changes because a diverse population is armed with more resources and kinds of knowledge than a homogeneous population (Zachary, 2003).

Welcoming Immigrants as an Economic Development Strategy

Moving beyond attempts to create a diversity advantage or creative milieu through attraction of immigrants alone, some cities are endeavouring to identify concrete programs and ways to integrate immigrants and their potential into the market and other arenas where their activity can be used as a springboard for citywide growth. Through proactive planning of the resources and tools needed to facilitate immigrant entrepreneurship, cities are calculating payoffs that can come from opening avenues of opportunity for a broader range of people. Thus some cities are advertising themselves as "pro-immigrant," "welcoming," and "immigrant-friendly."

This strategy is founded on studies that show the economic contribution immigrants make through purchases of goods and services and on the likelihood that these same immigrants will be self-employed (Fairlie, 2012). For example, in a study of three immigrant neighbourhoods in Boston, Borges-Mendez, Liu, and Watanabe (2005) found that immigrant entrepreneurs improve their neighbourhoods by

> Reviving commerce and investment in areas that had declined ... Incubating new businesses ... Attracting new customers. Providing some employment opportunities. Improving the physical quality and appearance of the buildings in which they operate and surrounding areas ... Enhancing public safety. (p. 3)

Immigrant activity is also being used to invigorate already thriving places, not just depressed areas. This injection of people, including but not limited to entrepreneurs, is seen as having the potential to expand markets beyond those of the immediate municipal vicinity. San Gabriel Square in San Gabriel, California, also known as "The great mall of China," services Chinese residents from the Los Angeles area in addition to being a recreational destination for others attracted to the many restaurants and specialty stores. The Square was initially funded by a Taiwanese bank and continues to be connected with the global economy, which boosts its economic potential (Cheng, 2010). Ethnic enclaves and economies can connect not just to local flows of money but to transnational pipelines as well, as noted by Zhou and Tseng:

> The Chinese community in Los Angeles exemplifies a new type of ethnic economy within which financial, labour, information, and commodity flows are international in scope, yet deeply intertwined and embedded within a local milieu of intense ethnic networking and entrepreneurship. (2001, p. 142)

These complex and rich transnational ties are what cities want to capture. Cities hope that immigrant entrepreneurs can jump-start economic development in neglected neighbourhoods, connect to globalizing markets in already thriving neighbourhoods, and create more opportunities in general for both the city and its residents through population growth. Of course, these opportunities are not without controversy, but immigrants are seen as a pool of residents that can potentially bring a bounty of assets to the city.

For New York, Los Angeles, and Chicago, maintaining and attracting immigrant population is a policy tied to their continuous history as gateways for immigration. Because immigrants wield economic and political power in these cities, the response has been to establish municipal immigrant affairs offices to address their specific needs (City of New York, n.d.; City of Los Angeles, n.d.). In many other cities, particularly rustbelt cities, it is fears of a diminishing population that impel a previously unheard of strategy for population and economic growth: embracing immigrants (Vogel, 2005; Preston, 2013). Cleveland, Dayton, Toledo, and Detroit have welcomed immigrants to promote economic prosperity and resilience (Boodhoo, 2012; McCord, 2012; R.L. Smith, 2013). These cities, with little recent experience of mass immigration, hope to strengthen their economic competiveness by retaining and attracting immigrants. Finally, in new immigrant destination cities, such as Boise, Idaho (City of Boise and Idaho Office for Refugees, 2010), becoming immigrant-friendly is a matter of survival. Inundated with immigrants and sometimes refugees, new destination communities often are ill prepared to address their needs.

Regardless of the motivations for creating immigrant-friendly plans, all of these cities are in the business of retaining immigrants. R.J. Smith (2013) identifies four immigrant recruitment policies: rhetorical, outreach, integration, and facilitation. Rhetorical policy includes passing pro-immigrant resolutions like Council Resolution No. 5073 in Eugene, Oregon: "A resolution in support of a statement of principles for immigrant integration" (City of Eugene, 2012). The coalition that pushed for that resolution wanted to create a "thoughtful public dialogue about immigration" (Russo, 2012). Outreach policies make information available to immigrants through translation, interpretation, and liaisons. Bloomington, Indiana, for example, offers a Spanish print publication and an e-newsletter to inform Latino residents about programs, services, and issues that pertain to them. Integration policies aim to help immigrants achieve citizenship by offering courses in English as a Second Language and in citizenship. Municipalities offer such classes through their public libraries or publish a list of local agencies that offer them. Facilitation policies encourage immigrants to participate in the civil and social life of a community even when they lack formal citizenship.

In 2012, Los Angeles became the largest city to provide undocumented immigrants with an identity card that gives access to services and makes them eligible for bank services (Miles, 2012). All these recruitment policies create an environment that acknowledges immigrants as members of the community.

Parallel with immigrant recruitment strategies and indispensable to their success are efforts to help native-born residents adjust to and appreciate immigrants. Welcoming America, a "national, grassroots-driven collaborative that works to promote mutual respect and cooperation between foreign-born and U.S.-born Americans" (Welcoming America Home Page, n.d.), started in Iowa in 2004 and has now expanded to many other states. In addition to its national and state campaigns, Welcoming America encourages municipalities "to support locally-driven efforts to create more welcoming, immigrant-friendly environments that maximize opportunities for economic growth and cultural vitality and position communities as globally competitive, 21st century leaders" (Welcoming Cities and Counties, n.d.). Welcoming America is supported by foundations such as Ashoka and Carnegie, and by corporations such as Starbucks and Western Union. The support by mainstream corporations suggests to the public that welcoming immigrants is logical, since global competition requires drawing on the resources of all who are competitive, whether citizens or not.

Planning for Immigrant-Friendly Cities in the Midwest

A growing immigrant population increases the demand for local goods and services and thereby stimulates new ethnic economies and boosts housing values, contributing to overall regional economic growth. Can cities plan to enlist this energy by attracting immigrants? Although some cities have implemented economic development strategies with immigration as one of their cornerstones, few have developed comprehensive community plans that welcome immigrants and – equally important – support their integration into the community. Leading such efforts in the Midwest are Dayton, Detroit, and Chicago.

While the responsibility for immigrant integration falls on a range of actors including the immigrants themselves, local government and community institutions are fundamental. The remainder of this chapter presents the immigrant integration strategies developed by the three major cities just named. The plans all first address immediate challenges of immigrants' social integration: improving their communication skills and providing access to health, education, legal, and municipal services. Each plan emphasizes the important economic benefits immigrants potentially offer. Key to each plan is a combination of local and neighbourhood-level strategies, such as marketing immigrant neighbourhoods, and offering small business a range of technical assistance combined with regional-level strategies that attempt to position the locality by developing personal networks, investment opportunities, and export connections in the global market.

Welcome Dayton

When we think of immigration, Dayton is not likely to come to mind; it does not fit the typical profile for an immigrant gateway. Yet this city has embarked on an ambitious plan to be immigrant-friendly. Dayton's population has been declining since the 1960s, from a peak of over 262,000 to just over 141,000 (US Census Bureau, 2010, 1998a). The slowdown in manufacturing and the movement to the suburbs explain much of the shrinkage.

Although Dayton's population – 52 per cent White, 43 per cent Black, and 3 per cent Hispanic (US Census Bureau, 2010) – is in decline, the immigrant population continues to grow. In 2010, 3.8 per cent of the population was foreign-born, a percentage similar to that of the state (US Census Bureau, 2010). That sliver of the total population includes immigrants and refugees from more than 100 countries, including Russia, Nigeria, India, Turkey, Libya, the Philippines, and Mexico (Welcome Dayton Plan, 2011).

In contrast to the many American localities that have organized against immigrants, Dayton city commission adopted the Welcome Dayton Plan on October 5, 2011. The plan lays out a framework for government, businesses, schools, and civic organizations to act inclusively, integrate immigrants into the community, and celebrate diversity (Welcome Dayton Plan, 2011). On its webpage the plan states:

> Welcome Dayton is a community initiative that reflects our country's core philosophy: people with diverse backgrounds, skills and experiences fuel our nation's success. The Welcome Dayton effort promotes immigrant integration into the greater Dayton region by encouraging business and economic development; providing access to education, government, health and social services; ensuring equity in the justice system; and promoting an appreciation of arts and culture. (Welcome Dayton Homepage, n.d.)

The Dayton plan has four sections: 1) Business and Economic Development; 2) Government and Justice System; 3) Social and Health Services; and 4) Community, Culture, Arts, and Education. For example, one initiative is to create an immigrant business district, an "international market place for immigrant entrepreneurship" (Welcome Dayton Plan, 2011, p. 7). The city will develop a marketing plan for the area, provide façade grants, and work with real estate agents to market the area. Other recommendations are to provide interpretation services to help immigrants to access services, and to increase opportunities to learn English. Sports are highlighted by a proposal to establish a Global Dayton Soccer event (see figure 10.1). The plan also calls for the creation of an advisory group of immigrants to work with the city and county commissioners.

The early impetus for the plan was work done by the Board of the Human Relations Council to address housing discrimination and anti-immigrant sentiment, especially against the undocumented. Later, the Welcome Dayton concept (see figure 10.2) grew out of discussions between the city manager, the director of the human relations council, and several city commissioners. A core team of city staff and a board member from the council then worked to garner commitment for the initiative (Welcome Dayton Plan, 2011).

As more actors got involved in the plan, its language shifted to include economic opportunities for immigrants. For example, Dayton Mayor Gary Leitzell commented, "Let's come up with something to attract immigrants, and therefore attract entrepreneurs" (Spary, 2011). The Dayton Area Chamber of Commerce has taken a similar stance, stating in a recent publication that "research has shown that immigrants are 2–3 times more likely to start a business than native-born residents. By attracting and retaining immigrants, the hope of the Dayton Region is to build our local economy and expand the diversity of our region" (Antonick et al., 2013, p. 6). Even the region's universities reflect this hope, funding international programs and initiatives to attract foreign students, especially Chinese (University of Dayton, China Institute, n.d.). The mantra is: Attracting immigrants will put Dayton back on the map as a competitor in the global economy (Castillo, 2011).

Global Detroit

Historically, immigrants and migrants have been central to Detroit's economy. European immigrants from France, England, Germany,

SEPTEMBER 13 - 14, 2013

Action Sports Center

1103 Gateway Drive

Dayton, Ohio 45404

Join us in promoting a stronger community for everyone represented in Dayton through recreational soccer. Get your youth or adult team together and come out and have fun as we educate each other about one another, increase cross cultural understanding, and be inclusive all through the game of soccer.

EVENT HIGHLIGHTS

Friday, September 13 — Opening Ceremonies and Police/Fire Scrimmage Game

Saturday, September 14 — All Youth and Adult Games; Multi-Cultural Food and Craft Fair

Teams: Each team must represent one country and at least one member of the team must be from that country.

Teams may be coed.

- YOUTH: 4 teams per age group
  6-7 yr olds; 8-9 yr olds; 10-12 yr olds
  13-14 yr olds; 15-18 yr olds

- ADULT: 8 teams per age group
  19-30 yr olds; 31 and over

Registration and Fees: By Saturday, August 3, 2013
$50 for youth teams $100 for adult teams

August 4 - August 17, 2013
$75 for youth teams, $125 for adult teams

The Tournament is free and open to the public to attend

For More Information:
Website: www.DaytonWorldSoccer.org
Email: worldsoccer@daytonohio.gov
Phone: 937-333-8400

YOU CAN CHANGE THE GAME...BE A VOLUNTEER, SPONSOR, FOOD OR CRAFT VENDOR OR ENTER YOUR TEAM!

Figure 10.1. Flyer for Dayton World Soccer Games. The event is used to show that the city is welcoming of immigrants and to promote cross-cultural understanding. Courtesy City of Dayton, Ohio.

Figure 10.2. Welcome Dayton logo. The logo also indicates that Dayton is immigrant-friendly. Courtesy City of Dayton, Ohio.

Ireland, and Poland created Detroit's early ethnic neighbourhoods. Later, Detroit also became home to immigrants from Eastern European and Middle Eastern countries as well as White and Black migrants from the American South who laboured in Detroit's great factories of the mid-twentieth century. After surging to more than 1.8 million, however, since 1950 the population has declined (US Census Bureau, 1998b). Today Detroit's population hovers around 700,000 and is 83 per cent Black (US Census Bureau, 2010). The population loss has left Detroit with a depleted tax base, vacant lots and properties, and run-down infrastructure, as well as with unemployment and crime. Yet, Detroit remains an important American city. Though struggling to gain a foothold in the new economy, it shows potential for recovery as downtown and historic neighbourhoods revitalize.

The growth in the immigrant population may have slowed Detroit's population loss, but perhaps a little late. According to the census, the foreign-born population increased between 1990 and 2010 from 3 to 5 per cent of Detroit's population, and the rate of decline for the foreign-born population, which since 1960 has been higher than for native-born residents, improved in that period (US Census Bureau, 2010, 1999). Michigan itself also shows a slow decline in population and a small but growing immigrant population that stood at 6 per cent in 2010 (US Census Bureau, 2010). Notably, Michigan now lags behind most US states in attracting foreign-born residents (US Department of Homeland Security, 2013). The growth in Detroit's immigrant population comprises primarily people born in Mexico, along with Middle Eastern immigrants from Iraq, Lebanon, and Yemen, and Asians from the Philippines, Pakistan, Korea, Vietnam, and Bangladesh (US Census Bureau, 2010).

Seeing the relatively untapped potential of the growing immigrant population, business groups created Global Detroit in 2010. Funded by the New Economy Initiative, the Detroit Regional Chamber of Commerce, and the Skillman Foundation, a study explored the impact that immigrants have on the Detroit regional economy. The final report, written by former Michigan Representative Steve Tobocman, draws from academic studies, interviews with national and local experts, accounts of best practices, and an advisory board comprising 38 regional leaders from businesses, universities, ethnic business chambers, foundations, and immigrant advocates (Overview of Global Detroit Initiative, 2011). Tobocman, now serving as the director of Global Detroit, wrote, "Metro Detroit is in crisis. If we don't take action to speed our own entry into the new economy, we will never return Detroit to a region of prosperity. Nothing can make a more powerful contribution to Detroit's rebirth than an affirmative immigrant-welcoming and global-connection building effort" (Tobocman, 2011).

Global Detroit aims to strengthen and expand Detroit's connections to the world by making the region more attractive to immigrants, international students, and business interests in investment and foreign trade. The report proposes 11 strategies to revitalize the regional economy. The first is the Welcoming America Campaign for Michigan, with the goal of helping non-immigrant residents adjust to the presence of immigrants. This includes promoting the positive impacts immigrants have already made in Detroit; see figure 10.3 for an example. Next is the creation of an EB-5 Investor Visa Center to facilitate permanent

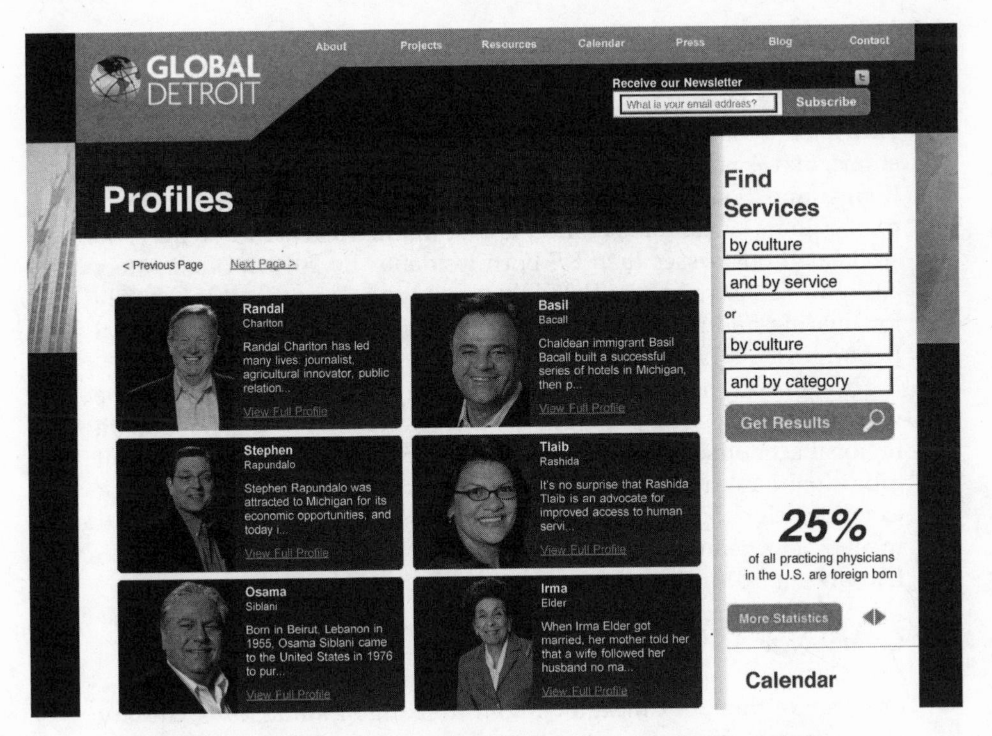

Figure 10.3. Global Detroit immigrant profiles. The profiles show that immigrants have both succeeded in the city and helped transform the city's economy. Courtesy City of Detroit, Michigan. http://www.globaldetroit.com/stories/

residency for any foreign investor (and family) who invests $1 million in an American business and creates 10 jobs.

Neighbourhood revitalization in immigrant communities is another Global Detroit strategy. It begins with "listening sessions" to learn about residents' views. In these neighbourhoods, ways are sought to make use of vacant land and buildings, for example by "homesteading" and by encouraging refugees in particular to settle in these areas. Homesteading involves giving away land or buildings at little or no cost. In exchange, the new owner must agree to improve the property; for example, by rebuilding or remodelling (Global Detroit Final Report, 2010). In 2012, Global Detroit helped to win a $2.4 million grant from the Kellogg Foundation to fund revitalization of immigrant neighbourhoods (Global Detroit neighborhood development collaborative gets $2.4 million for business development, 2012).

ProsperUS Detroit is an initiative that aims to train, support, and provide micro loans of up to $10,000 to low-income immigrant, refugee, ethnic, and African American entrepreneurs in the North End, Cody-Rouge, and Southwest Detroit neighbourhoods (ProsperUS Detroit, n.d.). The rationale for the grant is that immigrants are more likely to start small businesses than US-born residents. By 2013 approximately 20 entrepreneurs had completed the training and were eligible to apply for funding and receive one-on-one coaching (Global Detroit Annual Report, 2012, 2013).

The ultimate measure of Global Detroit's accomplishment will be whether Detroit sees population growth and more jobs along with regional economic growth. A hopeful sign is that even with the recent municipal bankruptcy, immigrants in Detroit continue to thrive. For example, Mexicantown, in southeast Detroit, continues to see investment dollars being put into new businesses and property rehabilitation (Advincula, 2013).

Chicago New Americans Plan

One cannot think of Chicago without thinking of immigrants; the city is a collage of distinctive ethnic neighbourhoods. Nineteenth-century immigrants came from Ireland, Germany, Sweden, England, and the Netherlands. Then in the early twentieth century, immigration and migration to Chicago boomed as unskilled workers flocked to its stockyards and mills from Eastern and Southern Europe, and rural Whites and Blacks migrated from the US South. The 1940s brought Latinos from Mexico, Puerto Rico, and Cuba. By 1950, Chicago's population had peaked at 3.6 million (US Census Bureau, 1998b). After 1965, immigration to the city from Asia, particularly China and India, increased. Immigrants from Central and South America followed in the 1980s. In the twenty-first century the region has continued to grow, but the city and some inner-ring suburbs have lost population since the 2000s (US Census Bureau, 2010). Chicago now has 2.7 million residents, 21 per cent of whom are foreign-born (US Census Bureau, 2010).

Chicago, the third largest city in the United States, is still a major destination for immigrants from all over the world, though the highest percentages today are from Mexico, Poland, and India (Paral, 2003). Like other Midwestern cities, the City of Big Shoulders emerged as a global city in the 1990s after decades of urban decline. However, the renaissance may be short lived, since a majority of immigrants now move directly to the region's suburbs and small towns.

Recognizing the importance of immigrants in Chicago's present as well as its past, in 2011 Mayor Rahm Emanuel created the Office of New Americans soon after taking office. Emanuel said, "I want to make Chicago the most immigrant-friendly city in the world and the Office of New Americans will ensure every law-abiding Chicagoan has access to the resources they need to become productive members of society and contribute to our thriving global city" (Emanuel, 2011). In addition, Emanuel put together the Office of New Americans Advisory Committee, 50 people representing different sectors, organizations, and communities, to identify issues for the new office.

The New Americans Plan now has 27 initiatives to ensure "that Chicago remains a place where diversity is welcomed and celebrated" (Emanuel, 2012). The initiatives fall into three divisions. 1) *Our Growth* lists initiatives that support immigrant-owned businesses and develop immigrant skills. 2) *Our Youth* focuses on the children of immigrants, their parents, and the schools. 3) *Our Communities* addresses public safety, access to services, and civic engagement. Examples, respectively, include increasing exports from immigrant-owned businesses and promoting tourism in immigrant neighbourhoods, supporting undocumented child arrivals, and reducing immigration services fraud (Chicago New Americans Plan, 2012). Given Chicago's steady flow of immigrants, the plan emphasizes retaining them by reaching out in ways targeted to parents, youth, businesses, students, and neighbourhoods. One such strategy targeted to business is to create a non-profit incubator and affordable kitchen spaces for restaurants and other food businesses.

The Chicago plan openly supports all immigrants regardless of legal status. The high level of organization in Chicago's immigrant communities creates venues where local politicians can advocate against the enforcement of federal immigration law at the local level. One result was the Chicago City Council's passage on September 12, 2012 of the Welcoming City Ordinance, an initiative of the New Americans Plan (Chicago city council passes "welcoming city ordinance" to protect undocumented immigrants, 2012). Although Chicago already had protective measures in place for undocumented immigrants, such as prohibiting city employees from asking about the status of people seeking services, this ordinance states that undocumented immigrants cannot

Table 10.1. Potential Economic Impact from the Chicago New Americans Plan Initiatives

When Chicago ...	... the City Will Benefit Greatly
Helps immigrant-owned businesses flourish	Immigrant owned businesses could create 100,000 to 200,000 more jobs if Chicago increased its job growth rate from small and medium-sized businesses to match the highest-performing US city.
Doubles the exports from immigrant-owned businesses	Doubling immigrant-owned business could produce an additional 24,000 to 30,000 jobs.
Becomes a more attractive city for high-skilled, foreign-born workers	For every 100 additional high-skilled foreign-born workers who work in science, technology, engineering, or math fields, 260 jobs could be created for US-born workers.
Increases graduation rates for high school immigrant students and helps more immigrants earn their GED certificate	For every additional high school diploma earned by Chicago students, gross state product could increase by $15,000.
Increases immigrant participation in early childhood programs	For every dollar invested in early childhood education, Chicago could save seven dollars in government spending.

Source: Excerpt from the Chicago New Americans Plan.

be detained unless they are the object of a criminal warrant (Chicago Municipal Code Chapter 2–173). The Chicago ordinance stands in stark contrast to Arizona and Alabama laws that enable local police to collaborate with federal immigration agents to deport undocumented immigrants. Table 10.1 provides a summary of the potential impacts that immigrants are expected to have in the city's economy.

Implications for Inclusive Planning Practice

Immigrants and immigration are becoming more prominent considerations in the planning of cities and neighbourhoods. All the community plans discussed in this chapter involved extensive outreach to schools, businesses, churches, service providers, and community-based organizations. Each planning team put together advisory committees and subcommittees that drew from organizations representing different neighbourhoods, ethnic communities, businesses, and service sectors. Planning processes included interviews and meetings with many individuals. The plans note the importance of creating coalitions and networks to share information, apply for grants, and tackle common problems together. Although it is not the scope of this chapter, several chapters in this volume, including chapter 9 by Nguyen, Gill, and Steephen, examine the importance of participation and the processes used in community plan making in multicultural cities. Specifically looking at challenges related to immigration, Main and Rojas in chapter 12 present ways to address participation in immigrant communities using visual arts. As discussed in the chapters, these types of planning processes can be long, arduous, and challenging on multiple levels; it is unlikely any one office alone would be able to address the many issues related to integration without the active participation and skill set of many different stakeholders and collaboration with multiple agencies and organizations.

Although comprehensive, the plans emphasize economic development, being driven politically by the economic imperative to stimulate growth that would improve their cities' competitiveness. The efforts in Dayton and Detroit, and to a lesser degree in Chicago, focus on immigrants as economic actors. This is not surprising given the recent economic downturn that left cities under tremendous pressure to create jobs. Strategies with such a focus are notably the most challenging, yet they are also the keys to success for these plans. Examples include creating opportunities for foreign investment (Dayton, Detroit), facilitating exports by immigrant-run businesses (Chicago), and retaining foreign students as residents after graduation from local universities (Detroit). Many of the strategies are place-based, targeting commercial strips and neighbourhoods where immigrants are concentrated (Detroit, Dayton, and Chicago).

The Dayton and Detroit plans notably emphasize documented immigrants, which helps to gain political support for their policies. The reality however, is that some families are made up of undocumented immigrants as well. The invisibility of the undocumented immigrant population in those plans not only is problematic for their scope but also is likely to reproduce social hierarchies. Averting the gaze from the realities of immigration obscures the power dynamics that give inequitable results for people's daily lives.

Dayton goes a step beyond Detroit in recognizing some of its plan's limitations and suggests that city staff work with legislators to change the regulations for service providers to allow them to be "friendly" to all immigrants. Many people and organizations in Dayton are behind its plan, but it has not escaped controversy. The Ohio Jobs and Justice

PAC predicts that the plan will turn Dayton into a sanctuary for "illegal immigrants" who will displace American workers. They refer to the plan as "Welcome Dayton! Illegal Alien Friendly City" (Welcome Dayton!, n.d.). To address the potential for backlash against integrating immigrants, both Detroit and Dayton employ public relations strategies to dispel the negative stereotypes about immigrants. Social media, billboards, flyers, presentations, and press releases have publicized the benefits of immigration to other residents and businesses.

Dayton, Detroit, and Chicago all identify ways to improve immigrants' access to city services and propose to create websites with information for them. A major element is translation of written materials and provision of volunteer and professional interpretation services (Dayton, Detroit). Cultural training to sensitize staff about the needs of immigrants (Dayton, Chicago) and hiring bilingual and multilingual staff (Dayton, Detroit) are called for in the plans. Chicago's plan identifies specific areas where increasing immigrant enrolment in early childhood programs and summer youth enrichment programs should be sought.

Whether these three immigrant-friendly plans make the political process more inclusive is an important consideration. All the plans acknowledge the need to promote immigrant leadership, especially in the immigrant business community. Dayton proposes to allow those who are not citizens to nevertheless sit on neighbourhood boards. Global Detroit proposes that key ethnic community leaders be invited to important official events and that training and spaces for nonprofit organizations and immigrant leaders be made available. Chicago proposes to promote immigrant participation in local school councils and park advisory councils and to assist immigrants with the naturalization process.

Although these initiatives will increase immigrant access not only to services but also to some local decision-making processes, the plans are limited in their efforts to integrate immigrants into the electoral political fabric. Chicago and Dayton identify areas for political incorporation – education, park, and neighbourhood boards and commissions – but more can be done. Historically, many US states allowed the foreign-born to vote in local, state, and federal elections, thus encouraging them to purchase land and settle (Hayduk, 2006). Today cities are again testing the waters for politically integrating immigrants through local voting rights, arguing that as residents of the community the foreign-born should have a legal say on local initiatives and in electing their local representatives (Coll, 2011). As cities extend voting rights, however, that brings attention to questions about the sovereignty of the state versus governance of the city when issues related to immigration arise at the local level (Coll, 2011; Varsanyi, 2005).

Conclusion

The intent of this chapter has been to present the content of three immigrant-friendly rustbelt community plans and to discuss their implications for inclusive planning. Future research should assess the effectiveness of such plans. Many questions remain: Do cities with more immigrant-friendly cultures actually attract and retain more immigrants, and as a result see economic growth? How do immigrants respond to immigrant-friendly initiatives? Do they feel welcome? Are these cities treating their immigrants any better than those without such plans do? Where do local institutions and organizations excel in meeting these challenges, and where do they need to improve? How do we transform policies and practices that are inherently anti-immigrant to be immigrant-friendly, particularly towards the undocumented? And finally, what is the role of municipal planning in such processes?

Who participates in city planning has broadened over the decades as we have come to recognize the work done by community members, organizations, and institutions, and also by government departments outside the municipal planning department. This is certainly the case for these immigrant-friendly community plans. However, the influence of municipal planners, particularly the land-use planner, is noticeably absent from the content of these plans. Although the three plans examined here have elements concerning physical development, often by way of creating business districts, the plans emphasize the investments in people and their economic capacity without considering the spatial regulations that shape much of what happens in each city. In chapter 8 Burayidi and Wiles point to the innovative work in physical planning and land use across the country, including, for example, accommodating multigenerational living, co-housing, and places of worship. In addition, none of the plans address the planning policies used to concentrate poverty and facilitate segregation. The obvious gaps indicate where a more encompassing vision of immigrants within the city is required.

Cities clearly should be welcoming to immigrants. But the heavy emphasis on the economic rationale should be expanded to full community inclusiveness. Valorizing one set of individuals while failing to justly incorporate others into the plans is often a shortcoming of plans that promote diversity (Peck, 2005; Peck, 2010). A policy that manages to welcome immigrants effectively into all aspects of the city can greatly benefit their future generations' success. Moreover, immigrants are only one of a city's subpopulations; questions about equity and justice for African Americans in cities has been one of planning's greatest challenges, and Doan in chapter 6 and Zaferatos in chapter 7 highlight other communities that are often neglected or omitted in discussions on city diversity. Focusing on inclusion enables planners and policy makers to recognize a range of exclusions from areas of growth that can disrupt planning's dominant economic development discourses. Planning for immigrants has arrived on the agenda. Much remains to be done, though, to make our cities inclusive.

REFERENCES

Advincula, A. (2013, Aug. 21). Detroit ethnic media tell different story – immigrants key to revival. *New America Media*, news feature. Retrieved from http://newamericamedia.org/2013/08/ethnic-population-in-detroit-remains-optimistic-despite-bankruptcy-filing.php.

Antonick, J., Barry, J., Harnish, D., Shannon, D., Stock, R., Weed, G., & Traynor, T. (2013). Economic outlook. Dayton Area Chamber of Commerce report. Retrieved from http://www.daytonchamber.org/index.cfm/linkservid/5D2A2D18-D090-74AD-DAEB3F421A1724ED/showMeta/0/.

Archibold, R.C. (2010, May 11). Phoenix counts Big Boycott Cost. *New York Times*. Retrieved from http://www.nytimes.com/2010/05/12/us/12phoenix.html.

Barron, J. (2007, May 9). Congregations to give haven to immigrants. *New York Times*. Retrieved from http://www.nytimes.com/2007/05/09/nyregion/09sanctuary.html.

Belson, K., & Capuzzo, J.P. (2007, Sept. 26). Towns rethink laws against illegal immigrants. *New York Times*. Retrieved from http://www.nytimes.com/2007/09/26/nyregion/26riverside.html.

Benjamin-Alvarado, J., DeSipio, L., & Montoya, C. (2009). Latino mobilization in new immigrant destinations: The anti-H.R. 4437 protest in Nebraska's cities. *Urban Affairs Review*, *44*(5), 718–735. http://dx.doi.org/10.1177/1078087408323380

Boodhoo, N. (2012, Feb. 22). Changing gears: Midwest embraces immigrants. *ideastream*. Retrieved from http://www.ideastream.org/news/feature/45289.

Borges-Mendez, R., Liu, M., & Watanabe, P. (2005, Dec. 1). Immigrant entrepreneurs and neighborhood revitalization: Studies of the Allston Village, East Boston and Fields Corner neighborhoods in Boston. *Institute for Asian American Studies Publications*. Paper 13. Retrieved from http://scholarworks.umb.edu/iaas_pubs/13.

Castillo, M. (2011, Oct. 8). Against the grain, Dayton, Ohio, embraces immigrants. *CNN U.S.* Retrieved from http://www.cnn.com/2011/10/08/us/ohio-dayton-pro-immigrant.

Cheng, W. (2010). "Diversity" on Main Street? Branding race and place in the new "majority-minority" suburbs. *Identities: Global Studies in Culture and Power*, *17*(5), 458–486. http://dx.doi.org/10.1080/1070289X.2010.526880

Chicago city council passes "welcoming city ordinance" to protect undocumented immigrants. (2012, Sept. 13). *Huffington Post Latino Voices*, 5. Retrieved from http://www.huffingtonpost.com/2012/09/13/chicago-welcoming-city-ordinance_n_1882115.html.

Chicago New Americans Plan. (2012). Retrieved from http://www.cityofchicago.org/city/en/depts/mayor/supp_info/chicago_new_americansplan.html.

City of Bloomington. (n.d.). *Latino programs: Services*. Retrieved from https://bloomington.in.gov/sections/viewSection.php?section_id=491.

City of Boise and Idaho Office for Refugees. (2010). *Refugee strategic community plan*. Retrieved from http://www.idahorefugees.org/home/refugee_community_plan/.

City of Eugene. (2012). Council Resolution No. 5073. Retrieved from www.eugene-or.gov/documentcenter/view/11013.

City of Los Angeles. (n.d.) *Mayor's Office of Immigrant Affairs*. Retrieved from http://www.lamayor.org/immigrants.

City of New York. (n.d.) *Mayor's Office of Immigrant Affairs*. Retrieved from http://www.nyc.gov/html/imm/html/home/home.shtml.

Coll, K. (2011). Citizenship acts and immigrant voting rights movements in the US. *Citizenship Studies*, *15*(8), 993–1009. http://dx.doi.org/10.1080/13621025.2011.627766

Cranford, C.J. (2005). Projects of solidarity and gender relations at work: Latina/o immigrant janitors in Los Angeles. *Conference Papers – American Sociological Association*, 1–25.

Crary, D. (2012, Apr. 29). Immigration victory in Arizona could spawn laws in other states. *Washington Post*, pp. A03.

Emanuel. (2011, July 19). Mayor Emanuel announces creation of Office of New Americans to support Chicago's immigrant communities and enhance their contributions to Chicago's economic, civic, and cultural life. News release.

Retrieved from http://www.cityofchicago.org/city/en/depts/mayor/provdrs/office_of_new_americans/news/2011/jul/mayor_emanuel_announcescreationofofficeofnewamericanstosupportch.html.

Emanuel. (2012, Dec. 4). Mayor Emanuel unveils first-ever Chicago New Americans plan. News release. Retrieved from http://www.cityofchicago.org/city/en/depts/mayor/provdrs/office_of_new_americans/news/2012/dec/mayor_emanuel_unveilsfirst-everchicagonewamericansplan.html.

Esses, V.M., Brochu, P.M., & Dickson, K.R. (2012). Economic costs, economic benefits, and attitudes toward immigrants and immigration. *Analyses of Social Issues and Public Policy (ASAP)*, *12*(1), 133–137. http://dx.doi.org/10.1111/j.1530-2415.2011.01269.x

Fairlie, R. (2012). Immigrant entrepreneurs and small business owners, and their access to financial capital. Commissioned by the Office of Advocacy, the United States Small Business Administration. Retrieved from http://www.sba.gov/sites/default/files/rs396tot.pdf.

Florida, R. (2002). *The rise of the creative class: And how it's transforming work, leisure, community and everyday life*. New York: Basic Books.

Florida, R. (2005). *Cities and the creative class: New York City*. New York: Routledge.

Freeland, G. (2010). Negotiating place, space and borders: The New Sanctuary movement. *Latino Studies*, *8*(4), 485–508. http://dx.doi.org/10.1057/lst.2010.53

Global Detroit Annual Report 2012. (2013, Jan.). Retrieved from http://www.globaldetroit.com/read-global-detroits-2012-end-of-the-year-report/.

Global Detroit Final Report (2010, Aug. 11). Retrieved from http://www.globaldetroit.com/wp-content/uploads/2014/10/Global_Detroit_Study.full_report.pdf.

Global Detroit neighborhood development collaborative gets $2.4 million for business development. (2012, Mar. 20). *Huffpost Detroit*. Retrieved from http://www.huffingtonpost.com/2012/03/20/global-detroit-neighborhorhood-development-collaborative_n_1367855.html.

Gomez, A., Wolf, R., Cauchon, D., & Raasch, C. (2012, June 26). "A mixed message": In Arizona's wake, states vow to forge ahead with new laws. *USA Today*.

Hackler, D., & Mayer, H. (2008). Diversity, entrepreneurship, and the urban environment. *Journal of Urban Affairs*, *30*(3), 273–307. http://dx.doi.org/10.1111/j.1467-9906.2008.00396.x

Hardwick, S.W. (2008). Toward a suburban immigrant nation. In A. Singer, S.W. Hardwick, & C.B. Brettell (Eds.), *Twenty-first century gateways:*

Immigrant incorporation in suburban America (pp. 30–52). Washington, DC: Brookings.

Hayduk, R. (2006). *Democracy for all: Restoring immigrant voting rights in the U.S.* New York: Routledge.

Jacobson, R.D. (2008). *The new nativism: Proposition 187 and the debate over immigration*. Minneapolis: University of Minnesota Press.

Kirchhoff, S. (2007, Mar. 13). Immigration debate squeezes some businesses: Cities, states passing rash of laws on illegal aliens. *USA Today.*

Leslie, D., & Catungal, J.P. (2012). Social justice and the creative city: Class, gender and racial inequalities. *Geography Compass*, *6*(3), 111–122. http://dx.doi.org/10.1111/j.1749-8198.2011.00472.x

Lennon, T.M. (1998). Proposition 187: A case study of race, nationalism, and democratic ideals. *Policy Studies Review*, *15*(2–3), 80–100. http://dx.doi.org/10.1111/j.1541-1338.1998.tb00780.x

Magaña, L. (2013). SB 1070 and negative social constructions of Latino immigrants in Arizona. *Aztlan*, *38*(2), 151–161.

Massey, D.S., & Sánchez, R.M. (2010). *Brokered boundaries: Creating immigrant identity in anti-immigrant times*. New York: Russell Sage Foundation.

McCord, E. (2012, Dec. 22). Immigrants welcomed: A city sees economic promise. *NPR Weekend Edition*. Retrieved from http://www.npr.org/2012/12/22/167797730/immigrants-welcomed-a-city-sees-economic-promise.

McKinley, J. (2009, Oct. 20). San Francisco alters when police must report immigrants. *New York Times*. Retrieved from http://www.nytimes.com/2009/10/21/us/21sanctuary.html.

Miles, K. (2012, Nov. 9). ID cards for undocumented immigrants in LA approved by council. *Huffington Post*. Retrieved from http://www.huffingtonpost.com/2012/11/09/id-cards-undocumented-immigrants-la_n_2102504.html.

Morgan-Besecker, T. (2013, July 27). Hazleton loses again in fight for illegal immigration act. *Times-Tribune.* Retrieved from http://thetimes-tribune.com/news/hazleton-loses-again-in-fight-for-illegal-immigration-act-1.1526626.

Muller, T. (1993). *Immigrants and the American city*. New York: New York University Press.

Overview of Global Detroit Initiative. (2011, Feb. 27). Retrieved from http://www.globaldetroit.com/wp-content/uploads/2014/10/Global_Detroit_Study.overview.pdf.

Paral, R. (2003, Sept.). Chicago's immigrants break old patterns. *Migration Information Source*. Retrieved from http://www.migrationpolicy.org/article/chicagos-immigrants-break-old-patterns/.

Paulson, A. (2006, May 3). Rallying immigrants look ahead. *Christian Science Monitor.*

Peck, J. (2005). Struggling with the creative class. *International Journal of Urban and Regional Research*, *29*(4), 740–770. http://dx.doi.org/10.1111/j.1468-2427.2005.00620.x

Peck, J. (2010). *Constructions of neoliberal reason*. Oxford: Oxford University Press. http://dx.doi.org/10.1093/acprof:oso/9780199580576.001.0001

Perla, H., & Coutin, S.B. (2009). Legacies and origins of the 1980s US–central American sanctuary movement. *Refuge: Canada's Periodical on Refugees*, *26*(1), 7–19.

Preston, J. (2007, July 27). Judge voids ordinance on illegal immigrants. *New York Times*. Retrieved from http://www.nytimes.com/2007/07/27/us/27hazelton.html.

Preston, J. (2010, Sept. 9). Court rejects a city's efforts to restrict immigrants. *New York Times*. Retrieved from http://www.nytimes.com/2010/09/10/us/10immig.html

Preston, J. (2012, June 27). Immigration ruling leaves issues unresolved. *New York Times*. Retrieved from http://www.nytimes.com/2012/06/27/us/immigration-ruling-leaves-issues-unresolved.html?pagewanted=all.

Preston, J. (2013, Oct. 6). Ailing Midwestern cities extend a welcoming hand to immigrants. *New York Times*. Retrieved from http://www.nytimes.com/2013/10/07/us/ailing-cities-extend-hand-to-immigrants.html.

Price, J.H. (2006, Sept. 21). Towns take a local approach to blocking illegal aliens. *Washington Times*. Retrieved from http://www.washingtontimes.com/news/2006/sep/21/20060921-110748-4227r/.

Price, M., & Benton-Short, L. (2007). Immigrants and world cities: From the hyper-diverse to the bypassed. *GeoJournal*, *68*(2–3), 103–117. http://dx.doi.org/10.1007/s10708-007-9076-x

ProsperUS Detroit. (n.d.). Retrieved from http://www.prosperusdetroit.org/.

Provine, D.M., & Sanchez, G. (2011). Suspecting immigrants: Exploring links between racialised anxieties and expanded police powers in Arizona. *Policing and Society*, *21*(4), 468–479. http://dx.doi.org/10.1080/10439463.2011.614098

Redmond, J. (2011, July 12). Georgia farmers to seek study of losses tied to labor shortages. *Atlanta Journal-Constitution*. Retrieved from http://www.ajc.com/news/georgiafarmers-to-seek-1012576.html.

Resolution would make Cook County "sanctuary." (2006, Sept. 8). *Chicago Tribune*. Retrieved from http://articles.chicagotribune.com/2006-09-08/news/0609080340_1_illegal-immigrants-maldonado-cook-county.

Russo, E. (2012, May 1). Coalition offers immigrants support: The new network pledges to work for immigration reform. *Register-Guard*. Retrieved from http://projects.registerguard.com/web/updates/27998196-55/network-immigrants-springfield-immigrant-immigration.html.csp.

Sassen, S. (1988). *The mobility of capital and labor: A study in international investment and labor flow*. Cambridge: Cambridge University Press. http://dx.doi.org/10.1017/CBO9780511598296

Sassen, S. (2002). *Global networks, linked cities*. New York: Routledge.

Singer, A. (2012, Mar. 15). Immigrant workers in the US labor force. *Brookings Institution Paper*. Retrieved from http://www.brookings.edu/~/media/research/files/papers/2012/3/15-immigrant-workers-singer/0315_immigrant_workers_singer.pdf.

Singer, A., Hardwick, S.W., & Brettell, C.B. (2008). *Twenty-first century gateways: Immigrant incorporation in suburban America*. Washington, DC: Brookings Institution Press.

Smith, R.J. (2013, Apr.). *Local government incorporation of immigrants in community economic development strategies*. Paper presented at the 2009 Annual Meeting of the American Sociological Association, San Francisco.

Smith, R.L. (2013, July 15). Welcoming America brings its pro-immigrant message to depopulated Cleveland. *Plain Dealer*. Retrieved from http://www.cleveland.com/business/index.ssf/2013/07/welcoming_america_brings_its_p.html.

Spary, C. (2011, Nov. 28). Hello world – welcome to Dayton. *Dayton Most Metro*. Retrieved from http://mostmetro.com/the-featured-articles/hello-world-welcome-to-dayton.html.

Steil, J., & Ridgley, J. (2012). "Small-town defenders": The production of citizenship and belonging in Hazleton, Pennsylvania. *Environment and Planning. D, Society & Space*, *30*(6), 1028–1045. http://dx.doi.org/10.1068/d0109

Sullivan, R. (2010). Organizing workers in the space between unions: Union-centric labor revitalization and the role of community-based organizations. *Critical Sociology*, *36*(6), 793–819. http://dx.doi.org/10.1177/0896920510376999

Sutterman, K. (2010, Dec. 17). Sanctuary city status sought for Iowa City and Cedar Rapids. *Gazette*. Retrieved from http://thegazette.com/2010/12/17/sanctuary-city-status-sought-for-iowa-city-and-cedar-rapids/.

Syrett, S., & Sepulveda, L. (2011). Realizing the diversity dividend: Population diversity and urban economic development. *Environment & Planning A*, *43*(2), 487–504. http://dx.doi.org/10.1068/a43185

Tobocman, S. (2011, Feb. 8). Immigration is key to building a global Detroit. *Model D Blog*. Retrieved from http://www.modeldmedia.com/features/tobocman060610.aspx.

University of Dayton, China Institute. (n.d.). Retrieved from https://www.udayton.edu/china_institute/.

US Census Bureau. (1998a). Population of the 100 largest urban places: 1960. https://www.census.gov/population/www/documentation/twps0027/tab19.txt

US Census Bureau. (1998b). Population of the 100 largest urban places: 1950. Retrieved from http://www.census.gov/population/www/documentation/twps0027/tab18.txt

US Census Bureau. (1999). Nativity of the population for urban places ever among the 50 largest urban places since 1870: 1850 to 1990 http://www.census.gov/population/www/documentation/twps0029/twps0029.html

US Census Bureau. (2010). QuickFacts for Chicago, Dayton and Detroit. http://quickfacts.census.gov/qfd/index.html

US Department of Homeland Security. (2013). *Yearbook of immigration statistics: 2012*. Washington, DC: US Department of Homeland Security, Office of Immigration Statistics.

Varsanyi, M.W. (2005). The rise and fall (and rise?) of non-citizen voting: Immigration and the shifting scales of citizenship and suffrage in the United States. *Space and Polity*, *9*(2), 113–134. http://dx.doi.org/10.1080/13562570500304956

Vogel, S. (2005, Aug. 16). Immigration and the shrinking city. *Model D Blog*. Retrieved from http://www.modeldmedia.com/features/shrinkage.aspx.

Welcome Dayton! (n.d.) Retrieved from http://www.ojjpac.org/Welcome_Dayton_Plan.asp.

Welcome Dayton Homepage. (n.d.). Retrieved from http://www.welcomedayton.org/.

Welcome Dayton Plan. (2011). Retrieved from http://www.welcomedayton.org/about/implementation-plan/.

Welcoming America Home Page. (n.d.) Retrieved from http://www.welcomingamerica.org/.

Welcoming Cities and Counties. (n.d.) Retrieved from http://www.welcomingamerica.org/get-involved/cities/.

Wilson, J.H., & Singer, A. (2011, Oct. 13). Immigrants in 2010 metropolitan America: A decade of change. *Brookings Institution Paper*. Retrieved from http://www.brookings.edu/~/media/research/files/papers/2011/10/13-immigration-wilson-singer/1013_immigration_wilson_singer.pdf.

Zachary, G.P. (2003). *The diversity advantage: Multicultural identity in the new world economy*. Boulder: Westview Press.

Zhou, Y., & Tseng, Y.F. (2001). Regrounding the "ungrounded empires": Localization as the geographical catalyst for transnationalism. *Global Networks*, *1*(2), 131–154. http://dx.doi.org/10.1111/1471-0374.00009

PART 4

Designing the Multicultural City

11 A Targeted Approach to Planning Socially Diverse Neighbourhoods

EMILY TALEN

Introduction

Socially diverse neighbourhoods are unlike most places in the US. They cannot be neatly characterized as poor, wealthy, suburban, inner-city, yuppie, Black, White, or Hispanic. They are instead distinguishable as places where residents are working out the day-to-day complexities that arise when different people occupy the same geographic space. They are not gentrified – although they often have strong gentrification pressures – and they are not suffering from high crime and disinvestment. But neither are their business districts and schools thriving. There is a constant tension arising from the ups and downs of paradoxical change: development pressures fighting against neighbourhood preservation efforts, older public facilities that deteriorate despite new investment, small businesses that are pushed out by chain stores, local institutions with far-flung constituencies, declining schools in the midst of rising wealth, old-timers who are displaced by young professionals, and real and imagined fears that gradually escalate. Eventually, such neighbourhoods often veer in one direction or another and lose their diversity.

What would a neighbourhood planning process devoted to sustaining the social diversity of such places be like? In this chapter, I argue that a targeted approach to neighbourhood planning is essential for accomplishing long-term neighbourhood stability. I overview a number of ways in which neighbourhood planning could be enlisted to support diversity, including public awareness, coding that supports diversity, and strategic public investment. The strategy relies on an integration of the three planning models that currently vie for planners' attention: communicative planning methods, New Urbanism, and the just city ideal. The three models, reviewed in a seminal paper by Susan Fainstein (2000), provide an ideal basis for a new neighbourhood planning approach specifically geared to socially diverse places. What socially diverse neighbourhoods need is not better dialogue, better form, or a commitment to social equity exclusively, but rather a combination of all three.

My focus is on neighbourhoods that are already diverse, where diversity is defined as the mix of different socioeconomic and ethnic groups within a neighbourhood. There is a fairly well established literature in planning that ties the goals of vitality, economic growth, tolerance, sustainability, and social justice to the mixing of races, ethnicities, ages, family types, and income levels (Talen, 2008). If planners are able to help sustain the diversity that exists, it is hoped, the number of diverse neighbourhoods will increase over time. This proactive approach is especially warranted given the documented evidence that well-serviced locations are losing their affordability and, thus, their diversity (Quigley, 2010; Pollack, Bluestone, & Billingham, 2010; Haughey & Sherriff, 2010).

Yet neighbourhood planning as currently practised does not offer a way to sustain or even account for mixed social structure. Neighbourhood planning is conceived as being neutral to population heterogeneity. The fact that there are no specific neighbourhood planning methods developed with diversity in mind is curious, given that a vibrant social mix is both a normative goal of planners and a significant challenge to neighbourhood stability.

Each of the dominant planning models reviewed by Fainstein has some applicability to the goal of constructing a neighbourhood planning approach that supports socially diverse places: the communicative model for its focus on inclusive planning processes; New Urbanism for its focus on the patterns and forms that support social diversity; and the "just city" model for its normative position that social benefits should be more equitably distributed. The support of socially diverse neighbourhoods will require an inclusive discourse, an encouragement of diversity-sustaining forms, and an effort to enlist both discourse and form in the support of a more just city.

The Diversity Goal in Planning

There is no explicit definition of the "socially diverse neighbourhood." Generally, people consider the mixing of people by race/ethnicity and by income level or wealth to be the most essential forms, though the mixing of age, family type, and household type is also important (Sarkissian, 1976). A diverse neighbourhood may have teenagers and elderly, married couples and singles, empty nesters and large families, waiters and teachers as well as professionals, affluent people, and people on fixed incomes, and people of varying racial, ethnic, and cultural backgrounds. In short, these neighbourhoods are places that harbour a full range of human complexity. Rough estimates put the number of neighbourhoods that could be characterized as racially/ethnically and economically diverse at anywhere from 5 per cent to 25 per cent of neighbourhoods in the US.[1]

Strategies for measuring the stability of diverse neighbourhoods include comparing how the proportion of one group changes over time relative to another (Galster, 1990), and then determining a threshold or tipping point that would indicate flight or displacement (Card, Mas, & Rothstein, 2008). In a comparative approach, changing proportions would be evaluated in function of overall citywide population shifts. In a market approach, stability would be a matter of assessing population proportions relative to housing affordability (Smith, 1998). Galster's (1998) "stock/flow model" defines stable integration on the basis of maintaining some diversity in the range of household flows in and out of the neighbourhood over a 10-year period.

The tools of planners have often undermined diversity by isolating urban functions into single-use districts, enforcing social segregation by zoning for separate housing unit types, and planning for monolithic, separating elements like expressways. The designation of "neighbourhood units" has excluded minorities (Silver, 1985), public works projects have displaced Blacks (Caro, 1974), and the push for "cold war utopias" in the form of peripheral, low-density development (Mennel, 2004) has fuelled segregation by income. Even the more recent attempts to explicitly incorporate diversity in planning have been known to falter (Day, 2003; Downs, 2000), reflecting a more generalized inability to deal appropriately with difference in land use decision making (Qadeer, 1997; Harwood, 2005).

The situation is disturbing and ironic, given that planners have a long history of valuing social diversity and even articulating in fairly explicit terms the urban forms and patterns that will support it (Talen, 2005). Decades ago, Lewis Mumford wrote about the importance of a social and economic mix, citing the "many-sided urban environment" as one with more possibilities for "the higher forms of human achievement"

(1938, p. 486). At mid-twentieth century, the rejection of suburbia by planners was based on the view that it lacked diversity and therefore was "anathema to intellectual and cultural advance" (Sarkissian, 1976, p. 240). Building up diversity through deliberate social mixing programs on the grounds that it would make communities more stable was a hallmark of 1950s social planning (Glazer, 1959), reconstituted through federal programs like HOPE VI, Choice Neighborhoods, and Moving to Opportunity. Planners continue to agree with Jacobs that urban diversity is "among our most precious economic assets" (1961, p. 219) and now endorse the notion that diversity promotes innovation and economic growth (Florida, 2002). The dominant view is that if diversity can be stabilized, neighbourhoods will be more resilient, limiting the ease with which they are abandoned and victimized by disinvestment.

In the environmental literature, the sustainable community is defined as "one in which diversity is tolerated and encouraged" and where "sharp spatial separation or isolation of income and racial groups" is non-existent (Beatley & Manning, 1997, p. 36). In smart growth, sustainable cities, and New Urbanist literatures, there is a wealth of ideas about how urban planning should be used to support social diversity: by mixing housing type and use, supporting locally owned business, promoting street connectivity, maximizing exchange through the public realm, and coding for compact, diverse, transit-oriented development (Jacobs, 1961; Talen, 2013). Planners are well versed in the negative implications of homogenized, privatized social worlds, where communal conflict is internalized or avoided rather than dealt with openly (Putnam, 2000; Baumgartner, 1991).

The key question, then, is: Why is a profession so often devoted to social diversity simultaneously connected to the policies, procedures, and tools that undermine it? The reasons are complex, involving not only the entrenched apparatus of land development to which planners are wedded but also larger structural issues involving political economy and social change (Van Kempen, 2002; Grigsby, Baratz, Galster, & MacLennan, 1987). Planners have to confront a wide range of counterforces in their quest to support diversity: individual behaviour and fear, consumer choice, discrimination in institutions and governance, and housing dynamics. There are the Tiebout hypothesis (social homogeneity is based on levels of taxes and services) and Schelling's hypothesis (households desiring at least some similar neighbours will result in widespread segregation) to contend with. For all of these reasons, planners are confronted with the reality of decreasing diversity in different realms: continued concentration of poverty, increased income segregation, widespread automobile dependence brought on by land use segregation, and increasing distances between where people live, work, and shop.

Attempts by planners to undo the damage of single-use zoning by enforcing mixed-use development have been problematic. As Jacobs warned, diversity that is "too feeble" can cause trouble by introducing variation in a place ill equipped to deal with strangers (Jacobs, 1961, p. 231). Already there is clear evidence that the movement of the middle-class and upper-class residents "back to the city" is not always indicative of healthy income diversity, but instead may only occur within the safe and exclusionary confines of gated or guarded enclaves.

A variety of other policies aimed at promoting racial diversity have been proposed, with mixed success. A few communities have undertaken proactive efforts to maintain racially integrated neighbourhoods, which have been referred to as "diversity-by-direction communities" (Peterman & Nyden, 2001). Such deliberateness tends to focus on Black-White integration. Planned diversity can either be a matter of government policy (e.g., Oak Park, Illinois, "Diversity Assurance Program") or a result of grass-roots effort (the diversity of Park Hill, Colorado, was fostered by an organization of church leaders). Juliet Saltman (1990) studied the neighbourhood stabilization efforts of five locales and found that the "fragile movement" was buoyed by progressive non-profit organizations, school boards, and neighbourhood groups, working in concert with pro-integrative governments. Other strategies for stabilizing diversity have included code enforcement, anti-blockbusting ordinances, bans on "for sale" signs, or even grants to individuals who work to support integration.

For planners, diversity might be encouraged by reversing the rules by which social segregation has occurred: allowing multi-family units where they have been excluded, and eliminating rigid building codes, minimum lot size, maximum density, minimum setbacks, and other barriers to infill development. Prior to these rule changes, diversity needs to be a more explicit target in neighbourhood-based planning efforts. Given planners' long-term interest in diversity, their failed track record in helping to sustain it, and their adoption of social mix strategies that may or may not have the desired effect, planners should have a key interest in exploring a variety of diversity-sustaining options. As a start, they could advocate planning strategies that are specifically tailored to supporting diverse urban neighbourhoods.

Three Diversity-Sustaining Planning Approaches

As it happens, a tailored diversity-enhancing neighbourhood planning approach could be best supported by drawing together the three planning models that dominate the planning field, as discussed by Fainstein (2000). First, the communicative planning approach would be a logical basis of support for neighbourhood planning that focuses on diversity. Methods falling under the communicative umbrella are often devoted to building inclusiveness, incorporating difference, reaching out to marginalized groups, and sensitizing planners to a wide variety of viewpoints and alternative ways of knowing (Harwood, 2005; Sandercock, 2000; Qadeer, 1997). These approaches rely on a more refined definition of what diverse urban places are and mean. Towards that goal, Sandercock has described the evolution of a "utopia with a difference" (1998, pp. 5, 119), which ties into Thomas's "unified diversity for social action" (1997, p. 258) and Friedmann's (2002) "open city" of diverse peoples supported by principles of ecology, citizenship, and regional governance. Lack of a good knowledge base about how to accommodate difference in planning has surely been a problem (Wallace & Milroy, 1999), but we have now been shown how "a thousand tiny empowerments" can constitute a new, socially transformative planning (Sandercock, 1997).

Yet the incorporation of difference in decision making is a different kind of goal from the goal of supporting neighbourhood-level social diversity. Planning for places currently composed of a complex range of people (diverse income levels, ethnicities, races, cultures, ages, and household types) requires something beyond inclusive exchange, or even a focus on empowering the underrepresented. Because engagement in the planning process always runs the risk of being motivated by a desire for group self-preservation – protecting one group from another in ways that are not mutually reinforcing – something more strategic is required. Those advocating various consensus-building approaches have recognized that dealing with diversity is likely to require something beyond merely democratizing and opening up public engagement, or simply having a more broad-minded view about the legitimacy of alternative cultural expressions. While theorists in the communicative planning mould have given us ideas about the mechanisms needed, and how the task of collaborative consensus building is to develop "'conversations' between stakeholders from different social worlds" (Healey, 1997, p. 219), we also need to know how that approach can be extended to the neighbourhood domain. How would innovation, "drama," and "a sense of play" as a way to "move the players and embed their learning deeply" (Innes & Booher, 1999, p. 19) help sustain social diversity in specific places?

What needs to be worked out is an approach that enlists ideas about consensus building and collaborative planning in ways that support socially diverse neighbourhoods specifically. The effort is not just a procurement of "collective agency" for the purpose of social transformation, political reform, and other radical planning ideals, which may in fact be ill suited to diverse neighbourhoods, but specifically to those neighbourhoods under pressure to transition towards non-diversity. Most importantly, the reconciliation of diverse points of view via creative group strategies has to be transferred to the realm of neighbourhood, where the goal of planning activity is not about a specific project or event but about the conscious support of diversity and the places that sustain it. Diversity is not accommodated, worked through, or simply tolerated; it is itself the object of support. The application of collaborative planning is likely to target ways of ensuring that diversity is fully represented throughout the neighbourhood planning process and that there is some concerted effort to keep the phenomenon of diversity and its positive implications as a central focus.

The second planning approach that has relevance for neighbourhood planning in diverse places is New Urbanism, whose physical prescriptions are in large part aimed at sustaining diversity. New Urbanists focus on providing a mix of housing types, building forms, and lot sizes, a connective street system, a supportive public realm, well-dispersed facilities and services, and codes that support these physical prescriptions, all of which are likely to be beneficial to the maintenance of social diversity. However, just as the communicative model does not necessarily result in diversity-supporting outcomes, diversity is unlikely to be supported by form and pattern alone. This situation is evident in the experience of New Urbanist development, which, paradoxical to its rhetorical goals, is able to command a high price for its amenity-rich buildings (Tu & Eppli, 2001; Talen, 2010). The laws of supply and demand, together with weak affordable housing subsidies, have ensured a lack of real diversity despite the inclusion of mixed housing types. In part because of a wide variety of market controls (Levine, 2005), neighbourhoods with exactly the kind of morphology that would seem to best support diversity (good connectivity, mix of uses, mix of housing type) are instead scarce, highly desired, and increasingly unaffordable to low- and middle-income groups. The inability to deliver affordable units within the context of walkable, mixed-income, quality environments has been viewed by some as a missed opportunity (Pyatok, 2002). New Urbanism was the movement that was going to do something about concentrated poverty by leveraging innovation in community design, but, without a more complex interweaving of policy tools, it falls short. Despite this inability to link design and policy, however, the physical ideas that New Urbanism offers about supporting diversity remain important.

Neither communicative planning nor New Urbanism can be relied upon as a model that alone sustains diversity. What they lack is a more complete adherence to the normative steps required – an unambiguous commitment to the realized "just city." They focus either on inclusive participation or correct form, devoid of specific strategies for arriving at social justice. As Fainstein (2000) argued, although the communicative model ensures an open exchange, it does not guard against unjust outcomes. In similar terms, New Urbanist models of good urban form do not ensure the existence of social diversity. Planning that supports the sharing of a neighbourhood by a diverse set of people is one realization of the just city ideal, and the maintenance of this diversity requires dedication to normative principles. Commitment to a fully integrated, inclusive society implies a level of involvement beyond dialogue and form, towards commitment to a type of planning that can support the long-term viability of socially diverse neighbourhoods.

A Targeted Approach to Neighbourhood Planning

From communicative models, we have inclusiveness, a focus on engaging excluded groups, and innovations in consensus building among fragmented groups. From New Urbanism, we have physical planning strategies aimed at facilitating diversity; most importantly, housing type mix, land use mix, community facilities within a quality public realm, and the need for codes that support social mix. From the just city model, we have a sense of commitment to employing a variety of redistributive mechanisms that help equalize access to resources. That different planning modes can and should exist simultaneously has been argued (Beard, 2003). How might neighbourhood planning integrate these three planning approaches to support socially diverse places?

One solution would be to structure the planning process itself around the idea of supporting diversity. The question then becomes: How can planners encourage a neighbourhood collectively to want to sustain the elements needed to support its diversity? Will a diverse group of people want to keep these sustaining elements or even incorporate new ones? In a society with such strong segregationist tendencies, where social mobility is equated with spatial mobility, something more specific than freeing people to engage in bottom-up, self-generative processes, building the right forms, or redistributing wealth may be required.

Conventional planning approaches do not offer much guidance. The intrinsic tension between the goal of increasing economic opportunity and the need to maintain social diversity is not a reality that informs planning method at the local level. The consequences of this neglect are easily recognized. Lees (2003, p. 623) showed how promoting "downtowns for people" and promoting "downtowns for opportunity" are contradictory, and that the resulting plans are not geared towards sustaining heterogeneity. Planners have to contend with not only land use regulation and a market that favour homogeneity but also with entrenched opposition to diversity, which is not uncommon even in neighbourhoods where diversity already exists (Nyden, Edlynn, & Davis, 2006). Fear of gentrification, fear of lower-income people degrading property values, fear of race mixing, fear of commercial uses encroaching on residential uses, fear of "non-consumers" occupying spaces of consumption – all of these familiar scenarios are in evidence in diverse neighbourhoods.

Unlike some conventional planning approaches, neighbourhood planning for social diversity cannot be value-neutral. It instead embraces difference and attempts to sustain it. It does this not by focusing on the empowerment of any particular group, but by ensuring that people are well informed about the diversity at hand, what that diversity means, and the positive ways in which it can be viewed. Tensions and conflicts that arise are not neutralized or assumed away, but confronted in a straightforward manner, largely by contextualizing them as conflicts that affect the social make-up and well-being of the neighbourhood. It is an approach that recognizes that the standard approach in which planners gather data, engage citizens in a far-reaching participatory process, and help formulate goals and objectives may not be effective for sustaining social diversity.

My suggested revisions to neighbourhood planning focus on the process of constructing a neighbourhood plan. I stop short of a full consideration of the plan's implementation, but the plan is intended to be a working document that incorporates specific implementation strategies directly. The overall approach starts with the identification of targeted, diverse neighbourhoods. Next, a citizens' planning group representing the diversity of the neighbourhood is appointed and charged with making sure that diversity is treated as an essential focus throughout the neighbourhood planning process. This is followed by a set of educational activities intended to inform neighbourhood residents about the diversity that exists around them, but that they might not be fully aware of. Finally, a neighbourhood plan is produced.

The neighbourhood plan that emerges out of this process must have specific requirements. A plan that supports diversity should, in particular, include three basic elements: 1) the establishment of an ongoing process of neighbourhood development review as a way of supporting the ongoing, shared management of neighbourhood change; 2) an urban code that ensures the provision of a supportive built environment; and 3) clear directives for targeted public investment that supports diversity, giving planners and residents the ability to push for expenditures that promote a "just city" by ensuring equitable access to resources for a diverse group of residents. These three elements mirror the three planning models discussed earlier.

The success of this effort rests on collaboration and engagement among a diverse group of residents – residents who reflect the diversity of the neighbourhood being planned for. Failure to actively engage this diversity will result in a failed process and an ineffective plan. Planners, however, do not play a neutral role in this process. The targeting of diverse areas and the effort to guide the planning process and the elements to be included in the plan reflect a normative position that values neighbourhood-level social diversity. It is a recognition too of the power planners have to advocate principles. Friedmann (2002) explained the manner in which planners champion specific positions: "They have the power to initiate a public discourse about the city. They have the power to involve the public and so engage them in mutual learning. They have the power to set agendas. And they have the power to advocate positions that have been carefully thought out in both ethical and practical terms."

Steps in the Targeted Neighbourhood Planning Process

i) Target Diverse Neighbourhoods

The first step is to decide which neighbourhoods should be targeted for a neighbourhood planning effort directed at sustaining diversity.

This is not the usual procedure. Most commonly, neighbourhoods are selected for planning work, followed by targeted investment based on level of distress or opportunity to stimulate private investment (see, for example, Richmond LISC, 2005, and McCarron, 2004).

Planners would need to propose measures to define diversity that make the most sense for their communities. There are a variety of possible approaches. For example, Khadduri and Martin (1997) used three categories of mix to define "mixed-income housing," while Smith (2002) proposed five categories based on the kind of income mix involved (e.g., "low-income inclusion" or "broad range of incomes"). Immergluck and Smith (2003) classified neighbourhoods in Chicago according to the internal mix of high- and low-income residents, ranging from "highly restrictive" to "highly diverse." Maly (2000) used a "neighborhood diversity index" to compare four population groups. Nyden et al. (1997) used a comparative approach (rather than an absolute definition of diversity), and defined a diverse census tract as one that approximated the racial and ethnic composition of the city as a whole. Galster (1998) defined a neighbourhood as "mixed" if no single group made up more than 75 per cent of the population. Ellen (1998) defined as "racially integrated" those neighbourhoods with a Black population between 10 and 50 per cent, which she justified as a compromise between the fact that Blacks make up a relatively small percentage of the population (roughly 13 per cent) and the idea that space should be shared equally.

Planners should also target neighbourhoods undergoing threats to existing diversity, the potential for instability (gentrification, displacement, disinvestment), and the likelihood of success (citizen interest, active and engaged local leadership). Depending on available resources, one or more diverse neighbourhoods can be targeted for special planning effort and focus.

ii) Appoint a Citizens' Planning Group

The second step is to assemble a citizens' planning group composed of local leaders who represent the diversity of the neighbourhood. The group would be enlisted to support the diversity-sustaining process being proposed. Planners could help the group work towards an appreciation of what diversity brings to the neighbourhood as a whole. Rather than being problematized, diversity would be considered as an asset and an essential part of the solution to community problems. Instead of being viewed as an imposed condition, whereby income and racial integration is forced by government fiat, diversity would be cast as something positive and unique. To accomplish this, interdependence must be established, whereby participants in the process "each have something others want and need something others can offer them" (Booher & Innes, 2002, p. 228). The challenge for the citizens' planning group is that, instead of applying this logic to a small group of participants, a sense of reciprocity and interdependence has to be established for the neighbourhood as a whole. The group could help articulate, for example, the ways in which the goals of diversity are interrelated: a neighbourhood that is open to a range of groups translates to improvements in neighbourhood services for all groups. A mix of uses that is good for the economic vitality of a neighbourhood adds interest and opportunity for every resident. Diversity may help stimulate an expanded set of locally based social networks, which may be viewed positively by many residents. There are also practical benefits: diverse people are more likely to have diverse schedules, thereby increasing the ability of the neighbourhood to patrol its streets informally at all times of the day.

iii) Build Public Awareness

One of the first tasks for the citizens' planning group would be to look for ways to increase public awareness of neighbourhood diversity. Specifically, there could be an effort to increase recognition and understanding of the kinds of diversity residents may only have a vague notion of. Activities to increase public awareness would need to take place during the initial months of the process. The ideas to be communicated would need to be simple, straightforward, and visually interesting, presented in a manner that is readily understood, and suitable for publication and exhibition throughout the neighbourhood. They should highlight racial, ethnic, income, age, and household diversity, and include some explanation about how the level of diversity has changed over time. Graphical output of various kinds could be exhibited in well-traversed public spaces and places including websites.

The citizens' planning group could tap the efforts of local schools to increase recognition and appreciation of diversity. Often, local schools and libraries are involved in projects devoted to celebrating diversity, although these efforts may remain detached from the work of neighbourhood planning. Children make art, write poetry, and find different ways to express their cultural identities and are encouraged to react to their differences from others in positive, affirming ways. Planners should work with the citizens' planning group to spotlight these efforts and integrate them into the neighbourhood planning process.

Planners could also help the citizens' planning group set up a website that spotlights diversity. The website could be used to expose the life stories of a wide range of residents, thereby taking a step towards building familiarity, tolerance, and social connection among people who otherwise would have no exposure to each other. It is widely recognized that a successful urban neighbourhood has a variety of spaces for social and cultural transaction. Planners can make use of technology as a "place" for exchange – an alternative mechanism for increasing the transaction base of a diverse neighbourhood.

Technology has been used in a variety of ways to increase community capacity building, and planning for a diverse neighbourhood would likely benefit from such approaches. For example, groups connected to the Community Technology Centers' Network have been working to provide access to information technology to strengthen the social, economic, and cultural life of communities (Hampton & Wellman, 2003). Organizations like Beyondmedia Education seek to give access to media tools so that people can "document and communicate their stories, serve as educators and role models for others, influence public policy, and generate social transformation." The book *Community Building on the Web* (Kim, 2000) documents the ways in which websites can be used as catalysts for community interaction.

iv) Produce a Neighbourhood Plan

With a citizens' planning group in place, and the establishment of preplan activities focused on building public awareness of neighbourhood diversity, the next step is the creation of a neighbourhood plan. The production of a neighbourhood plan puts the idea of a shared future in concrete terms. The act of organizing the various ideas that emerge into a neighbourhood plan is a way of focusing the discussion. The plan can be used as a framework to channel individual ideas towards something tangible – collectively realized, positive outcomes for the diverse neighbourhood.

The provision of such a framework may be especially important in a socially diverse neighbourhood where lower-valued homes and businesses are seen as a threat to higher-valued ones. One study showed that the neighbourhood plan was critical for garnering support for affordable infill housing, because it embedded the infill within a larger context, i.e., the affordable housing was contextualized and planned for (Deitrick & Ellis, 2004). The plan takes on the expanded goal of ensuring that it is possible to "envision each building, each development project, in relation to a positive ideal" (Brain, 2005, p. 32). Doing away with negative feelings about subsidized housing, small businesses, or social service agencies in favour of positive feelings about diversity is going to require strong conceptualization of the neighbourhood, and the plan is an essential means for accomplishing that.

Planners argue that the process of plan making itself is valuable because it promotes broader notions of the common good and encourages residents to break from the constraints of self-interest (Sirianni & Friedland, 2001). Some neighbourhood plans have been deemed effective because citizens acquired increased feelings of "community and efficacy" (Fainstein & Hirst, 1996). In the case of neighbourhood planning in diverse places, plan making can be a way of ensuring that the diversity of a neighbourhood is understood and supported.

During the creation of a neighbourhood plan in a socially diverse area, it may be especially important for participatory methods to avoid "the tyranny of structurelessness" (Sirianni & Friedland, 2001, p. 24). A variety of methods are possible, from week-long intensive charrettes to months-long visioning efforts involving a variety of public participation methods. Whatever process is used, the key point for planning that supports social diversity is that neighbours be kept informed of the tradeoffs involved in whatever is being proposed – how the proposals they advocate have consequences for other people currently in their neighbourhood as well as the neighbourhoods around them. If they propose, for example, to downzone portions of the neighbourhood so that multi-family units are disallowed, or to limit the ability to add accessory units, they need to be made aware of the effect this might have on reducing diversity, the ability to retain essential services they might deem important, the ability to sustain a walkable environment, the increase of traffic arterials through the neighbourhood, or the cost of housing.

In addition to their support of a process that keeps diversity front and centre, planners should be well informed about the various design strategies that could be used to stabilize and promote diversity: ideas about sustaining social mix, connecting a diverse population, and maintaining security. In particular, they can illuminate the importance of housing mix, business mix, facilities that support connection, and the potential benefits of breaking through residential enclaves. They can suggest locations for nodes of new non-residential growth – strategically targeted in function of the housing diversity that surrounds them – that can be used as a basis of support. They can suggest ways to intersperse different housing types, and show how codes can be important for supporting the mix. They can show how strategically placed public investment can support diversity. Planners can be an essential resource on design ideas, pointing out when and where specific ideas being proposed may have the effect of undermining diversity.

In these ways, a plan aimed at supporting socially diverse neighbourhoods would be different from a conventional neighbourhood plan. And it would likely include three main elements, in line with its diversity-supporting purpose: a) establishment of a process for shared management of the built environment as a neighbourhood-stabilizing strategy; b) regulatory reform including new types of codes that encourage a coherent yet flexible guide for the built environment; and c) recommendations for public investment that stimulates positive change, giving the neighbourhood the kinds of improvements it needs without undermining its diversity.

Shared Management

A neighbourhood plan should not be an end-state project, which is in any case considered to be at odds with diversity (Jacobs, 1961). Jane Jacobs wrote, "A successful city neighborhood is a place that keeps sufficiently abreast of its problems so it is not destroyed by them" (p. 112). Keeping abreast implies some form of ongoing neighbourhood governance. A neighbourhood plan that supports diversity should lay out a process of collaboration whereby the diverse nature of the neighbourhood is borne in mind as residents respond to and address various proposals for change. The emerging "collective intelligence" of recurring responses to recurring problems should not be permitted to be about co-opting space for control by one group or another. People need to think positively about the process and each other to ensure their neighbourhood sustains the diversity that exists. Planners have the power to shape these kinds of processes.

A diverse people with a unifying, shared notion of themselves (via the neighbourhood and articulated in the neighbourhood plan) cannot escape the very real problems of living together in proximities that require the equitable sharing of resources and the sharing of responsibility for the continued maintenance of those resources. There needs to be an operating system that focuses on diversity preservation through a shared commitment to place. Sociologists have documented the ability of diverse people to come together to work on neighbourhood revitalization efforts, especially where the focus has been on bringing rich and poor people together to address affordable housing, schools, safety, and public services (Warren, 2001). Planning researchers have shown how diverse groups of people, if engaged in "authentic dialogue," have the ability to find innovative and nuanced solutions to problems involving "shared and complmentary identities" (Booher & Innes, 2002, p. 227). It is this kind of collective effort that is essential for the long-term viability of the socially diverse neighbourhood.

Often, it is through civic institutions of various kinds that shared management of the neighbourhood takes hold. Civic institutions are generally defined as non-governmental, faith-based, and/or community-based organizations. Through them, collective, grassroots citizenship can be leveraged to sustain essential neighbourhood facilities, which are likely to be a primary focus of the plan. Like all neighbourhoods, diverse neighbourhoods should be well serviced, with good parks and schools, and spaces and streets that are safe. The success and safety of these public places depend to some degree on the health of the civic institutions needed to sustain them. While this is critically important in disadvantaged neighbourhoods, as documented in books like Stephen Goldsmith's *Putting Faith in Neighbourhoods: Making Cities Work through Grassroots Citizenship*, it may be just as important in the socially diverse neighbourhood, where bonds among residents are likely to be stressed for reasons having to do with cultural, social, and economic difference. The tie into existing neighbourhood institutions will help residents assume management quickly, fostering the establishment of a proactive, resident-led approach to neighbourhood management. The neighbourhood plan is essential for setting the rules for this kind of shared institutional approach.

Shared management can take different forms. A citizens' review committee could be established to gain more control of property development in the neighbourhood. Diverse neighbourhoods might also be open to community policing efforts, whereby residents work in tandem with police enforcement to take control of neighbourhood security. While policing partnerships are generally aimed at reform of failed methods, they are likely to be essential in places where residents are particularly anxious and fearful of others. In a diverse neighbourhood, locally driven efforts are not about transforming the neighbourhood, but about connecting diverse interests in a way that is sustaining. Again, community-based institutions – schools, neighbourhood associations, faith-based organizations, and the like – are essential for promoting the level of participation and responsibility necessary for this kind of involvement.

The capacity of neighbours to work together to address issues like crime – their collective efficacy – is crucial. But the achievement of collective efficacy requires a certain level of social cohesion. As one sociologist put it: "Socially cohesive neighborhoods will prove the most fertile contexts for the realization of informal social control" (Sampson, Raudenbush, & Earls, 1997, p. 919). In diverse neighbourhoods, a different kind of effort may thus be required – one focused on increasing the capacity of neighbours to work together. Residents in diverse neighbourhoods need to be engaged, and a focus on shared management of everyday issues, spelled out in the plan and using civic institutions of all kinds, may be the way to achieve this. Residents of diverse neighbourhoods should be entreated to participate in activities that sustain the qualities that make their neighbourhood unique and vital.

Code Reform

Another appropriate focus of the plan would be regulations that guide neighbourhood development. Land use regulations and building codes have a direct bearing on the ability of neighbourhoods to sustain diversity. Codes of all kinds can either stifle and undermine diversity, or help it flourish. Certain physical features are important for sustaining a diverse neighbourhood, and codes are an important way to introduce or maintain them. Most obviously, diverse neighbourhoods require a range of housing choices. They must simultaneously support home ownership and rental housing, integrate a range of housing types, densities, and levels of affordability, and foster a mix of uses, services, and facilities. There must be tangible opportunities for residents to "move up" without necessarily moving out. Strategies for mixing housing in otherwise single-family neighbourhoods include allowing corner duplexes, walk-up apartments on side streets, duplexes that look like single-family homes, and accessory units over garages. Part of the effort to reform codes may involve counteracting regulations that decrease affordability: excessive environmental regulations, restrictions on growth, exclusionary zoning, overbearing subdivision controls, impact fees, and burdensome, outdated codes of all types (Galster, Tatian, Santiago, Pettit, & Smith, 2003). There may be a focus on reversing the rules by which social segregation occurred: allowing multi-family units where they have been excluded, and eliminating minimum lot size, maximum density, minimum setbacks, and other rules that work to prevent housing unit diversity.

But a diverse neighbourhood is not just a matter of throwing a diverse range of housing types together. It may also be necessary to find ways for housing diversity to blend and therefore minimize negative perceptions. To do that, there needs to be a good understanding of the tradeoff between heterogeneity and uniformity in housing design. While some have argued that there should be no difference in the design and quality of housing for different income categories, architectural variety is also seen as a way of promoting diverse cultures. In any case, supporting diversity through housing design will require balancing cultural and individual expression with an integration that works.

Corresponding to housing type diversity, there needs to be a "diversity of forms and facilities" that responds to the needs of a diverse population (Qadeer, 1997). This may require greater flexibility in codes, increasing "undirected hetero-zone spaces," including locations for work-play, live-work, or play-live (Boyer, 1990; Ellin, 1996). Diversity of uses also requires a "fine grain" in the urban texture (Jacobs, 1961). Small lot sizes and small block sizes are more likely to encourage a diversity of building types and uses, and this morphology is something codes could incorporate and/or protect. There may need to be some design control (via form-based codes) accompanying the increase in permitted uses, to ensure neighbourhood compatibility. In addition, the design of public space, including streets, parks, and sidewalks, is an essential factor in the ability to successfully mix residential, commercial, and industrial uses (Angotti & Hanhardt, 2001).

Targeted Public Investment

Often, public investment in neighbourhoods is motivated by a desire to stimulate private investment. Alexander Garvin argues throughout his book *The American City: What Works, What Doesn't* that planners ought to support "strategically planned investment" rather than "routine capital spending" because the former can "spark further investment by private businesses, financial institutions, property owners, and developers" (Garvin, 2002, p. 4). This may make sense for public investment intended to revitalize and transform depressed areas, but it may not always be the most important strategy for supporting diverse neighbourhoods.

For one thing, diverse neighbourhoods need an infrastructure that supports positive social connection, and that means paying particular attention to public space – not necessarily by increasing opportunities for private development. Parks, plazas, streets, and other elements of the public realm act in particular as both mitigators and generators of diversity. Such spaces may sustain diversity by offering shared space as opposed to privatized residential space, therefore providing a better chance for informal, collective control and a sense of shared responsibility. Since diversity decreases distances between elements whose compatibility is questioned, investment in the public realm may be essential for holding disparate elements together. If residents do not value the external context in which increased proximities occur but instead focus on the isolated value of the individual dwelling, social diversity may eventually lose support.

Sustaining mixed housing type in a diverse neighbourhood is also dependent on strategic investment because diverse neighbourhoods are prone to gentrification and displacement pressures. The design requirements of diverse neighbourhoods make them more desirable places, thus threatening the very diversity they were meant to protect. Because diverse neighbourhoods become victims of their own success, a proactive response to the affordability problem is essential. Targeted public investment should be used to ensure that diverse neighbourhoods are not permitted to appreciate beyond the means of middle and lower income groups. If neighbourhoods of diversity are also to be well planned and serviced, there will need to be a political commitment to maintaining their social diversity. This commitment becomes real if such policies are embodied in the neighbourhood plan.

Those working to provide affordable housing, like community development corporations, need to be supported by including their efforts to build housing in diverse places directly in the plan. Non-profits are essential in diverse neighbourhoods because they act as agents of social renewal for all residents, in addition to building affordable housing and helping to deliver social services (Briggs, Mueller, & Sullivan, 1997). These and other organizations working to promote diversity by developing infill assisted housing need to be enlisted as partners in the formulation of the neighbourhood plan, including suggestions for strategic, diversity-stabilizing public investment. Increased private investment in the built environment is often considered an important strategy for distressed communities (Lamore, Link, & Blackmond, 2006), but the case could also be made that diverse areas need strategic investment that serves the purpose of stabilizing diversity.

Development projects that are likely to be most supportive of diversity are those that are mixed in use and mixed in housing type, and either type requires financial backing from multiple public and private sources. A neighbourhood plan with the goal of supporting diversity could include strategies to increase the viability of mixed-use, mixed-income projects. Within the context of a well-conceived plan, supported by an inclusive neighbourhood planning process, the desire to support diversity-sustaining development projects is likely to engender greater support. Planners could assist residents in laying out strategies for provision of a variety of supports, from technical assistance to tax incentives and grant monies, to make it easier for developers working in diverse neighbourhoods to combine funding in effective ways. The plan could also support funding for small, independently run businesses in the form of micro-lending, the practice of making small loans to budding entrepreneurs who would otherwise lack access to capital.

Conclusion

Cities in the US are defined by high levels of spatial sorting. Through transportation, land use zoning, housing, mortgage lending, energy, school finance, and many other types of policies, social groups and economic functions have become increasingly segregated. Such systems of separation are interlinked, one feeding into and sustaining the other. This is why it is essential that planners proactively support neighbourhoods that, despite the odds, maintain a level of social diversity.

The challenges involved in sustaining such neighbourhoods are significant. Some argue that Americans are generally loath to have neighbours of lower status than themselves, a "dirty little secret buried in the shelves of social science poverty studies" (Heclo, 1994, p. 422). Ethnographers have found that working-class residents feel especially vulnerable living close to "lower-class populations" (Kefalas, 2003, p. 100). Yet planners can play a role in helping residents of diverse neighbourhoods better appreciate the diversity that defines their neighbourhood. They can support socially diverse neighbourhoods by initiating a neighbourhood planning process geared specifically to embracing the diversity around them in positive, stabilizing ways. They can make it easier for people who share a diverse neighbourhood to collaborate, to participate in "re-working the meaning of a place-based political community" (Healey, 1997, p. 124).

Neighbourhood planning for socially diverse areas involves converging the three main models of urban planning and using them to support the long-term viability of socially diverse places. Planners can work from communicative models of inclusive discourse that is nevertheless well informed about the diversity that exists, they can marshal principles of good urban form that are likely to be supportive of social diversity, and they can advocate an important aspect of the just-city model by working to stabilize places of social diversity through targeted public investment. The combined model incorporates inclusive participation and good urban form in service to the just city.

Above all, planning for neighbourhoods that are already socially diverse requires a shift in emphasis. Planners often define urban planning as being about *change*: encouraging change that is desired and discouraging change that is not (Garvin, 2002). In the case of planning for social diversity, however, it might be more apt to define urban planning as being about *stability*: how to keep a place diverse and prevent it from being taken over by one particular social group or one particular land use. In such cases, the goal of urban planning may be to encourage change that supports a stable heterogeneity – the continued presence of diversity – while discouraging change that undermines it. This requires strategic thinking, since support of a diverse neighbourhood runs the risk that any targeted planning effort will ultimately undermine the very diversity planners hope to protect.

NOTE

1 Phil Nyden estimated 5–10 per cent at a panel discussion on diversity, which took place in Chicago on June 9, 2006, at the I-Space Gallery, 230 W. Superior St.

REFERENCES

Angotti, T., & Hanhardt, E. (2001). Problems and prospects for healthy mixed-use communities in New York City. *Planning Practice and Research, 16*(2), 145–154. http://dx.doi.org/10.1080/02697450120077352

Baumgartner, M.P. (1991). *The moral order of a suburb*. New York: Oxford University Press.

Beard, V.A. (2003). Learning radical planning: The power of collective action. *Planning Theory*, 2(1), 13–35. http://dx.doi.org/10.1177/1473095203002001004

Beatley, T., & Manning, K. (1997). *The ecology of place: Planning for environment, economy, and community*. Washington, DC: Island Press.

Booher, D.E., & Innes, J.E. (2002). Network power in collaborative planning. *Journal of the American Planning Association*, *21*, 221–236.

Boyer, M.C. (1990). Erected against the city: The contemporary discourses of architecture and planning. Center 2: Architecture vs. Planning: 36–43.

Brain, D. (2005). From good neighborhoods to sustainable cities: Social science and the social agenda of the New Urbanism. *International Regional Science Review*, *28*(2), 217–238. http://dx.doi.org/10.1177/0160017605275161

Briggs, X. de S., Mueller, E.J., and Sullivan, M.L. (1997). *From neighborhood to community: Evidence on the social effects of community development*. New York: New School for Social Research.

Card, D., Mas, A., & Rothstein, J. (2008). *Are mixed neighborhoods always unstable? Two-sided and one-sided tipping*. Washington, DC: National Bureau of Economic Research. http://dx.doi.org/10.3386/w14470

Caro, R.A. (1974). *The power broker: Robert Moses and the fall of New York*. New York: Alfred A. Knopf.

Day, K. (2003). New Urbanism and the challenge of designing for diversity. *Journal of Planning Education and Research*, 23(1), 83–95. http://dx.doi.org/10.1177/0739456X03255424

Deitrick, S., & Ellis, C. (2004). New Urbanism in the inner city: A case study of Pittsburgh. *Journal of the American Planning Association*, *70*(4), 426–442. http://dx.doi.org/10.1080/01944360408976392

Downs, A. (2000). How city planning practices affect metropolitan-area housing markets, and vice versa. In L. Rodwin & B. Sanyal (Eds.), *The profession of city planning* (pp. 117–130). New Brunswick, NJ: Center for Urban Policy Research.

Ellen, I.G. (1998). Stable racial integration in the contemporary United States: An empirical overview. *Journal of Urban Affairs*, *20*(1), 27–42. http://dx.doi.org/10.1111/j.1467-9906.1998.tb00408.x

Ellin, N. (1996). *Postmodern urbanism*. New York: Princeton Architectural Press.

Fainstein, S. (2000). New directions in planning theory. *Urban Affairs Review*, *35*(4), 451–478. http://dx.doi.org/10.1177/10780870022184480

Fainstein, S.S., & Hirst, C. (1996). Neighborhood organizations and community planning: The Minneapolis Neighborhood Revitalization Program. In W.D. Keating, N. Krumholz, and P. Star (Eds.), *Revitalizing Urban Neighborhoods* (pp. 96–111). Lawrence: University Press of Kansas. .

Florida, R. (2002). *The rise of the creative class*. New York: Basic Books.

Friedmann, J. (2002). City of fear or open city? *Journal of the American Planning Association*, *68*(3), 237–243. http://dx.doi.org/10.1080/01944360208976270

Galster, G.C. (1990). Neighborhood racial change, segregationist sentiments, and affirmative marketing policies. *Journal of Urban Economics*, *27*(3), 344–361. http://dx.doi.org/10.1016/0094-1190(90)90006-9

Galster, G.C. (1998). A stock/flow model of defining racially integrated neighborhoods. *Journal of Urban Affairs*, *20*(1), 43–51. http://dx.doi.org/10.1111/j.1467-9906.1998.tb00409.x

Galster, G.C., Tatian, P.A., Santiago, A.M., Pettit, K.L.S., & Smith, R.E. (2003). *Why not in my backyard? Neighborhood impacts of deconcentrating assisted housing*. New Brunswick, NJ: Center for Urban Policy Research.

Garvin, A. (2002). *The American city: What works, what doesn't*. New York: McGraw-Hill.

Glazer, N. (1959). The school as an instrument in planning. *Journal of the American Institute of Planners*, *25*(4), 191–196. http://dx.doi.org/10.1080/01944365908978333

Grigsby, W., Baratz, M., Galster, G., & MacLennan, D. (1987). *The dynamics of neighbourhood change and decline*. Oxford: Pergamon.

Hampton, K., & Wellman, B. (2003). Neighboring in Netville: How the internet supports community and social capital in a wired suburb. *City & Community*, 2(4), 277–311. http://dx.doi.org/10.1046/j.1535-6841.2003.00057.x

Harwood, S.A. (2005). Struggling to embrace difference in land-use decision making in multicultural communities. *Planning Practice and Research*, *20*(4), 355–371. http://dx.doi.org/10.1080/02697450600766746

Haughey, R., & Sherriff, R. (2010). Challenges and policy options for creating and preserving affordable housing near transit and other location-efficient areas. Washington, DC: What Works Collaborative, Center for Housing Policy, and National Housing Conference.

Healey, P. (1997). *Collaborative planning – shaping places in fragmented societies*. London: Macmillan Press.

Heclo, H. (1994). Poverty politics. In S. Danziger, G. Sandefur, & D. Weinberg (Eds.), *Confronting poverty: Prescriptions for change* (pp. 396–437). New York: Russell Sage Foundation.

Immergluck, D., & Smith, G. (2003). Measuring neighborhood diversity and stability in home-buying: Examining patterns by race and income in a robust housing market. *Journal of Urban Affairs*, *25*(4), 473–491. http://dx.doi.org/10.1111/1467-9906.00173

Innes, J.E., & Booher, D.E. (1999). Consensus building as role playing and bricolage: Toward a theory of collaborative planning. *Journal of the American Planning Association*, *65*(1), 9–26. http://dx.doi.org/10.1080/01944369908976031

Jacobs, J. (1961). *The death and life of great American cities*. New York: Vintage Books.

Kefalas, M. (2003). *Working-class heroes: Protecting home, community, and nation in a Chicago neighborhood*. Berkeley: University of California Press.

Khadduri, J., & Martin, M. (1997). Mixed-income housing in the HUD multifamily housing stock. *Cityscape: A Journal of Policy Development and Research*, *3*(2), 33–69.

Kim, A.J. (2000). *Community building on the web*. Berkeley, CA: PeachPit Press.

Lamore, R., Link, T., & Blackmond, T. (2006). Renewing people and places: Institutional investment policies that enhance social capital and improve the built environment of distressed communities. *Journal of Urban Affairs*, *28*(5), 429–442. http://dx.doi.org/10.1111/j.1467-9906.2006.00308.x

Lees, L. (2003). Super-gentrification: The case of Brooklyn Heights, New York City. *Urban Studies (Edinburgh, Scotland)*, *40*(12), 2487–2509. http://dx.doi.org/10.1080/0042098032000136174

Levine, J. (2005). Zoned Out: Regulation, markets, and choices in transportation and metropolitan land use. Washington, DC: RFF Press.

Maly, M.T. (2000). The neighborhood diversity index: A complementary measure of racial residential settlement. *Journal of Urban Affairs*, *22*(1), 37–47. http://dx.doi.org/10.1111/0735-2166.00038

McCarron, J. (2004, July). The power of sticky dots. *Planning Magazine*, 10–13.

Mennel, T. (2004). Victor Gruen and the construction of Cold War utopias. *Journal of Planning History*, *3*(2), 116–50. doi:10.1177/1538513204264755.

Nyden, P., Edlynn, E., & Davis, J. (2006). *The differential impact of gentrification on communities in Chicago*. Chicago: Loyola University Chicago Center for Urban Research and Learning.

Nyden, P., Maly, M., & Lukehart, J. (1997). The emergence of stable racially and ethnically diverse urban communities: A case study of nine U.S. cities. *Housing Policy Debate*, *8*(2), 491–534. http://dx.doi.org/10.1080/10511482.1997.9521262

Peterman, W., & Nyden, P. (2001). Creating stable racially and ethnically diverse communities in the United States: A model for the future. *Social*

Policy and Administration, *35*(1), 32–47. http://dx.doi.org/10.1111/1467-9515.00218

Pollack, S., Bluestone, B., & Billingham, C. (2010). *Maintaining diversity in America's transit-rich neighborhoods: Tools for equitable neighborhood change.* Boston: Dukakis Center for Urban and Regional Policy at Northeastern University.

Putnam, R.D. (2000). *Bowling alone: The collapse and revival of American community*. New York: Simon and Schuster.

Pyatok, M. (2002). The narrow base of the New Urbanists. *Progressive Planning Magazine*. Spring. http://www.plannersnetwork.org/2002/04/the-narrow-base-of-the-new-urbanists/

Qadeer, M.A. (1997). Pluralistic planning for multicultural cities: The Canadian practice. *Journal of the American Planning Association*, *63*(4), 481–494. http://dx.doi.org/10.1080/01944369708975941

Quigley, L. (Ed.). (2010). *Preserving affordable housing near transit: Case studies from Atlanta, Denver, Seattle and Washington, D.C.* Columbia, MD: Enterprise Community Partners.

Richmond LISC. (2005). *The ripple effect: Economic impacts of targeted community investments*. Richmond, VA: Richmond LISC. http://www.lisc.org/content/publications/detail/762/.

Saltman, J. (1990). *A fragile movement*. New York: Greenwood.

Sampson, R.J., Raudenbush, S., & Earls, F. (1997). Neighborhoods and violent crime: A multilevel study of collective efficacy. *Science*, *277*(5328), 918–924. http://dx.doi.org/10.1126/science.277.5328.918

Sandercock, L. (1997). *Towards Cosmopolis: Planning for multicultural cities.* Chichester: John Wiley & Sons.

Sandercock, L. (2000). Cities of (in)difference and the challenge for planning. *DISP*, *140*, 7–15. http://www.nsl.ethz.ch/index.php/en/content/view/full/96

Sarkissian, S. (1976). The idea of social mix in town planning: An historical overview. *Urban Studies (Edinburgh, Scotland)*, *13*(3), 231–246. http://dx.doi.org/10.1080/00420987620080521

Silver, C. (1985). Neighborhood planning in historical perspective. *Journal of the American Planning Association*, *51*(2), 161–174. http://dx.doi.org/10.1080/01944368508976207

Sirianni, C., & Friedland, L. (2001). *Civic innovation in America: Community empowerment, public policy, and the movement for civic renewal.* Berkeley: University of California Press.

Smith, A. (2002). Mixed-income housing developments: Promise and reality. Neighborhood Reinvestment Corporation: Joint Center for Housing Studies of Harvard University. Online at http://www.academia.edu/5192888/The_Promise_of_Mixed-Income_Housing_for_Poverty_Amelioration.

Smith, R.A. (1998). Discovering stable racial integration. *Journal of Urban Affairs, 20*(1), 1–25. http://dx.doi.org/10.1111/j.1467-9906.1998.tb00407.x

Talen, E. (2005). *New Urbanism and American planning: The conflict of cultures*. London: Routledge.

Talen, E. (2008). *Design for diversity: Exploring socially mixed neighborhoods*. London: Architectural Press (Elsevier).

Talen, E. (2010). Affordability in New Urbanist development: Principle, practice, and strategy. *Journal of Urban Affairs, 32*(4), 489–510. http://dx.doi.org/10.1111/j.1467-9906.2010.00518.x

Talen, E. (Ed.). (2013). *Charter of the New Urbanism* (2nd ed.). New York: McGraw-Hill.

Tu, C.C., & Eppli, M.J. (2001). An empirical examination of traditional neighborhood development. *Real Estate Economics, 29*(3), 485–501. http://dx.doi.org/10.1111/1080-8620.00019

Van Kempen, R. (2002). The academic formulations: Explanations for the partitioned city. In P. Marcuse & R. van Kempen (Eds.), *Of state and cities: The partitioning of urban space* (pp. 35–56). Oxford: Oxford University Press.

Wallace, M., & Milroy, B. (1999). Intersecting claims: Planning in Canada's cities. In T. Fenster (Ed.), *Gender, planning and human rights* (pp. 55–73). London: Routledge.

Warren, M.R. (2001). *Dry bones rattling: Community building to revitalize American democracy*. Princeton: Princeton University Press.

12 Imagining the Future: The Role of Visual Arts in the Planning of Multicultural Communities

KELLY MAIN AND JAMES ROJAS

Introduction

Challenges to community participation in diverse and marginalized neighbourhoods are both well documented and an ongoing concern in the field of planning (Checkoway, 2009; Cushing, Wexler Love, & van Vliet, 2012; Forester, 1999; Healey, 1997; Abramson, Manzo, & Hou, 2006; Umemoto, 2001; Umemoto & Igarashi, 2009). How do we get the full spectrum of a community's residents to participate in the process? How can the process address conflicts between these residents and produce meaningful solutions to planning problems? These tasks are difficult enough to accomplish within longtime homogeneous communities comfortable with interacting with local government, but when it comes to multicultural immigrant neighbourhoods, the job can prove especially daunting. According to the 2010 US Census, of the roughly 40 million immigrants living in the United States, more than half have come to live here since 1990, with about one-third entering the country since 2000 or later. While 20 million of these immigrants rate themselves as speaking English "very well," the other half report that they struggle with the language to one degree or another. Language issues are often compounded by the all-too-common problems in multicultural immigrant communities of little or no computer access, suspicion of government officials, planners' cultural biases, and cultural divisions between cultures within the communities (US Census Bureau, 2010). How, then, is it possible to create inclusive, cross-cultural, and productive public outreach/participation processes in such communities?

Current planning outreach/participatory practices that rely heavily on verbal communication can reinforce the growing gap between those who speak the technical language of a planner and those who do not. Moreover, cultural conflicts between communities over the appropriate use of city space can be heightened by language differences that make communication and understanding during outreach extremely challenging. This chapter argues that public outreach/participation in multicultural communities can benefit from employing the arts (approaches to understanding and communicating about place that de-emphasize language), to both enhance cultural expression and expand cross-cultural communication and problem solving regarding planning issues. These methods also offer an opportunity to help communities access their imagination to develop innovative perspectives upon, and solutions to, local problems and to create compelling stories regarding these perspectives and solutions.

To illustrate the potential of using the arts for public outreach/participation, this chapter will cover case studies of two methods – photovoice and Place It! Photovoice is a process in which participants explore and express their ideas about the built environment through photography, ultimately creating photos and stories that both empower the participants and can be used to influence planning policy and projects. Place It! – spontaneous modelling – is a process in which participants are given approximately 20 minutes, a blank piece of paper, and a huge assortment of recycled and random objects to choose from to address a planning question related to the city (e.g., How would you design your ideal city?). After they have created these tableaus, they tell their stories in very brief and simple language.

Before reviewing descriptions and case studies of photovoice and Place It!, including our experiences in using them, we review the literature on the challenges to participation in diverse/marginalized communities and arts-based social change. We argue that these two methods address a significant number of the challenges associated with working in communities who feel uncomfortable, unwelcome, and/or unable to participate in more traditional participation efforts. Because these methods initiate (Place It!) or involve community members in extensive active learning (photovoice) about each other, they raise the opportunity for meaningful community building. Because both methods include the interactive creation of visually compelling desired futures, they provide the possibility of empowerment of marginalized communities.

Challenges to Community Participation in Diverse Communities

Rapid globalization has produced communities of diverse cultural and ethnic backgrounds in the US. A recent estimate of foreign-born residents (both documented and undocumented) places the number at 39 million (Lichter, 2013). Since 1990, "majority-minority" communities in the US have more than doubled, from approximately 7 to 14 per cent[1]. In 2010, "minority" populations made up approximately 37 per cent (112 million) of the US population, and the US is predicted to be a "majority-minority" society by 2043. Many of these communities experience social and economic marginalization that, in turn, negatively impacts civic participation (Derr, Chawla, Mintzer, Cushing, & van Vliet, 2013; Hum, 2010; Umemoto & Igarashi, 2009). Concerns about cultural and political fragmentation are present in popular literature (Buchanan, 2011; Huffington, 2010), as well as academic discussions of the constraints of deliberative planning in a multicultural society (Checkoway, 2009; Healey, 1997; Abramson, Manzo, & Hou, 2006; Umemoto & Igarashi, 2009). Increasing diversity is "no guarantee that people of different cultural backgrounds will share the same physical and social spaces" (Lichter, 2013, p. 361), both literally and figuratively. It would seem then that in our near future, there will be a growing need to address challenges to community participation in multicultural communities.

Challenges to community participation in diverse and marginalized neighbourhoods range from language differences to more deep-seated fears of, or inability to understand, "the other." In marginalized communities and communities with foreign-born populations, these challenges may be further exacerbated by limited access/resources/time to participate and unfamiliarity with participation processes. Furthermore, the subject matter of planning, itself, may be framed in such culturally specific ways as to cloud the relevancy for many community members. All of these challenges have led to calls for increased cultural competency and training for planners (Burayidi, 2003) and the development of strategies/tools for working in multicultural communities (Qadeer, 2009).

One of the most obvious and significant challenges to community participation in multicultural communities is language, which "carries with it the power to discourage or encourage, repress or release, legitimize or degrade" (Umemoto, 2001, p. 23). Language differences inhibit communication not only between planners and their communities but also, in highly diverse communities, between community members themselves. To address language differences, planners have been encouraged to employ translators, particularly during public engagement processes (Qadeer, 2009). However, language differences belie two other significant challenges to communication between cultural groups: 1) that of understanding "the other" and 2) cultural differences in ways of knowing and interpreting the world (Umemoto, 2001; Sandercock, 1998).

Theoretical criticisms of rational and modernist planning (Harvey, 1989; Soja, 1989) and conceptualizations and case studies of both successful and problematic deliberative planning process (Cushing et al., 2012; Derr et al., 2013; Forester, 1998; Hum, 2010; Jojola, 1998; Qadeer, 1997; Umemoto & Igarashi, 2009) illustrate the importance of culture and ethnicity to situated knowledge – what we know – and, in fact, the nature and grounds of knowledge – how we come to know or make sense of the universe (Umemoto, 2001; Umemoto & Igarashi, 2009; Sandercock, 1998). Translation does not necessarily guarantee an accurate communication of ideas, as translators must be able to do more than express the same words in a different language; they must interpret and accurately represent the meaning of words and ideas. Words acquire meaning through lived experiences and cultural practices (Umemoto, 2001), and, thus, meaningful communication between groups requires more than the translation of words; it requires a fuller understanding of the lived experiences and cultural practices of others.

Challenges to meaningful participation can be further exacerbated in diverse communities that have experienced some marginalization, whether economic or social. In economically marginalized communities, access to venues for participation and time to participate can be problematic. In a study of Latino and/or foreign-born park users in Chicago, Stodolska and Almeida Santos (2006) demonstrated that access to and cost of transportation and the number of working hours were factors commonly listed by participants as limiting leisure activities. Similarly, in a study of civic participation by Latino youths in Boulder, Colorado, Derr et al. (2013) found that after-school jobs and living at the periphery of communities without transportation options inhibited Latino youths' civic participation. Discrimination may also make community members feel unable or unwelcome to participate, particularly in processes conducted by agencies associated with the discrimination or in locations associated with these agencies. For instance, consider the legacy of local governments that spearheaded the significant number of redevelopment efforts in the US that bulldozed neighbourhoods, such as in Boston (Gans, 1959) or Los Angeles (Hines, 1982), or more recent "revitalization" efforts that have led to the displacement of existing communities of colour (Smith, 1996) or the militarization of public spaces (Davis, 1992). In diverse communities that include migrant populations, the challenges associated with language, access, time, and discomfort may be exacerbated.

Finally, broader criticisms related to the relevancy of planning concerns may be particularly applicable in diverse and marginalized communities. In a discussion of challenges to public participation in San Francisco, Prowler (2007) asks: "How would you explain to a single mom in the Tenderloin, a teenager in the projects, a couple starting to look for a place to buy, or a grocer what planning can do to make their daily life better or worse? It's too abstract" (p. 2). This potential "irrelevancy" may be the result of the narrowness of the issues covered and/or the way in which these issues are framed or input is sought in the participation process. In their study of Latino youth in Boulder, Derr et al. (2013) found that the youth were more likely to participate when they could affect the relevancy of issues being discussed by framing the issues themselves, such as by being asked to "reimagine a civic area." Moreover, the youth were more likely to participate when the methods were "relevant" to them, such as through art, digital media, and photography.

Taken together, the challenges associated with community participation in culturally diverse and marginalized communities may produce feelings of disempowerment. Will I be able to participate? Will my concerns be understood and acknowledged? Will my participation change anything? If the answer to any or all of these questions is no, then for a potential participant, overcoming the significant challenges to participate may not appear to be worth the effort. Thus, to design genuine and meaningful participation processes in culturally diverse communities, concerns related to access, time, capacity, welcomeness, language, understanding, and relevancy need to be considered. The following sections cover the potential for the arts, and more specifically two processes – Photovoice and Place It! – to address these challenges to community participation in diverse and/or marginalized communities.

The Arts and Civic Participation

In the last several years, a growing movement of community artists has been working with planners to improve civic participation

(Catherwood-Ginn & Leonard, 2012; Walter, 2013). This kind of movement – arts-based social change/social practice art/participatory art – has demonstrated the power of the arts to address several of the challenges to community participation in diverse and marginalized communities. From the perspective of the artist, the viewer becomes the creator, and audiences are no longer passive in the production of art (Kennedy, 2013). This active participation in the creation of art meshes well with planners' concerns for more meaningful civic participation (Walter, 2013). More recently, planners have turned their attention to the arts as an economic development tool. Less frequently explored is the potential of art to build community and achieve change (Yonder, 2005). The openness fundamental to the creative process makes it possible for community members, as artists, to define their concerns and issues, and this, in turn, increases the relevancy of the discussion and the motivation to participate (Derr et al., 2013).

William Cleveland, director for the Center for the Study of Art and Community Development in the state of Washington, describes the purpose of arts-based community development as

> Arts-centered activity that contributes to the sustained advancement of human dignity, health, and/or productivity within a community. These include arts-based activities that:
>
> - EDUCATE and INFORM us about ourselves and the world
> - INSPIRE and MOBILIZE individuals or groups
> - NURTURE and HEAL people and/or communities
> - BUILD and IMPROVE community capacity and/or infrastructure. (Cleveland, 2011, p. 4)

The purposes of educating, informing, nurturing, and healing align with concerns that participation improves understanding between groups (planner and communities, communities and communities) and address fragmentation and marginalization. Some of this work draws on Paulo Freire's (1974) education reform theory, a key concept of which is "critical consciousness," learning through action and self-reflection (Cushing et al., 2012). Freire asserted the power of identifying forces in society that impede one's ability to take action and then entering into the active process of constructing or reshaping reality. An example of one such project is Crossroads Charlotte. In 2003, having identified significant levels of interracial and social distrust among community groups in the Charlotte-Mecklenburg area of North Carolina, a local steering committee, put together by the Foundation for the Carolinas, concluded that the mistrust was a result of feelings about access, inclusion, and equity. Community members wrote four stories centred on these issues, stories that were used to prompt dialogue between communities. However, it was not until artists were engaged to perform live poetic responses that Crossroads Charlotte became popular. Since 2003, associated projects including poetry, film, visual art, and theatre have continued to inspire public participation (Catherwood-Ginn & Leonard, 2012) in the form of grassroots movements to address other social justice issues.[2]

With a focus on community and public art and capacity building (economic and social), many artists have placed their projects at sites that are much more accessible and convenient for community participation (Cleveland, 2011; Catherwood-Ginn & Leonard, 2012; Walter, 2013). For example, Brooklyn's Laundromat Project's tag line is "Wash Clothes, Make Art, Build Community," and its stated intention is to:

> [Bring] engaging, community-responsive art and artists into local coinops across Greater New York City. By helping turn imaginations into creative fuel, we empower communities of color living on modest incomes to dream new visions for their own neighborhoods.[3]

The non-profit group holds free drop-in art classes that focus on a variety of issues: "The Art of Protest," the creation of protest posters to address social justice issues; "One garden-to-go, Please!," the creation of small terrarium gardens; and Harlem Is ... "The Personal Symbol Project," which they describe as:

> Draw. Cut paper. Roll ink. Layer. Repeat. While your clothes are on the spin cycle, you'll learn how to make a meaningful symbol representing your neighborhood – Harlem – that you can also use as a stamp to repeat onto any surface. Using materials you can find at home and some poems for inspiration, you'll be making your own collograph print in no time.[4]

In the following section, we describe two processes that employ the arts to engage culturally diverse community members on community planning issues: Photovoice, employed as a participatory process in a college studio project, the purpose of which was the amendment of a downtown-specific plan (Main); and Place It!, created and used in more than 300 workshops (Rojas). In describing the two processes and examples of their use, it is our intent to point out the strengths and weaknesses of these processes to address the challenges to community participation in culturally diverse and marginalized communities.

Photovoice

Photovoice is a technique developed in 1992 by C.C. Wang at the University of Michigan and M.A. Burris, a program director for public health at the Ford Foundation. The method is designed to empower community members through their own development of compelling stories about their lives and needs, using photos they have taken and narratives they have developed (Wang & Burris, 1994). Through self-reflection and sharing, participants gain both a personal and a collective knowledge of their situation. With these visual and written stories, participants may influence decision makers. Photovoice's use as a tool for community involvement has been growing since it was first employed in the Yunnan province in China to help rural women influence government actions and policies affecting them. Used most frequently in the field of public health, photovoice projects have been completed on a diverse variety of issues such as women's health, maternal and child health, individuals living with HIV, and the influence of immigration on Latino adolescents.

The process used in photovoice is meant to be both capacity-building and influential. Participants are first trained in specific skills – basic photography techniques, learning to take photos for maximum impact – and then asked to take pictures that reflect a particular topic. Topics can be openly defined – What is important to you? – or more narrowly defined – How can we make a more pedestrian-friendly neighbourhood? Once participants have taken their photos, they return to discuss them and their meaning. Group discussion is an important component of the process, because it is through discussion of the photos that participants become reflective about themselves and understanding of their collective concerns. Next, participants are typically asked to create an exhibit of their photos, stories, and suggestions for change. This involves further discussion and capacity building – learning how to develop a narrative/prose, telling a story using photos, demonstrating the importance of an issue, identifying structural problems, learning about traditional solutions, and developing their own solutions. Finally, venues are sought out, by participants and/or organizers, where participants can exhibit their photos and stories. While not strictly conceived of as arts-based social change by Wang and Burris, photovoice provides an experience in which participants can experience Freire's critical consciousness and express themselves through visual arts and narrative.

Examples of the power of photovoice and its use in community planning can be found in recent photovoice projects focused on the relationship between public health and the built environment. Six communities participating in the Healthy Eating, Active Communities Program in California used photovoice to get input from local youth about how young people could lead more active and healthier lives (Healthy Eating, Active Communities Program, 2009). After taking photos related to this concern, students returned to discuss the photos with adult activists. They discussed the problems the participants thought the photos portrayed and how these problems might be addressed. The adult organizers stress the importance of participants developing their own ideas and solutions, so that they feel ownership of the process. Ultimately participants in this program presented their photos and narratives to local community leaders and business owners, such as members of the parks and recreation commission and local store owners, to influence local community plans and the food available in local markets.

In 2011–12, Keith Woodcock[5] and I (Main) used photovoice in Santa Paula, California, as part of a community participation process for an update of the city's Downtown Improvement Plan. Established in 1872, primarily as a consequence of the area's petroleum reserves, the town lies 65 miles northeast of Los Angeles, nestled in a bucolic inland valley 12 miles from the California coast. Today Santa Paula, which has dubbed itself "The Citrus Capital of the World," is made up of approximately 30,000 residents, with Latinos being the majority. The city's economy, just as its moniker suggests, is now centred on agriculture. But while oil production has taken a back seat to oranges and lemons, Santa Paula, perhaps owing to its relative isolation, has managed to retain its small-town character, unlike the large metropolitan areas that surround it. Hollywood is fond of filming here, in fact, though mostly in period pieces where Santa Paula is meant to represent quintessential Middle America. How best to maintain the essence of Santa Paula's historic downtown while ensuring its economic survival and a healthy community was a focus of the planning project. College students in the community planning laboratory of the City and Regional Planning

Program at California Polytechnic State University undertook the update of the Downtown Improvement Plan for two quarters.

We used photovoice to find out what youth in Santa Paula thought about their community and health.[6] The photovoice project was sponsored by STRIDE (Science through Translational Research in Diet and Exercise), an interdisciplinary research centre at Cal Poly that promotes healthier living.[7] The chair of the English Department at the high school, Nicola Lamb, agreed to incorporate the photovoice project into two senior-level high school English classes. Initially, students were asked to take photos that addressed four questions: What do you like about your city? What would you like to change? How does the city (built environment) support health? How could the city be healthier? The high school students received disposable cameras and instructions about how to tell a story using photos and narratives. For almost three months, with guidance from Lamb and input from Woodcock, the students photographed, discussed, and developed a photo exhibit about their city.

Once Lamb's students completed the project, the high school students presented their exhibit to the Cal Poly college students, who were responsible for identifying issues raised by the high schoolers and addressing those in the Downtown Improvement Plan. It's interesting to note that high school students echoed many of the same concerns heard from other Santa Paulans who participated in the project – the importance of preserving the historic and small-town feeling of Santa Paula, the need for more variety in the businesses on Main Street, and the improvements needed to streets and parks. High school students also raised their own unique issues: concerns about gang graffiti, how much they loved the new bike path in town, how they wanted more activities to be healthy, how many of them had two working parents and needed something safe and constructive to do in the afternoon, how proud they were of their community's farmworker heritage and the new monument in town to farmworkers. The photos and narratives allowed students to tell stories that captured their feelings about their community and about the places in Santa Paula that were special to them, needed improvement, and affected their activity level. Some of the statements made by students are captured in figures 12.1 through 12.4.

Our photovoice project had several limitations. First, because of the time frame of our community planning laboratory, our students, rather than the high school students, were tasked with developing

Figure 12.1. Mural depicting the variety of people who live in Santa Paula and their rich culture.

"This town should reflect its own true beauty with its people, attractive architecture, and gorgeous landscape that are depicted through our murals around Santa Paula" – Yesenia and Maria

the solutions to the problems raised by the high school students.[8] It is desirable, as in the Healthy Eating, Active Communities (HEAC) Program efforts, that photovoice participants not only participate in but also, preferably, lead the process of developing solutions. And, while policies and programs addressing the high school students' needs were included in the plan, none of these ideas has been implemented. We have moved on to our next studio, and Ms Lamb has also moved on to teaching her next classes. In other words, at this moment there are no advocates in the community to help the students ensure the implementation of programs addressing their concerns. Currently, we are working, from a distance, with community members to find a venue for the photographs and to find other ways for students to bring

Figure 12.2. Vacant lot providing an opportunity for community space downtown.

"Having a butterfly farm [on this lot] would be a community activity with 'Santa Paula Beautiful' and the agricultural program at the high school. It would bring the community together with volunteer work and love for one's neighbourhood and each other." – Sandra and Camille

the photovoice project back to the attention of the city council members and others who can affect implementation.

When Santa Paula High School students made their presentation to us, we spent some time asking them what they thought of the photovoice project.

> "Our town just usually doesn't get the recognition it should and that's what this project [photovoice] did for us." – Gabriel, Santa Paula High School.

Figure 12.3. Stop sign showing graffiti and a need for the city to tackle quality of life issues.

"The tagging on the stop sign is … the type of graffiti that makes it seem like a neighbourhood is a certain gang's territory or turf and gives that area a bad reputation. Tagging in such obvious places like this shows the neglect of maintenance, because such places are not hard to find but rather are out in the open. So maybe what we should do is band together as a city, do what the sign states, and help stop the graffiti." – Elmer

> "I'd recommend this project [photovoice] to other schools 'cause they can go out and see their city the way it is … what's wrong and what's good about the city, and then learn from it and show others the city, and the city can be improved." – Rosalino, Santa Paula High School

> "I think it made them [the high school students] better citizens … It kind of gave them confidence in themselves that they are, in fact, citizens of this town, and they have a voice." – Nicola Lamb, English department chair, Santa Paula High School

Figure 12.4. Streets with holes and cracks showing the need for infrastructure improvements to enhance quality of life for residents.

"Santa Paula is a safe community; however, there are some areas that can be improved. Our town is very beautiful but streets with holes, cracks, and potholes catch the eye of many people. Our town is full of little streets, and most of them are terribly destroyed. Fixing things like streets will keep Santa Paula the same beautiful town it always has been." – Diana

These statements provide some evidence that photovoice had accomplished one of its intended consequences – a voice and sense of empowerment had been found, at least temporarily, by those who usually don't have them.

Place It!

In 2007, while working for Los Angeles' Metropolitan Transportation Authority, I (Rojas) developed Place It! as an engagement tool to use in

Figure 12.5. The Place It! technique used at a laundromat in New York City.

Latino communities. During the process, participants are organized into small groups, asked a simple question (e.g., What is your ideal city? or What was your favourite childhood place? or How would you increase walking in your community?) and then given up to 30 minutes to develop a three-dimensional model that addresses the question. The only requirements of the process are that every participant create a model – a small vignette of urban, suburban, or rural life – and then briefly explain their ideas to the rest of the group using their model. Because, in my experience, traditional, centralized approaches of outreach aren't working, and particularly those held in traditional, centralized locations, I've held workshops where people were already present: on the street, in parking lots, on sidewalks, in parks and vacant lots, in a laundromat and a horse stable.

In 2012, I collaborated with artists Elizabeth Hamby and Hatuey Ramos-Fermin on a project called "Mind the Gap/La Brecha," as part of the

Laundromat Project's Create Change Artist Residency. This residency places artists in their local laundromats to do community-based art works with their neighbours. Mind the Gap/La Brecha explored issues related to waterfront access in the South Bronx. I worked with the artists and the community to create interactive 3D models, exploring neighbours' perceptions of the water and waterfronts, and talking with them about the Harlem River and the Bronx Kill – two waterways that are physically proximate, but because of highways, industry, and land speculation, feel thousands of miles away.

To build their models, participants use thousands of small, colourful objects that are piled in the middle of tables in the laundromat. They can use what they like based on texture, colour, form, or idea. For example, green yarn can become grass, blue poker chips can represent the ocean, and hair rollers can represent apartments or office buildings. Some participants almost immediately begin constructing something that answers the question posed, while others start in a more reflective state. Many participants start with a few materials and continue to add throughout the process to finish off their models, which become increasingly elaborate. After the flurry of activity slows down, participants frequently began to talk with each other and look around at the models created by their neighbours. It is not unusual for participants to pull out their cell phones to take pictures of their models.

After developing their models, everyone, regardless of age or language, is required to present what they've produced to the rest of the group. In their language of choice, participants are given one minute to state their name and tell about their idea/plan to the group. Others at the table, or the facilitator, help by translating for the rest of the audience. A one-minute time frame for presentations keeps the pace of the meeting moving and reduces the pressure on participants to create elaborate descriptions. The participants are asked to stand up and gather around each table to see each other's work and hear the presentations. There are no formal planning terms or concepts discussed during the meeting, and, thus, in describing their models, participants interject their own personal experiences, memories, and random thoughts about places, both real and imagined. Participants' descriptions are often animated, and they present their ideas with conviction, enthusiasm, and/or concern for themselves, family, friends, community, and environment. Some designs address principles: "I think every office building should have a grammar school attached to it." Some address very specific place-based problems or challenges: "These aerial bike ways are going to improve transportation while preserving the environment" or "I added a swing set to create

Figure 12.6. Residents of Tijuana, Mexico, use Place It! technique to design a new park. Colourful everyday objects such as hair rollers, green yarn, and blue poker chips are used to depict land uses such as grass and buildings.

a child-friendly plaza." Some designs capture an experience: "Jefferson Street is too noisy, so I added trees to tone down the noise." Still other designs refer to emotional states: "All these buttons are happy people living in the city." After each presentation, the audience applauds.

Once a person has completed his or her one-minute presentation, the facilitator quickly synthesizes the information for the audience, using both the terminology of the presenter and urban planning terms to begin to familiarize the participants with community planning concepts and to validate that they both understand and can develop ideas related to city planning. Up to this point, Place It! typically takes less than an hour, depending on the number of participants. If each participant is going to build a model, I recommend that the group be no larger than

30 participants. This keeps the exercise under an hour, which serves several purposes. It makes participation much more manageable for children and seniors, who may tire of the activities, and for anyone who has just an hour to devote to the task.

If consensus building is a desired outcome for the process, after participants have completed their individual tasks, they are asked to work together and create a group model synthesizing all their ideas into one diorama. At this point, they may be asked to pick a specific location in the community where their model should be realized. During this phase of the exercise, the participants tend to combine models, only to realize that they must compromise, altering their ideals for the greater good of the community. The time for the consensus-building activity is typically 15 minutes. Once each team completes its group model, participants are given two minutes to explain their model to the rest of the audience.

Two workshops I recently completed in Southern California are examples of how Place It! can be used to understand common concerns and develop collective ideas and/or build capacity for future action. The workshops were held for Pacoima Beautiful, a non-profit located in a low-income Latino neighbourhood in Northeast San Fernando Valley. Staff recognized that significant barriers exist for some residents in regard to civic engagement, so every year they organize the People's Planning School for their Spanish-speaking constituents to help build their planning capacity (see figure 12.7).[9] One purpose of the workshops was to explore ways to further engage marginalized Latinos in Northeast San Fernando Valley on urban planning issues.

During the first part of each workshop, participants were asked: "How would you create your ideal community?" Dozens of ideas emerged from the individual models; however, many reflected urban design elements from Latin America such as plazas, shrines, soccer fields, jungles, or ranchos. Even though the participants lived in the US, they imagined a life here based on their memories. Through this part of the exercise participants got to know each other better, in some sense sharing who they were by sharing where they came from and how they would like to live. Once everyone presented, many participants commented on the similarities in their ideas, which appeared in their group designs during the second part of the workshop, when participants were placed in teams and asked to identify a specific problem in their neighbourhood and then design a solution to that problem.

Workshops recently held in Eugene, Oregon,[10] revealed slightly darker concerns regarding life for Latinos in the US. The purpose of the

Figure 12.7. Pacoima Beautiful People's Planning School. The technique provides a way for people with limited English or professional planning skills to participate in the planning process.

workshops was to engage marginalized Latinos in Eugene/Springfield on urban planning issues. To get people to participate, the effort partnered with community-based organizations (Huerto de la Familia and Pilas) which had already developed a sense of trust in the community. The first step was to identify where Latinos were already participating and then approach them in safe and comfortable ways. Teachers in local schools were contacted, and they sought out the parents of Latino children who were enrolled in their schools. Inviting children to participate in the workshops enabled parents to spend time with their children. It was also a priority that children be given a voice in the city planning process, so they could learn how community engagement can be fun and to ease the seriousness of the planning process. Thus the workshops became more like family gatherings than intimidating planning sessions.

Two workshops were held, with a significant turnout. There were 90 Latino participants, many of whose voices are often left out of the planning process because they do not typically participate in more formal local processes in Eugene/Springfield. Participants were asked to create their ideas of community. As they explained their models, it became clear that they were concerned about their safety. One of the most common responses heard at both workshops was that participants longed for security but didn't feel safe in the public spaces of their communities. They also felt left out, whether because of ethnic discrimination, immigrant status, or other reasons. Instead of asking for social services, better housing, or transportation, they spoke about wanting a better sense of belonging and the ability to enjoy public spaces. From the emotions and ideas expressed in the workshops, it was clear the participants' feedback was deeply felt.

Place It!'s brevity is both its limitation and strength in terms of building community ties. As a one-time event, Place It! is an icebreaker. In making it possible for disparate groups to share their values and ideas within an hour and in a playful manner, Place It! enables participants to experience their commonalities before they have time to focus on their dissimilarities and grievances with each other. However, in one hour, it is unlikely that group members will have time to fully explore their common impediments, desires, and needs, to develop what Freire would call a critical consciousness.

Conclusion

Literature on community participation in culturally diverse communities – particularly communities that have been marginalized – identifies access, time, and language differences as key challenges to participation. Participation venues must be conveniently located in places accessible to community members. Ideally locations are where community members already gather, such as public (parks, streets, sidewalks, and schools) and community spaces (local businesses, religious facilities). One of the strengths of Place It! is that, because of its simple materials and relatively quick pace, it can be (and has been) conducted easily and in a wide variety of places. City planners can include this technique in the public workshops and community group meetings that they ordinarily conduct. More importantly, city planners should use Place It! wherever community members can already be found. Photovoice, too, can be conducted in locations where people already gather (schools, community centres, religious facilities, laundromats), but the extended time required to complete the effort makes it slightly less flexible for both the participants and facilitators. One way to make photovoice much more convenient is to include it within the curriculum of existing learning environments (schools, community centres, religious facilities). With community partners, such as teachers, health officials, and non-profits, working on capacity building, city planners can utilize photovoice to address community conflicts and to build capacity among community members. By de-emphasizing the role of discussion and prose early in the process, both methods make it easier for community members with limited English skills to engage in the process. Place It!'s one-minute time limit for presentations and purposeful disuse of planning terminology are effective ways of addressing language challenges. Many participants have an easier time building a physical solution – using their hands and minds to create and reimagine their idea of an ideal city, street, public space, or building – than talking about it. The later phases of photovoice require significant language skills. This impediment is addressed through facilitators who help the participants develop their storytelling capacity, both in verbal and written forms, and thus address language issues through capacity building.

Lack of capacity also presents a challenge to participation in diverse and marginalized communities. The activities required to complete a photovoice project make it an excellent tool to build capacity. Because of photovoice's learning component – photography, storytelling, the built environment (both problems and solutions) – it has significant capacity-building potential. Through self-reflection generated by their photographs, participants can access memories, emotions, and ideas to develop a new consciousness regarding their relationships with the built environment. Place It! is a one-time event. During this event, participants explore their relationship with the built environment using very simple objects. Because of their familiarity, simple everyday objects like hair rollers, buttons, plastic bottle caps, and game pieces put people at ease. The quick time frame requires that people think "on their feet" using their existing capacity. Because it is a creative exercise to explore ideas, there is no right or wrong answer. Place It! does make it possible for people to realize their ability to understand and build their neighbourhoods and communities.

Perhaps the greatest challenges to participation in diverse and marginalized communities are the stereotypes and misunderstandings that exist between individuals and groups and planners and the communities they serve – the need to cross cultural bridges, better understand each other, and develop community ties. The interactive processes that are undertaken in photovoice make possible significant self-reflection and collective learning. By sharing their personal understanding in group discussion, participants can develop a deeper understanding of each other and a collective consciousness regarding shared impediments to improvements in their neighbourhoods. As an icebreaker, Place It! has the potential to initiate understanding between participants; however, its brevity makes it a more limited tool for building understanding.

Ultimately, photovoice and Place It! are useful participation tools in multicultural and marginalized communities only if they empower these communities to make changes that are relevant to them. As a creative exercise, photovoice has been designed to give participants the freedom to explore concerns and to address needs. As an exercise that produces photos and stories that are compelling to local community members, government agencies, and policy makers and have resulted in changes in plans, policies, and the physical environment, photovoice has been a valuable tool for change. To be effective, photovoice projects do require commitment to the advocacy and implementation portion of the process. Exhibits must reach the community members, government agencies, and policy makers who can facilitate change. The three-dimensional models created in the Place It! process are typically disassembled at the end of the hour. Only the narrative remains, in the form of a synthesis of the ideas, and is passed on to agencies and policy makers. The compelling sharing that goes on during a Place It! session is, therefore, difficult to preserve and may limit its potential to enable significant community change on its own. However, ideas developed during the Place It! sessions have been incorporated into community and project plans, and it is likely that these sessions influenced the dialogue.

While planners have begun to explore the potential of the arts to improve community participation in diverse and marginalized communities, many of the claims regarding community building and efficacy remain anecdotal. Although anecdotal evidence suggests that singular artistic endeavours and relatively new methods to planning, such as Place It!, are effective tools in these communities, their efficacy is relatively untested in planning research. Several empirical studies have documented the immediate sense of empowerment and community experienced by participants in the photovoice process, as well as the changes to the physical environment that have resulted from the compelling exhibits and advocacy produced. However, the long-term effects of these projects on community ties and empowerment go relatively unexplored. The capacity of the arts to fire a community's imagination and empower its residents to achieve lasting change is promising. This subject deserves further attention from both practitioners and researchers if these methods' potential for multicultural and marginalized communities is to be realized and their limitations addressed.

NOTES

1 From 13.5 per cent to 31.7 per cent of principal cities, 6.9 per cent to 14.7 per cent of suburbs, and 7.0 per cent to 12.6 per cent of non-metropolitan places.
2 Such as Amendment One, a law defining marriage between one man and one woman as the only legal form of marriage, and the 2012 murder of 17-year old Trayvon Martin (p. 14).
3 This description of the Laundromat Project can be found on its home page at http://laundromatproject.org, retrieved August 15, 2013.
4 Description appears as part of their "Works in Progress" program, http://laundromatproject.org/nontsikelelo-mutiti-elizabeth-hamby-and-hatuey-ramos-fermin/, retrieved May 27, 2015.
5 A lecturer at Cal Poly with professional community planning experience throughout the state of California.
6 Initially, we partnered with the local Boys and Girls Club to involve children who attend the club after school. With our support, Boys and Girls Club staff were to supervise the efforts and help the children review their photographs and develop their narratives.
7 Dr Ann McDermott, STRIDE's director, strongly supported the idea of exploring high school students' perceptions of the relationship between public health and the built environment.
8 Following the presentation by high school students, community planning laboratory students incorporated policies, programs, and ideas into the updated Downtown Improvement Plan to address the issues and concerns raised by Santa Paula High School students, including historic preservation, park improvements, street improvements, an entertainment district, and a children's public art program. In addition, a recommendation for the creation of a Youth Commission was included in the plan, to lay the foundation for continued participation by Santa Paula's students. The photos and narratives of the students have been included in the Downtown Improvement Plan and will be exhibited in both San Luis Obispo and in Santa Paula.

9 Two workshops were held in the evening after work. Pacoima Beautiful had already developed a sense of trust in the community, so we were able to hold the meetings at the local library and at Pacoima Beautiful Offices. Food was provided, and children were invited to attend.

10 This effort was completed with Professor Gerardo Sandoval of the Planning, Public Policy and Management Department at the University of Oregon as part of a HUD Sustainable Communities Grant. With the assistance of the Lane Livability Consortium, which comprises local agencies seeking to engage the entire community in planning efforts while at the same time recognizing that significant barriers exist for some community residents, the Regional Planning Grant has provided an opportunity to explore and better understand these barriers in ways that will lead to better engagement and planning processes. Work through the Consortium commenced in 2011 and concluded in 2014.

REFERENCES

Abramson, D.B., Manzo, L.C., & Hou, J. (2006). From ethnic enclave to multiethnic translocal community: Contested identities and urban design in Seattle's Chinatown-International District. *Journal of Architectural and Planning Research*, 23(4), 341–364.

Buchanan, P.J. (2011). *Suicide of a superpower: Will America survive to 2025?* New York: Thomas Dunne Books.

Burayidi, M.A. (2003). The multicultural city as planners' enigma. *Planning Theory & Practice*, 4(3), 259–273.

Catherwood-Ginn, J., & Leonard, B. (2012). Playing for the public good: The arts in planning and government. Retrieved August 29, 2013 from http://animatingdemocracy.org/resource/playing-public-good-arts-planning-and-government.

Checkoway, B. (2009). Community change for diverse democracy. *Community Development Journal: An International Forum*, 44(1), 5–21. http://dx.doi.org/10.1093/cdj/bsm018

Cleveland, W. (2011). *Arts-based community development: Mapping the terrain.* Animating Democracy. Retrieved May 27, 2015, from http://www.lacountyarts.org/UserFiles/File/CivicArt/Civic%20Engagment%20Arts%20Based%20Community%20Develop%20BCleveland%20Paper1%20Key.pdf.

Cushing, D.F., Wexler Love, E., & van Vliet, W. (2012). Through the viewfinder: Using multimedia techniques to engage Latino youth in community planning. In L. Vazquez and M. Rios (Eds.), *Diálogos: Placemaking in Latino communities* (pp. 172–185). London: Taylor and Francis.

Davis, M. (1992). Fortress Los Angeles: The militarization of urban space. In M. Sorkin (Ed.), *Variations on a theme park: The new American city and the end of public space* (pp. 154–180). New York: Hill & Wang.

Derr, V., Chawla, L., Mintzer, M., Cushing, D., & van Vliet, W. (2013). A city for all citizens: Integrating children and youth from marginalized populations into city planning. *Buildings*, 3(3), 482–505. http://dx.doi.org/10.3390/buildings3030482

Forester, J. (1998). Rationality, dialogue and learning: What community and environmental mediators can teach us about the practice of civil society. In M. Douglas and J. Friedman (Eds.), *Cities for citizens: Planning and the rise of civil society in a global age* (pp. 213–225). Chichester: John Wiley.

Forester, J. (1999). *The deliberative practitioner: Encouraging participatory planning processes*. Cambridge, MA: MIT Press.

Freire, P. (1973). *Education for critical consciousness*. New York: Seabury Press.

Gans, H.J. (1959). The human implications of current redevelopment and relocation planning. *Journal of the American Institute of Planners*, 25(1), 15–26. http://dx.doi.org/10.1080/01944365908978294

Harvey, D. (1989). *The condition of postmodernity: An enquiry into the origins of cultural change* (vol. 14). Oxford: Blackwell.

Healey, P. (1997). *Collaborative planning: Shaping places in fragmented societies*. London: Macmillan.

Healthy Eating, Active Communities Program. (2009). *Photovoice as a tool for youth policy advocacy*. Oakland, CA: Public Health Institute.

Hines, T.S. (1982). Housing, baseball, and creeping socialism: The battle of Chavez Ravine, Los Angeles, 1949–1959. *Journal of Urban History*, 8(2), 123–143. http://dx.doi.org/10.1177/009614428200800201

Huffington, A. (2010). *Third World America: How our politicians are abandoning the middle class and betraying the American dream*. New York: Crown Publishers.

Hum, T. (2010). Planning in neighborhoods with multiple publics: Opportunities and challenges for community-based nonprofit organizations. *Journal of Planning Education and Research*, 29(4), 461–477. http://dx.doi.org/10.1177/0739456X10368700

Jojola, T.S. (1998). Indigenous planning: Clans, intertribal confederations, and the history of the All Indian Pueblo Council. In L. Sandercock (Ed.), *Making the invisible visible: A multicultural planning history* (pp. 17–31). Berkeley: University of California Press.

Kennedy, R. (2013, Mar. 20). Outside the citadel, social practice art is intended to nurture. *New York Times*. Retrieved August 3, 2013 from http://www.nytimes.com/2013/03/24/arts/design/outside-the-citadel-social-practice-art-is-intended-to-nurture.html

Lichter, D.T. (2013). Integration or fragmentation? Racial diversity and the American future. *Demography*, *50*, 359–391.

Prowler, D. (2007, Jan. 1). Form foils function: How our process prevents real planning – and what we can do about it. *BeyondChron: The voice of the rest*. Retrieved May 26, 2015, from http://quartz.he.net/~beyondch/news/index.php?itemid=5803.

Qadeer, M. (1997). Pluralistic planning for multicultural cities: Canadian practice. *Journal of the American Planning Association*, *63*(4), 481–494. http://dx.doi.org/10.1080/01944369708975941

Qadeer, M.A. (2009). What is this thing called multicultural planning? *Plan Canada: Special Edition: Welcoming Communities: Planning for Diverse Populations*, 10–13.

Sandercock, L. (Ed.). (1998). *Making the invisible visible: A multicultural planning history* (vol. 2). Berkeley: University of California Press.

Sandercock, L. (2000). When strangers become neighbours: Managing cities of difference. *Planning Theory & Practice*, *1*(1), 13–30. http://dx.doi.org/10.1080/14649350050135176

Smith, N. (1996). *The new urban frontier: Gentrification and the revanchist city*. London: Routledge.

Soja, E.W. (1989). *Postmodern geographies: The reassertion of space in critical social theory*. London: Verso.

Stodolska, M., & Almeida Santos, C. (2006). Transnationalism and leisure: Mexican temporary migrants in the US. *Journal of Leisure Research*, *38*(2), 143–167.

Walter, D.S.A. (2013). Urban design interventions: An emerging strategy of arts-based social change. Master's thesis, University of Oregon.

Wang, C., & Burris, M.A. (1994). Empowerment through photo novella: Portraits of participation. *Health Education & Behavior*, *21*(2), 171–186. http://dx.doi.org/10.1177/109019819402100204

Umemoto, K. (2001). Walking in another's shoes: Epistemological challenges in participatory planning. *Journal of Planning Education and Research*, *21*(1), 17–31. http://dx.doi.org/10.1177/0739456X0102100102

Umemoto, K., & Igarashi, H. (2009). Deliberative planning in a multicultural milieu. *Journal of Planning Education and Research*, *29*(1), 39–53. http://dx.doi.org/10.1177/0739456X09338160

US Census Bureau. (2010). American Community Survey, 2006-2010. Retrieved August 5, 2013 from http://factfinder.census.gov/faces/tableservices/jsf/pages/productview.xhtml?pid=ACS_10_5YR_B05002&prodType=table.

Yonder, A. (Ed.) (2005). Arts, culture, and community. *Planners Network: The Magazine of Progressive Planning*, 165. Retrieved August 7, 2013 from http://www.plannersnetwork.org/2005/10/fall-2005-arts-culture-and-community/.

13 Religious Clusters and Interfaith Dialogue

SANDEEP K. AGRAWAL

Introduction

Canada is a multicultural and multi-religious society committed to religious pluralism. Consistent with changing immigration patterns, growing proportions of the population report religious affiliations other than Christian, including Muslim, Hindu, Sikh, and Buddhist.[1] The changing composition of immigrants resulting in increased religious pluralism in Canada is further evidenced by the proliferation of non-Christian places of worship. This is most visible in Greater Toronto and Greater Vancouver, where almost half of all immigrants live. Greater Toronto hosts the "largest conglomeration of ethnic minority places of worship" in Canada (Agrawal, 2009, p. 64). Indeed, as Beyer (2005) notes, "multi-faith religiousness in Canada is overwhelmingly a development that affects the large urban agglomerations [with] 90 percent of those who identify with non-Christian world religions [living] in the six largest metropolitan areas," especially Toronto and Vancouver (pp. 168–69).

The growth of religious plurality and geographic proximity raises important questions about religious tolerance and conflicts (Keaten & Soukup, 2009). In the US, for example, the location of mosques has been a controversial land use issue, especially following the terrorist attacks of 9/11. The planning profession is challenged to accommodate religious places of worship, especially those of new immigrants whose religions differ from the Judeo-Christian faiths (Agrawal, 2009). Neighbourhood resistance to the location of such places of worship, among other factors, has been a major stumbling block. In a bid to solve this problem, planners in Canadian metropolitan areas have resorted to siting religious places in vacant lands originally zoned as industrial or agricultural (Agrawal, 2009). Over time, such planning actions have led to the formation of religious clusters in a few places where multiple institutions of different faith groups reside in close proximity.

Religious land uses take many forms, sizes, and configurations, from start-up temples or mosques in rented or purchased industrial, commercial, or residential units, renovated warehouses, churches, or other buildings to major works of religious architecture, large mixed-use religious campuses, or clusters of places of worship (Hoernig, 2009). Faith-based neighbourhoods, growing around places of worship, are another form of religious development (Agrawal, 2008). This chapter explores the role of planning policies and the physical proximity of places of worship in facilitating interfaith dialogue. The research focused on two religious clusters, perhaps unique in North America: one in the Greater Toronto Area, which emerged due to incremental zoning changes over time; and the other in Greater Vancouver, which came about by a deliberate policy of the local government. Using key informant interviews as the method, and Allport's contact hypothesis as the guiding theoretical framework, the study explored the effects of proximity and contact (interaction or encounters) on intergroup relations. We first review Allport's contact hypothesis as it relates to difference before returning to a discussion of the findings from the study.

Interfaith Dialogue and Allport's Contact Hypothesis

The clustering of religious institutions has led to the geographic proximity of different religious institutions in space. While this proximity may enhance religious encounter and dialogue, it is by no means a guarantee. Interfaith dialogue is defined as a discourse between individuals or groups who hold differing religious beliefs and convictions for the purposes of understanding their faiths (Mojezs, 1998). It is a "conscious process, through which deliberate efforts toward understanding the religious stranger are implied," which differs from a simple encounter in that the latter is not rooted in a *deliberate* interaction (Urbano, 2012, p. 150). Interfaith dialogue emphasizes cooperation and the exchange of theological beliefs. It is an organized formal dialogue to "promote respect for difference, encourage cooperation, and overcome conflict" (Seljak, 2009, p. 22) by exploring values and challenges common among participants as well as exposing prejudices (Smock, 2003).

Interfaith dialogue has been proposed as a way to counter hostile attitudes held towards religious "others"; it is a "conscious effort to step out of our comfort zone and to find inspiration from those who come from a world different than our own" (Mews, 2006, p. 80).

In this chapter, interfaith dialogue is defined as an *intentional* encounter between individuals who adhere to differing religious beliefs and practices in an effort to foster respect and cooperation among these groups through organized dialogue. Allport's contact hypothesis is employed as the guiding framework of discussion. Contact hypothesis pays particular attention to diversity, communication, and interaction – the principles on which interfaith dialogue is based. The hypothesis emerged after the Second World War in an effort to alter negative stereotypes of racial groups, shaped by the belief that ignorance gives rise to prejudice and that if individuals across diverse "group lines" became better acquainted with each other, they would discover their common humanity (Pettigrew & Tropp, 2006).

Gordon Allport (1954), in his seminal work exploring the nature of prejudice, introduced contact hypothesis as a mode of analysis to investigate the role of contact in reducing prejudice and minimizing conflict. According to his hypothesis, four key conditions are necessary for a positive attitude shift by intergroup members to reduce prejudice and increase cohesion. These are equal status, common goals, intergroup cooperation, and the support of authorities, laws, and customs.

While contact hypothesis was developed largely to explore and explain segregation and anti-Black prejudice, its scope has widened over the years. Today, it "remains the main theoretical framework for mixing and desegregation" (Wood & Landry, 2008, p. 107). Within its larger application, this hypothesis has most commonly been used to explore the relations between diverse racial, ethnic, and cultural groups (Laurence, 2011). Nonetheless, only a few studies of interfaith encounters have applied the contact framework of analysis (Hemming, 2011; Ipgrave, 2002; Connolly, 2000).

Migration affects the religious landscape in receiving countries, changes the cultural landscape, and results in encounters between host communities and diasporic religious communities (Kong, 2010; Peach, 2002). Yet, there is a dearth of planning literature on religious relations in plural societies. Yiftachel's (1992) study is one of the few comprehensive works that utilizes ethnic and religious relations as a framework for analysis in planning. Winkler's (2006) work in Johannesburg, South Africa, shows that the secular values of planning theory and practice are often indifferent to – or at times in conflict with – the religious beliefs of the majority of citizens on behalf of whom planners make decisions.

A recent study by Dwyer, Gilbert, and Shah (2012) draws attention to the way the sacred has been constructed as an incongruous use in the suburbs. The authors point out that the construction of religious buildings in suburban landscapes is "out of place," a discourse situated within the historical theorization and articulation of the suburbs as a site of modernization, materialism, and secularization. Dwyer et al. (2012) use this incongruity as a starting point to address the complexities of relationships between religion and suburban space. Gale and Naylor (2002) also address the inaptness of minority places of worship in urban landscapes, exploring the intersection of planning, policy, and individuals of difference.

A few works (such as Agrawal, 2008; Kong, 2010; Ley, 2008; and Nye, 1993, as cited in Kong, 2001) attest to the significance of religious minority places of worship in (sub)urban settings as markers of permanence and belonging, as well as an assertion of immigrants' claim to space. Other studies on the nexus of religion, immigration, and "discriminatory" planning policies are limited to the sites on which places of worship are built and the land use and design problems associated with their construction. The challenges posed by multiple religions and faiths to contemporary urban planning first received attention in Great Britain (Thomas & Krishnarayan, 1994). In his subsequent publications (Hutchings & Thomas, 2005; Thomas, 2008), Thomas draws attention to the importance of faith groups in the planning process.

In Canada, studies have mostly focused on difficulties encountered by places of worship in the planning approval process (Agrawal, 2009; Beattie & Ley, 2001; Germain & Gagnon, 2003; Isin & Siemiatycki, 1999). Gagnon, Dansereau, and Germain's (2004) work on Hasidic Jews in metropolitan Montreal draws on case studies to illuminate the complex relations between people of difference who occupy the same social space, and the role of urban planners in mitigating such land use conflicts.

Dwyer, Tse, and Ley's (2013) research uses No. 5 Road in Richmond, British Columbia, one of the two clusters in this chapter, as a case study to explore the role of religious institutions in immigrant integration. The study concludes that these religious institutions contribute to the integration of immigrants in multiple ways. They facilitate transnational activities such as international migration and the transfer of funds between countries and intra-communal services. They also provide local networking opportunities for recent immigrants. The authors suggest that No. 5 Road provided an educational resource for city residents about diverse religions and cultures. They also point to City Museum's efforts to host exhibitions as a good example of telling the story of No. 5 Road: it "is not simply about preserving a historical record; it is also grounding discussions about religion, ethnicity, and integration in concrete events, instead of letting the conversation remain at an abstract level" (p. 49).

The two studies closest to the topic at hand are those of Hoernig (2009) and Agrawal (2008). Hoernig's study explored various forms of religious development in Canada and their implications for urban planning. Agrawal (2008) in his study explored whether religion influences neighbourhood location decisions and whether it contributes to the social capital within so-called faith-based neighbourhoods in the Greater Toronto Area. His study suggests that faith is not an all-encompassing characteristic of these neighbourhoods and has a weak influence in promoting neighbourliness. Another related study, Hackworth and Stein (2012), documented the rapid expansion of places of worship in Toronto's industrial lands and the mounting tension between faith and economic development.

Religious Clusters and Interfaith Dialogue

This chapter focuses on two religious clusters as case studies. These clusters originated under different planning regimes: one in a suburban municipality in the Greater Toronto Area and the other in a suburban municipality in Greater Vancouver. These constellations of religious institutions are likely the only ones of their kind in North America and thus present an opportunity to test their usefulness in promoting interfaith understanding.

The religious cluster in the Toronto area is located on a quiet street – Professional Court – off a major arterial road in the city of Mississauga, Ontario (see figure 13.1). The area is zoned light industrial, and six places of worship are located there. The places of worship came about through a zoning designation change from industrial to institutional on a case-by-case basis over a period of 20 years. The six ethnic places of worship include a Korean church, an Orthodox Indian church, a Jamaican Pentecostal church, two Indian temples, and a mosque. All of these places of worship except the Korean church participated in the study.

A cluster of about 24 religious assemblies is located in the city of Richmond, British Columbia, along one side of No. 5 Road, a three-kilometre

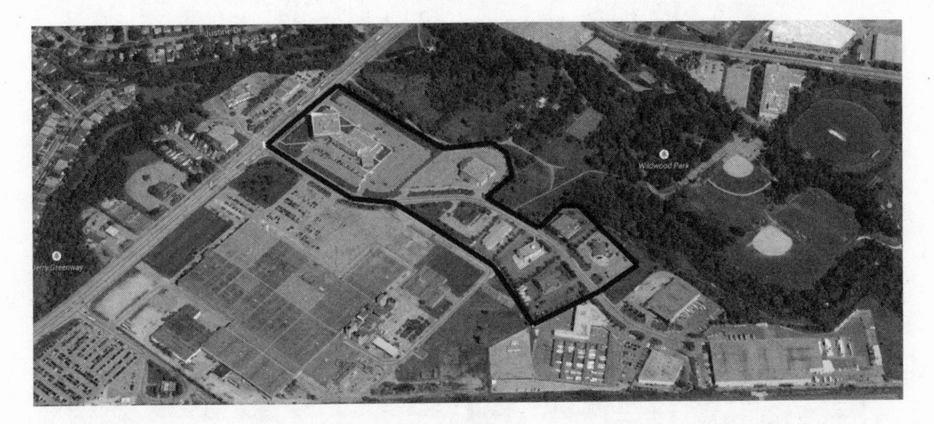

Figure 13.1. The religious cluster along Professional Court, City of Mississauga. The buildings within the black outline show the religious uses on the site. Map data: Google.

stretch between Blundell Road and Steveston Highway (see figure 13.2). This linear cluster is colloquially referred to as "Highway to Heaven." In 1990, in the absence of any available developable land within the city limits and given the increasing demand for new places of worship, the city of Richmond, in consultation with the province's Agricultural Land Commission, established a back lands policy which rezoned lands along the eastern side of No. 5 Road from agricultural to assembly use, including places of worship, schools, and other public use facilities. However, the assembly use of the land was limited to a depth of about 360 feet from No. 5 Road and the remaining back portion of the property was to be dedicated to active farming. A variety of faith groups are now represented there: Christian, Jewish, Buddhist, Muslim, Sikh, and Hindu. The site includes two mosques, eight Christian churches (six of which are Chinese congregations), three Buddhist temples, two Hindu temples, a Sikh gurudwara, and six religious schools.

This study employed a qualitative methodological approach to gather data through semi-structured interviews with key informants. The interviews elicited insight into the informants' experiences of worshipping in close physical proximity with diverse faith groups as well as their involvement in and perception of interfaith activities.

In the Toronto area, eight key informants were interviewed. Three of them were interfaith advocates and practitioners who helped to

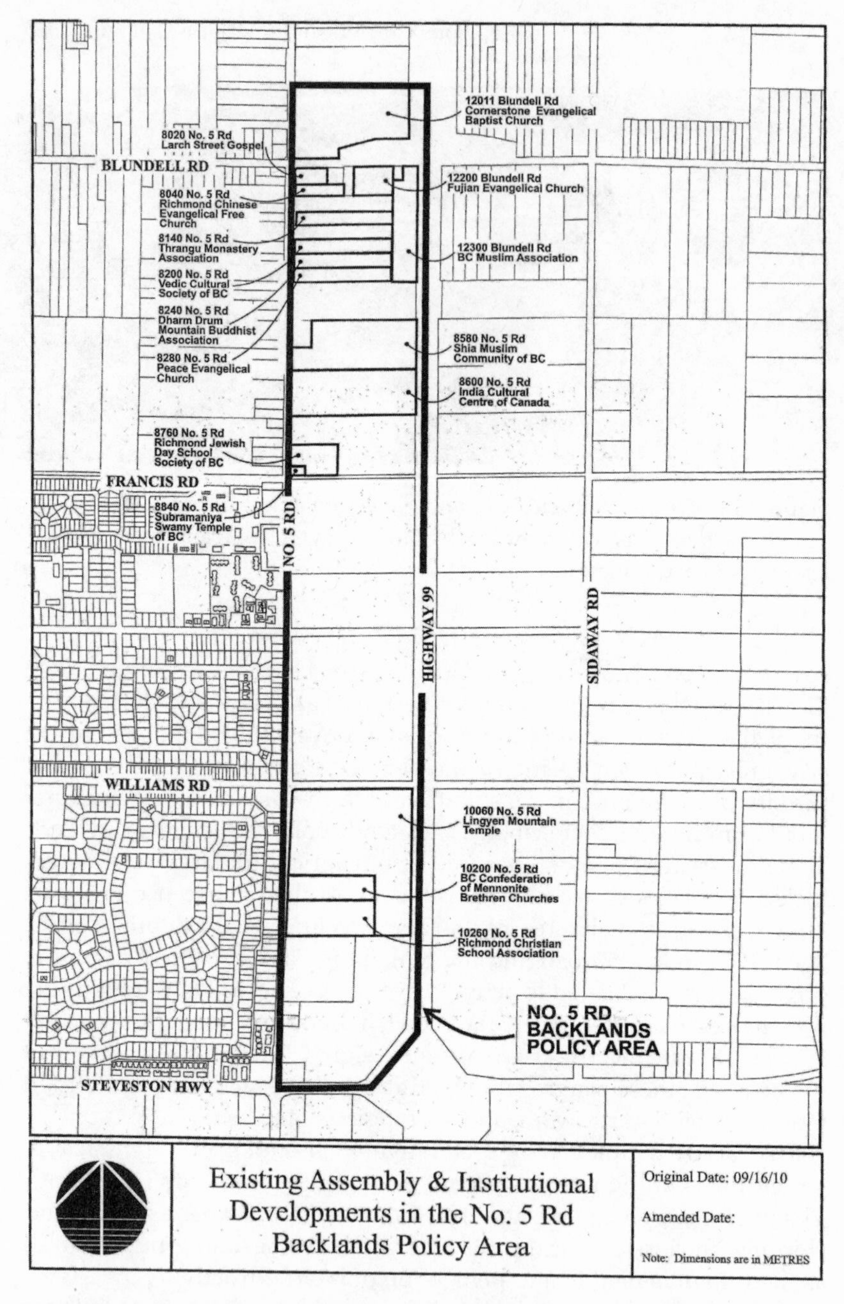

Figure 13.2. The religious cluster along No. 5 Road in the Greater Vancouver area. Note the different cultural uses, including the India Cultural Centre of Canada, the BC Muslim Association, the Thrangu Monastery Association, and the Neighbourhood Christian School. Courtesy City of Richmond, British Columbia, Canada.

document their experience in interfaith work, explore whether geographical proximity was a factor in interfaith dialogue, and identify the positive benefits or tangible outcomes of participating in interfaith dialogue. The five other participants comprised members of boards of directors and representatives of the religious institutions on Professional Court. These five provided insight into the effects on interfaith dialogue of physical proximity among diverse places of worship in the cluster.

In the case of Greater Vancouver, nine individuals were interviewed. Seven of them were leaders or representatives of religious institutions and schools on No. 5 Road; the other two were Richmond officials who were knowledgeable about the area. The members of the religious institutions helped document their experience of being in geographic proximity to other places of worship and identify benefits, if any, in promoting interfaith dialogue. The city officials provided the background and the policy context for the emergence of the cluster.

Three major findings emerged from the study: 1) community services such as parking and municipal projects provide an opportunity for informal encounters between members of different religious groups; 2) diverse religious groups cooperated on matters of common interest such as disaster relief; and 3) government programs are not a preferred way to promote interfaith dialogue. We discuss each of these findings in turn.

Points of Interaction

The literature suggests that spaces that are shared by individuals of difference could offer an environment in which encounters can occur. These are encounters between individuals who may not otherwise interact. By being in close geographical proximity, the places of worship within the cluster create a unique opportunity for people to engage with the religious "other." Community service, parking, and municipal projects provided the points of interaction for the faith groups.

i) Community Service

The existence of common social goals between two or more religious groups provides an opportunity for intergroup contact. Common goals can foster mutual understanding and cooperation between diverse groups who may otherwise not interact or work together on projects.

In turn, this contact and joint effort could produce an environment in which prejudice is reduced and relationships are established.

Non-denominational community service surfaced as an important point of collaboration among the faith groups studied. All five religious establishments on Mississauga's Professional Court were involved in community service work and so were the institutions on No. 5 Road in Richmond. This took one of two forms: international relief work or local community outreach activities. Adherents engaged in international work such as organizing to send relief aid for natural disasters; for instance, when the tsunami struck Japan or when large parts of India were affected by floods. Local community involvement often took the form of distribution of food through a food bank operated by the place of worship, helping the homeless or others in need beyond their own congregation and faith, or contributing to charity events for a sick kids' hospital or a cancer foundation.

All five of the religious leaders on Professional Court commented on the possibility of collaboration for social service work. Ujwal,[2] a member of the board of directors of one of the religious institutions, recognized the existence of common goals: "Often enough there are similarities between our operations and our missions." Pradyumn, a representative of a different religious organization, explained the collaboration between his temple and the neighbouring Orthodox Church:

> And then we also worked together during the tsunami … We [collected] some food, donated to the appropriate association, and recently we also raised [money] during December for homeless people. [We supported] the soup kitchen [by organizing] a drive to donate all the canned food. We [did] that together.

This informal collaboration, albeit at a low level, is also evident at the mosque, according to Mohammad, who represents the local mosque: "Even with the church next door, they have a food bank also, so in the event they should be short of anything … we can help them out with that and vice versa." Mohammad noted that these charity drives are two "individual things," that is to say that there is no *official* collaboration between the faith groups. However, it is the physical proximity of the church that made this exchange possible. Collaboration between the faith groups located within the cluster is an example of "bridging social capital" (Laurence, 2011), which could lead to improved attitudes between the groups.

The sentiments expressed by Richmond's No. 5 Road respondents were no different, but most of their community work seemed directed towards their own congregations with very little or no collaboration with others. One institution did, however, highlight making donations for a cancer foundation and bringing guest speakers to learn about other religions as a way to serve the larger common good.

When asked whether working for the larger social good could be used as a point of collaboration, a pastor of a religious institution on No. 5 Road responded: "It is possible to come together for a higher social good – goodwill towards human kind. You can rally people around that idea. For instance, the issue of abortion can bring Muslims and Christians together ... If something threatens our existence, a non-religious matter, then that can bring us together. For instance, if there is a threat to the practice or expression of religion, it will bring us together." Another pastor from the same institution, however, contradicted this: "We cannot take out religion, even for the social good. Social good comes out of our theological values."

From these statements, we can conclude that the emergence of a common vision can lead to collaboration between diverse faith groups in a manner that is not necessarily related to their faith but is rooted in meeting a social need. Also, religious groups that differ in their beliefs are willing to discuss matters that may threaten their common existence, such as restrictions of their religious freedom. This buttresses the point made by Harris and Young (2009) that in order to bridge the gap between people of diverse backgrounds, it is necessary to focus on shared visions and tasks.

ii) Parking

Parking was an issue that was mentioned by almost all religious leaders in both clusters and was identified as a common problem experienced by many places of worship. Consider Mohammad's statement with respect to Professional Court:

> We have a very, very good relationship in terms of not necessarily meeting frequently or having frequent dialogues as such, but because one of the issues we have in many places of worship is parking ... Because our events happen to not coincide at the same time, we have an understanding with the other community groups to utilize their parking facility in the event we have an overflow ... At the very beginning, all stuck together and said this is okay.

Mohammad's statement makes it clear that parking offers an opportunity for interaction. While mundane in nature, concerns over parking created a space for the diverse faith leaders to connect and dialogue as neighbours. Thompson, a representative of a place of worship on Professional Court, noted that parking quickly became an issue which the religious leaders realized they could work together to address:

> Well, it's a mutual thing, because I approached [the mosque] first ... I introduced myself [and] where I am coming from. And we decided to work together. He said, "No problem, you can use my parking." This is how we got to know each other.

At No. 5 Road, Shalom, the principal of a religious school, responded similarly: "Our relationship with our neighbour goes as far as sharing parking. It just started last year when I walked into the temple and introduced myself." Similar examples were cited on key religious festivals when neighbouring institutions offered their parking lots for worshippers of different faiths. Using each other's parking lots was evidence, as one interviewee told us, of the level of trust between the different religious communities.

iii) Infrastructure Service

Another major instance of collaboration on No. 5 Road was the installation of a main sewage pipe initiated by the city in the early 2000s, which required the neighbouring religious communities to contribute funds towards its provision. One respondent mentioned how his establishment allowed his immediate neighbour to temporarily flow the sewage through their property until the sewer lines were put in place. However, not everyone agreed with their neighbours, and the cooperation lasted only a short time. Disagreements arose over whether financial contributions by each institution for the sewer line should be based on the frontage of the lot, which is the usual practice, or the size of the facility.

Other expressions of collaboration existed in the form of sharing private infrastructure such as a gymnasium facility. An early city initiative for the construction of a highway interchange brought institutions of different religions together but was also short-lived (Dwyer et al., 2013). An early initiative on the interchange linked a number of institutions together under the leadership of the British Columbia Muslim

Association, which was severely affected by the project, but they could not convince other immediate neighbours to join them.

iv) Disaster Management

Responses to local disaster management planning also prompted institutions to collaborate and cooperate. One of the religious schools on No. 5 Road pointed to a nearby religious institution as part of its disaster management plan. In the event a natural disaster strikes, the students will leave the school and take refuge there. The cooperating religious institution was quick to emphasize this arrangement as a sign of cooperation among neighbours. A puzzling part, however, was that the school representative referred to this institution by an incorrect name and an incorrect religious affiliation. Perhaps it was an honest mistake or perhaps it is indicative of how little these neighbours know each other.

v) Development and Expansion Plans

An ongoing issue concerning the role of the religious institutions along Richmond's No. 5 Road is safeguarding the agricultural land reserve and restricting the expansion of existing development. The back lands policy required the institutions to actively farm the land to the back of their properties. In some cases this has been a source of conflict between the religious institutions and the city. Many institutions are successfully growing fruits and are engaged in active farming. However, not all of the institutions have fulfilled the expectation of using their land for agricultural purposes.

For instance, one of the institutions submitted an application to expand its facility deeper into its agriculturally zoned land and erect a statue. This would have defied the height constraints in the area, especially because it is so close to the Vancouver airport. Although the application was subsequently withdrawn, it gives us an interesting peek into neighbourhood politics and the reasons why sometimes neighbours come together. The opposition to the application came mainly from the residential neighbourhoods on the opposite side of No. 5 Road and not from the religious institutions, which would have been equally affected by the development. The members of a nearby religious institution admitted to supporting the application despite its excessive size and height because they thought that, if the application was approved, it would help them get their own expansion plans cleared by the city.

The above discussion provides examples of shared "functional" space and infrastructure that are used by religious institutions in the neighbourhood. The necessity for increased parking, sharing the cost of infrastructure, and preparation for a natural disaster led to informal collaboration between the groups. These encounters were not premised on religious exchange. Nonetheless, the examples illustrate the way in which the sharing of space and infrastructure – public or private – brings about interactions and encounters between members of diverse ethnoreligious groups. The cases of sewage lines, road improvements, and development applications demonstrate that such interfaith collaborations were often one-off projects and that their failures and successes were usually not the product of theological conflicts. Shared parking in both clusters is thus far the only example of sustained engagement by the religious institutions.

Seeing the "Other" as Human

The contact hypothesis articulates the power of contact and interaction to alter existing prejudices, particularly towards out-groups. This study also showed that encounters and interactions among the different faith groups have a "humanizing" effect. Mohammad touched on how the physical proximity of places of worship allows him to see his diverse faith neighbours as humans and not just as religious individuals.

> Having us on the same block, so to speak, allows me to see how our neighbours interact outside of their place of worship, and so even without us even saying anything, it's something you can see, right? And to me, this is something probably a bit more beneficial than arranging a meeting to be talking all the time.

Pradyumn also noted a common humanity to which all belong:

> As you know or learn about another religion your thinking about that religion changes, and the same thing happened to me as well. My thinking, rather my respect for Islam and Christianity, has really gone up. Basically, what I learned is that though they are Muslims, Christians, or Hindus, the first thing is that we are all human beings. And as human beings, we all

> three different organizations are helping another human being ... I didn't have that kind of understanding before.

It is important to note that the above comments were made by respondents at Mississauga's Professional Court. Respondents at No. 5 Road made it clear that they respected their neighbours as human beings and as people of other faiths, but that the views they held had nothing to do with the proximity to the other religious groups.

Religion as a Point of Dissonance

One finding that emerged from this study is the pessimism held by leaders of the places of worship regarding interfaith dialogue. Participants echoed concerns that religion can be a possible point of contention and contestation, which may lead to hostile relations between the groups. There appeared to be consensus that religion is a "hot" topic that can divide people more than unite them. Participants spoke about religious intolerance and conflict on a national and global level and drew attention to the difficulties that might arise if a meeting between faith groups was premised on understanding each other's religion. Consider, for example, Pradyumn's explanation of the relationships between his temple and the neighbouring places of worship on Professional Court:

> You know, I really feel proud when I see this kind of thing [collaboration between the neighbouring faith groups over parking] happening here, that two different religions or three different religions are working so closely with each other. I mean ... there is lots of tension going on across the world ... mainly because of religion.

Paul, an interfaith advocate, echoed similar sentiments, showing apprehension about the discussion of the various religious tenets: "But when it comes to religion ... most of the bad things happening in our world is because of religious belief ... so we don't want to get involved in that."

On Richmond's No. 5 Road, Pastor Mike said, "The problem with people not coming together is religion ... It is religion that is keeping people in the silos." John, another pastor, agreed with Pastor Mike that religion keeps people in silos: "Interfaith activity diminishes the identity of a religion." Along the same lines, Lama, a Buddhist master, opined, "We do not need interfaith dialogue. Religion is not like shopping to choose what you like. It's better not to compare, but to respect each other."

While some religious leaders in both clusters were open to further interaction with their faith neighbours, they were not certain that this encounter should occur under the auspices of a formal interfaith dialogue. As one religious leader put it, "Religiously, you're not going to see eye to eye ... but socially ... it is okay, something that involves the community."

These responses illustrate a lack of receptivity on the part of religious leaders to engage with individuals of difference on matters of faith. This would most closely align with the exclusivist model of interfaith dialogue, which is characterized by claims of the religious uniqueness of one's own revelation and "oneness and onlyness" as the language of identity (Eck, 2003; McCarthy, 1998). According to this model of interfaith dialogue, individuals maintain a conviction in their own path, while rejecting and claiming the inadequacy of all other paths. This approach is the least conducive to dialogue, as the exclusivist member will engage in order to convert rather than to understand (Abu-Nimer, Khoury, & Welty, 2007).

In this context, it would seem unlikely that the participants would be receptive to dialogue on religious matters. Thus, the religious leaders in this study demonstrated an interest in furthering their knowledge about religious out-groups but tended to engage in dialogue for non-religious purposes.

The Role of Government

The position that religious leaders maintain on government involvement in interfaith work is of interest, especially in the context of interfaith dialogue as a formal, organized initiative. When asked about government's involvement in interfaith dialogue initiatives, many on Professional Court thought the role of government should be minimal. One respondent said, "Politics and religion don't mix." Another stated, "It [government's involvement] becomes too formal, too political, too much red tape." Two other religious leaders favoured a limited role for government, which would largely involve providing funding and support for interfaith initiatives.

An official from Richmond, who was responsible for coordinating the efforts of the city council and faith groups, argued that government funding and government initiative led to interfaith dialogue on shared issues such as perception, religious intolerance, discrimination, and racism. Funding through a government program, Embrace BC (a provincial fund for anti-racism initiatives including interfaith bridging), helped create an Interfaith Committee in Richmond, BC, in 2009, which initiated a number of programs for No. 5 Road. However, when the provincial funding dried up, the committee became defunct and with it died the interfaith initiatives.

To illustrate his point, the city official cited development of Temple Tours of the institutions on No. 5 Road as evidence of the work of the Interfaith Committee. He also mentioned the city-organized Doors Open Richmond program, the exhibition about the religious heritage on No. 5 Road at the City Museum, and the plans to promote No. 5 Road as a tourist destination. He believed there was a willingness among No. 5 Road institutions to participate in the interfaith dialogue and activities, but it required sustainable government funding along with the involvement of a secular coordinating agency to organize and promote interfaith dialogue.

Interestingly, in response to the question of government involvement, while a few religious leaders on No. 5 Road remembered the works of the Interfaith Committee, others did not acknowledge it or mentioned it in passing in discussions on interfaith dialogue. Many rejected outright the idea of having a committee to promote interfaith dialogue.

Participants' assertions that external involvement should be minimal are countered by Seljak (2009), who argues that Canada is not keeping pace with other Western nations that receive high levels of religious minorities and that policy makers should take heed of this issue. He suggested that one function of an interfaith council is to "promote understanding and cooperation among various religious groups" (p. 28). Seljak's work is of interest in this conversation because he recognizes the importance of interfaith initiatives in cultivating relations between individuals and promoting religious identity; but this study suggests that this kind of encounter is already occurring, although informally. There seems to be some support, albeit limited, among the religious leaders of the two clusters for an institutionalized interfaith organization.

Lessons from Religious Cluster and Interfaith Dialogue

The study shows that the effect of geographic proximity through clustering on interfaith dialogue is weak. Proximity does not appear to be a significant factor in the establishment of "organized" interfaith initiatives.

Planning actions can therefore only go so far in promoting interfaith dialogue. Proximity does, however, play a role in increasing contact between faith groups, and it appears that these encounters are influential in improving interfaith relations, as evidenced by increased interaction and cordial relations, and in altering attitudes about the religious "other." Proximity is also a factor in promoting interfaith encounters and interactions. These encounters are influential in altering attitudes and perspectives about an out-group.

Proximity then has the potential to facilitate the process of breaking down barriers and building bridges between diverse communities, including faith communities. Proximity can also minimize the segregation and avoidance between groups living in diverse communities, a concept put forth by Cantle (2005). Despite the increasing diversity in multicultural, cosmopolitan urban centres, there is "evidence of considerable levels of separation to the extent that the relationship between communities might also be characterized as one of 'parallel lives'" (Cantle, 2005, p. 14).

The faith groups ran a number of similar yet distinct social and community programs and worked together infrequently and at a low level. Municipal projects and plans did corral the neighbours together but only for the short duration of the projects. Parking seemed to be a major area of sustained collaboration between the places of worship within the clusters, suggesting that the sharing of functional space can provide a point of encounter between the faith groups. This type of encounter can be situated within the discourse on proximity in the context of a modern (sub)urban neighbourhood. As Qadeer and Kumar (2006) note, "the contemporary neighbourhood is a community of polite but limited social relations. People normally have a largely nodding acquaintance with other residents of their street" (p. 15).

The social capital in a faith-based institutional cluster seems no different from that of any suburban residential neighbourhood. While participants mentioned that their close physical proximity with the diverse religious organizations enabled them to see their neighbours as humans rather than just the "religious other," it did not seem to foster greater bonds between religious leaders or adherents. None of the participants mentioned developing friendships with members of another faith community, and simultaneously they expressed little interest in or need or desire for interfaith dialogue.

Finally, the study indicates that the religious leaders did not fully support interfaith dialogue or any institutionalized interfaith initiatives.

Participants in this study envisioned an encounter rooted in community service and premised upon a clearly defined social goal, rather than one related to religion or theology. This finding is significant because studies show that out-groups working together towards a common goal are linked by mutual rapprochement, which contributes to a decrease in prejudiced attitudes or hostile relations (Amir, Ben-Ari, & Bizman, 1985, p. 213). The presence of common goals and values unrelated to a theology is significant in that it is a domain of community cohesion (Cantle, 2005).

From a planning perspective, it is important to note that neither of the two clusters emerged with the intention of promoting interfaith dialogue. The planning policy that created No. 5 Road was primarily intended to solve the problem of exponentially increasing demand for new places of worship and the threat of development pressure on precious agricultural land. In other words, No. 5 Road was not unplanned but was intentionally created, albeit accidentally, by a distinctive planning policy. However, the consequence is a landscape of religious diversity, where buildings from different faith communities stand somewhat incongruously side by side. Unlike No. 5 Road, the clustering along Professional Court did not happen by a deliberate policy, but occurred through site-by-site zoning changes over time.

All in all, development projects such as road and sewer line improvements and disaster management plans brought the neighbours together to work towards a common cause. The contacts and interactions developed through these projects could be channelled into a much more concrete and sustainable engagement to stimulate interfaith dialogue. Planning actions can thus provide harmonious multifaith neighbourhoods if they are combined with interfaith initiatives coordinated by a government-funded program run by a secular civil society organization.

This study demonstrates that while the original intention was not the creation of multifaith communities, much less a vehicle for interfaith dialogue, both places have potential for building an environment for such dialogue. However, to maximize its utility, sustained funding for coordinating the interfaith activities is needed. Planners can encourage participation of faith groups in public hearings and discussions on matters that are common or concerning to all groups. Working together on such projects may increase appreciation for persons in different faith groups without interfering with their individual religious identities. Clustering places of worship as a deliberate planning activity could promote interfaith encounters that could decrease prejudice and increase appreciation for persons of other religious faiths, even if it does not result in interfaith dialogue.

NOTES

1 In 2011, according to the National Household Survey, the largest faith in Canada was Christianity. About two-thirds of Canada's population (67.3 per cent) reported that they were affiliated with a Christian religion. Roman Catholics were the largest Christian religious group in 2011. About 38.7 per cent of the population identified themselves as Roman Catholic, down from 43 per cent a decade ago; 7.2 per cent of Canada's population reported affiliations with Muslim, Hindu, Sikh, or Buddhist religions, up from 4.9 per cent a decade ago. People who identified themselves as Muslim made up 3.2 per cent of the population (2 per cent in 2001), Hindu 1.5 per cent (1 per cent in 2001), Sikh 1.4 per cent (0.9 per cent in 2001), Buddhist 1.1 per cent (1 per cent in 2001), and Jewish 1.0 per cent (1.1 per cent in 2001). Each of them saw an increase except for Jewish.

2 Pseudonyms are used to protect the privacy of individuals interviewed.

REFERENCES

Abu-Nimer, M., Khoury, A., & Welty, E. (2007). *Unity in diversity: Interfaith dialogue in the Middle East*. Washington, DC: US Institute of Peace.

Agrawal, S. (2008). Faith-based ethnic residential communities and neighbourliness in Canada. *Planning Practice and Research*, *23*(1), 41–56. http://dx.doi.org/10.1080/02697450802076431

Agrawal, S. (2009). New ethnic places of worship and planning challenges. *Plan Canada: Special Edition: Welcoming Communities: Planning for Diverse Populations*, 64–67.

Allport, G.W. (1954). *The nature of prejudice*. New York: Addison-Wesley Publishing.

Amir, Y., Ben-Ari, R., & Bizman, A. (1985). Prospects of intergroup relations in an intense conflict situation: Jews and Arabs in Israel. *Journal of Asian and African Studies*, *20*(3), 203–217. http://dx.doi.org/10.1177/002190968502000307

Beattie, L., & Ley, D. (2001) *The German Immigrant Church in Vancouver: Service provision and identity formation*. Working Paper Series, October, Research on Immigration and Integration in the Metropolis, Vancouver.

Beyer, P. (2005). The future of non-Christian religions in Canada: Patterns of religious identification among recent immigrants and their second generation, 1981–2001. *Journal of Religion/Sciences*, *34*(2), 165–196. http://dx.doi.org/10.1177/000842980503400202

Cantle, T. (2005). *Community cohesion: A new framework for race and diversity*. New York: Palgrave. http://dx.doi.org/10.1057/9780230508712

Connolly, P. (2000). What now for the contact hypothesis? Towards a new research agenda. *Race, Ethnicity and Education*, *3*(2), 169–193. http://dx.doi.org/10.1080/13613320050074023

Dwyer, C., Gilbert, D., & Shah, B. (2012). Faith and suburbia: Secularisation, modernity and the changing geographies of religion in London suburbs. *Transactions of the Institute of British Geographers*. Abstract retrieved from Wiley Online Library DOI: http://dx.doi.org/10.1111/j.1475-5661.2012.00521.x.

Dwyer, C., Tse, J.K.H., & Ley, D. (2013). Immigrant integration and religious transnationalism: The case of the "Highway to Heaven" in Richmond, BC. Metropolis British Columbia Centre of Excellence for Research on Immigration and Diversity, Working Paper Series 13–06.

Eck, D. (2003). *Encountering God: A spiritual journey from Bozeman to Banaras*. Boston: Beacon Press.

Gagnon, J., Dansereau, F., & Germain, A. (2004). "Ethnic" dilemmas? Religion, diversity and multicultural planning in Montreal. *Canadian Ethnic Studies Journal*, *36*(2), 51–75.

Gale, R., & Naylor, S. (2002). Religion, planning and the city: The spatial politics of ethnic minority expression in British cities and towns. *Ethnicities*, 2(3), 387–409. http://dx.doi.org/10.1177/14687968020020030601

Germain, A., & Gagnon, J. (2003). Minority places of worship and zoning dilemmas in Montreal. *Planning Theory & Practice*, *4*(3), 295–318. http://dx.doi.org/10.1080/1464935032000118652

Hackworth, J., & Stein, K. (2012). The collision of faith and economic development in Toronto's inner suburban industrial districts. *Urban Affairs Review*, *48*(1), 37–63. http://dx.doi.org/10.1177/1078087411420374

Harris, M., & Young, P. (2009). Developing community and social cohesion through grassroots bridge-building. *Policy and Politics*, *37*(4), 517–534. http://dx.doi.org/10.1332/030557309X435529

Hemming, P.J. (2011). Meaningful encounters? Religion and social cohesion in the English primary school. *Social & Cultural Geography*, *12*(1), 63–81. http://dx.doi.org/10.1080/14649360903514384

Hoernig, H.J. (2009). Planning amidst cultural diversity: Lessons from religious development. In S. Agrawal, C. Andrew, & J. Biles (Eds.),

Plan Canada: Special Edition: Welcoming Communities: Planning for Diverse Populations, 55–59.

Hutchings, E., & Thomas, H. (2005). The business case for equality and diversity: A UK case study of private consultancy and race equality. *Planning Practice and Research*, *20*(3), 263–278. http://dx.doi.org/10.1080/02697450600568605

Ipgrave, J. (2002). Interfaith encounter and religious understanding in an inner-city primary school. PhD thesis. University of Warwick, Coventry.

Isin, E.F., & Siemiatycki, M. (1999). *Fate and faith: Claiming urban citizenship in immigrant Toronto*. CERIS Working paper. Retrieved November 12, 2012 from http://ceris.metropolis.net/wp-content/uploads/pdf/research_publication/working_papers/wp8.pdf.

Keaten, J.A., & Soukup, C. (2009). Dialogue and religious otherness: Toward a model of pluralistic interfaith dialogue. *Journal of Intercultural Communication*, *2*(2), 168–187. http://dx.doi.org/10.1080/17513050902759504

Kong, L. (2001). Mapping "new" geographies of religion: Politics and poetics in modernity. *Progress in Human Geography*, *25*(2), 211–233. http://dx.doi.org/10.1191/030913201678580485

Kong, L. (2010). Global shifts, theoretical shifts: Changing geographies of religion. *Progress in Human Geography*, *34*(6), 755–776. http://dx.doi.org/10.1177/0309132510362602

Laurence, J. (2011). The effect of ethnic diversity and community disadvantage on social cohesion: A multi-level analysis of social capital and interethnic relations in UK communities. *European Sociological Review*, *27*(1), 70–89. http://dx.doi.org/10.1093/esr/jcp057

Ley, D. (2008). The immigrant church as an urban service hub. *Urban Studies (Edinburgh, Scotland)*, *45*(10), 2057–2074. http://dx.doi.org/10.1177/0042098008094873

McCarthy, K. (1998). Reckoning with religious differences: Models of interreligious moral dialogue. In S. Twiss & B. Grelle (Eds.), *Explorations of global ethics* (pp. 73–117). Boulder: Westview Press.

Mews, C.J. (2006). The possibilities of interfaith dialogue. *Meanjin*, *65*(4), 78–80.

Mojezs, P. (1998). The who and how of dialogue. In D. Bryant & F. Flinn (Eds.), *Interreligious dialogue: Voice from a frontier* (pp. 199–206). New York: International Religious Foundation.

Nye, M. (1993). Temple congregations and communities: Hindu constructions in Edinburgh. *New Community*, *19*, 201–15.

Peach, C. (2002). Social geography: New religions and ethnoburbs – contrasts with cultural geography. *Progress in Human Geography, 26*(2), 252–260. http://dx.doi.org/10.1191/0309132502ph368pr

Pettigrew, T.F., & Tropp, L.R. (2006). A meta-analytic test of intergroup contact theory. *Journal of Personality and Social Psychology, 90*(5), 751–783. http://dx.doi.org/10.1037/0022-3514.90.5.751

Qadeer, M., & Kumar, S. (2006). Ethnic enclaves and social cohesion. *Canadian Journal of Urban Research, 15*(2), 1–17.

Seljak, D. (2009). Dialogue among the religions in Canada. *Horizons, 10*(2), 22–32.

Smock, D.R. (2003). *Building interreligious trust in a climate of fear: An Abrahamic trialogue*. Washington, DC: United States Institute of Peace.

Thomas, H. (2008). Race equality and planning: A changing agenda. *Planning Practice and Research, 23*(1), 1–17. http://dx.doi.org/10.1080/02697450802076407

Thomas, H., & Krishnarayan, V. (1994). *Race, equality and planning: Policies and procedures*. Aldershot: Brookfield and Avebury.

Urbano, R. (2012). Levinas and inter-faith dialogue. *Heythrop Journal, 53*(1), 148–161. http://dx.doi.org/10.1111/j.1468-2265.2010.00635.x

Winkler, T. (2006). Reimagining inner-city regeneration in Hillbrow, Johannesburg: Identifying a role for faith-based community development. *Planning Theory & Practice, 17*(1), 80–92.

Wood, P., & Landry, C. (2008). *The intercultural city: Planning for diversity advantage*. London: Earthscan.

Yiftachel, O. (1992). *Planning a mixed region in Israel: The political geography of Arab-Jewish relations in the Galilee*. Aldershot: Avebury Press.

PART 5

Enhancing the Cultural Competence of Planners

14 Negotiating Culture: Towards Greater Competency in Planning

MICHAEL RIOS

Introduction

Today, we find ourselves in a world defined by transnational and transregional migration, owing to natural and human-induced disasters, human conflict, economic restructuring, and the intended and unintended consequences of government policies and actions. Struggles over place materialize at the intersections of these multi-scalar forces and are experienced as everyday contests over the rights to space and place. It is against this backdrop that place and culture collide in ways that raise critical questions for the future of planning practice. For example, little thought has been given to issues arising from an increasingly multiethnic and multiracial world, and this results in a superficial treatment of a plural public realm. The limits to current approaches are evident inasmuch as planners stay within the narrow confines of professional norms and fail to address claims by new social groups. This complicity defines the current crisis of urbanism – an ambivalence about which public is being served, the reproduction of existing social structures that perpetuate exclusion and social inequality, and the absence of ethico-political considerations.

In this chapter, I highlight the importance of place as the space where new forms of citizenship are being produced as marginalized populations struggle to build community and gain social and political standing. I draw attention to the difference that culture makes for these communities and, specifically, how the convergence of place and culture instigates negotiations of belonging, authorship, and power to establish what groups can expect of one another. Negotiations are the basis for agreements and provide shared experiences that maintain relationships into the future. These "cultural contracts" measure the degree to which values and commitments are exchanged between groups – including professionals and the publics they purport to serve. Unlike the problems of sprawl and environmental degradation, there are no easy technical solutions to working more effectively with culturally diverse communities. While cultural diversity is a growing part of planning discourse, most schools do not offer courses on the topic as part of core curricula, nor are students required to demonstrate their ability to work effectively across cultures. One implication for planning education is the need for a greater focus on cultural competency, as measured by the level of cross-cultural communication between individuals and among different social groups to determine why place matters, for whom, and with what results.

The Changing Public Realm, Difference, and the Culture of Planning

As citizenship is inextricably linked to membership in a political community (Marshall, 1977), one issue concerns how citizens – as an ensemble of individuals and groups – seek to organize and order their coexistence. Membership in a political community does not exclude the potential for competition, conflict, and disagreement among adversaries, but rather it is guided by identifying the space between these differences to arrive at more democratic solutions (Hillier, 2003). To address current limitations in the planning field to a changing public sphere, a focus on the polity draws attention to how political relationships between different social groups and institutional actors are constituted in planning policy and practice, the challenges therein, and how strategic alliances are forged across cultural differences.

The main method in planning to address cultural differences in the polity is public participation. From a historical perspective, the use of citizen involvement to form a political community is not new. Many methods and techniques have been developed to involve different groups in public decision making (Sanoff, 2000). However, despite the best efforts of practitioners, the planning profession has not kept pace with the changing conditions that surround public processes. Participatory methods are often used in mainstream practice as a form of placation and performance to manufacture consensus rather than a means to enter into meaningful dialogue and decision making about conflicts and differences among participants, professionals, politicians, and other stakeholders (Hajer, 2005). Furthermore, participatory technologies often privilege technical and institutional instrumentality that has the effect of silencing already marginalized groups (Lake, 1994). This problem is compounded by the fact that the types of citizens involved in the public process and motivations for participation have changed significantly since the beginning of the citizen participation movement over a generation ago. A growing number of cities in the United States and elsewhere are now home to transnational populations whose cultural practices differ not only from the majority but also from prior generations of immigrants. These demographic changes have implications for planners who facilitate public processes in settings in which there are diverse participants. The growing number of majority-minority cities discussed by Burayidi and Wiles in chapter 8 underscores that we are truly at a cultural crossroads and sheds light on a number of problems in the planning arena, which include a lack of participation in planning decisions, a cultural divide that exists between practitioners and communities, and urban designs that do not meet cultural needs and preferences. These are the primary planning concerns that came out of a series of *diálogos*, or dialogues, sponsored by the Latinos in Planning Division of the American Planning Association. In 2005, professionals were asked to identify and prioritize key planning and design challenges during nine forums held in several states. These important efforts highlight the growing gap between planners and marginalized communities and the need for a critical re-examination of the public realm. However, the findings also call for problematizing the culture of planning itself.

Over the past two decades, a number of scholars have critiqued planning from a multicultural perspective. Much of this work is a direct challenge to modernist conceptions of planning defined by rationalist and universal approaches to city building and culture, and overlaps with the work of other scholars who are critical of public engagement methods, arguing that planners are complicit in reproducing social relationships of power. Regardless of the orientation, a number of themes have emerged. Some of these include:

- Respecting difference and highlighting creativity in insurgent forms of citizenship and planning,
- Drawing attention to an increasingly multiethnic and transnational public due to migration, the fluidity of borders, and the existence of hybrid identities,

- Identifying informal planning processes outside of state establishments and the increasing role of institutions representing the interest of marginalized communities, and
- The need for alternative approaches to planning in multicultural communities that emphasize relational and dialogical approaches.

Of these perspectives, Leonie Sandercock (2003; 1998) has been the most prominent in theorizing the prospects for multicultural cities. Sandercock has employed the term "cosmopolis" as an imagined future – a utopia – that epitomizes how multiethnic populations live together, how planning needs to be theorized anew in anticipation of this inevitability, and what planning practices need to change to allow strangers to "co-exist with one another." Sandercock challenges the precepts of modernist planning and normative practice that privilege scientific rationality, comprehensiveness, a focus on the state, and universal notions of the public interest, and replaces this emphasis with a focus on social justice, community, citizenship, and multiple publics.

Sandercock poses the question, important to the essays in this book, "How can stroppy strangers live together without doing each other too much violence?" (2003, p. 3). Central to Sandercock's response are two reinforcing positions: one is problematizing planning practice and the need to recognize cultural difference; the other is a normative vision of planning's future that is defined by a commitment to cross-cultural exchange. Sandercock argues that this transformation is only achievable through democratic politics, coalition building, and a conception of citizenship that is best understood in the context of everyday lives. However, while she identifies the importance of cultural difference and calls for a new "urban politics of interculturalism," her argument falls within a set of binary social relationships between the powerful and the marginalized, expert and non-expert planners, and leaves little room for imagining cross-cultural formation beyond hierarchically established boundaries. The point here is not to minimize Sandercock's contribution to theorizing the multicultural city, but rather to suggest that there is always something more, and beyond, the recognition of difference or management of coexistence. Moreover, Sandercock's postmodern perspective on structural change, albeit important, privileges cultural recognition and symbolism over more *emplaced* practices that are defined in material and locational terms.

Place is a setting to realize rights that are related to inhabitation, use, and access, but are only made possible through the formation of place-based alliances. Some have argued that the "power to act" resides locally, but that this power also incorporates extra-local forces aimed at producing a progressive or extroverted sense of place in global, local, and trans-local processes (Agnew, 1987; Amin, 2004; Massey, 1991). As Doreen Massey (2005) has pointed out, place necessitates invention and negotiation to make sense out of the complex power geometry of space and time. This "practising of place" is inherently political and requires action between people – re-acting and pro-acting, drawing together local circumstances with external forces at play. A challenging but important role for planning is to assist communities by linking, bridging, and mobilizing multiple understandings and visions of place that take into account extra-local factors.

Identity, Belonging, and Citizenship

The territorial basis for cultural and political claims begins with a sense of belonging – in a place, within a culture, and as part of a community. The formation of identity is often a product of what Harvey (1996) has called an "imaginary of belonging": a foundation to achieve social cohesion and solidarity. Like place, identity is not static but is fluid and contingent and depends on its historical context and on trajectories that are yet to be realized. For many diaspora communities, identity takes the form of a common sense of territorial association for members who identify, either through birth, ancestry, or social imagination, with a homeland other than the one they currently inhabit. Paradoxically, the diaspora experience is often about settling down, but having roots elsewhere (Brah, 1996; Blunt & Dowling, 2006; Kalra, Kaur, & Hutnyk, 2005). The diaspora experience questions the fixity of boundaries of here and there, inside and outside, and what it means to dwell in place.

According to Castells (1997), there are three forms and origins of identity building in a contemporary networked society: 1) a legitimizing identity that is introduced by dominant institutions of society, including the state and civil society organizations that promote a narrow nationalistic agenda; 2) a resistance identity (such as religious fundamentalism, ethnically based nationalism, and inward-looking communities) that is generated by actors in positions of little or no power vis-à-vis dominant institutions; and 3) a project identity (such as feminism) that draws on cultural resources to build a new identity in order to redefine one's position in society and transform social relations.

All three of these identities are present in place-making processes. However, of the three, only project identities hold the potential for transformative social relations through relational cultural practices. This assertion does not suggest that resistance identities close off the possibilities for project identities. Castells argues that "project identity, if it develops at all, grows from communal resistance" (p. 11). His point is that projects, which begin with resistance, need to allow space for new imaginaries in order to move from a culturally defensive position to one that holds open the possibility for broader institutional and societal change.

Theorizing identity vis-à-vis group claims is also a key focus of citizenship studies. Themes include the evolution of rights in contemporary society, debates between universalism and particularism, and the identification of new spaces of citizenship outside of formal state structures. Some have argued that citizenship is a multilayered process because of the interpenetration of public institutions into private affairs and vice versa, transnational migration, and the rise of new social movements that challenge state governments (Yuval-Davis, 1999). Given these larger trends, some are calling for the deepening of democracy and a fundamental redefinition of political processes (Mouffe, 2000). New forms of political discourse, representation, and accountability have been theorized to redress structural relationships of race, class, and gender, among other forms of inequality (Young, 2000). This redress includes reconceptualizing the public realm and citizenship in terms of difference and the intersectionality (see Doan's discussion in chapter 6) of multiple and layered identities (Young, 1990; 1996; Yuval-Davis, 2003). In addition, other arguments have been made for pluralism and a multiplicity of political communities as a challenge to universal and deliberative conceptualizations of the public sphere that are inherently undemocratic owing to the lack of mutual recognition in decision-making spaces (Fraser, 1992; Sanders, 1997). Consistent with a conception of multiplicity in the public sphere are the claims by social groups that interact with different and overlapping polities, as is the case with diasporic and transnational communities (Laguerre, 2005; 2006).

Within anthropology, the use of cultural resources (such as customs, inventions, institutions, and knowledge) to claim rights and create new polities and spaces has often been described as *cultural citizenship*. Cultural forms of citizenship represent "the ways people organize their values, their beliefs about their rights, and their practices based on their sense of cultural belonging rather than on their formal status as citizens" (Silvestrini, 1997, p. 44). Cultural citizenship focuses on the political in shaping cultural identity and rights within "zones of difference within and between cultures" (Rosaldo, 1989, p. 28). As an analytical lens, cultural citizenship highlights the important role of culture in struggles for a democratic society aimed at changing structural relationships of power such as racism and neoliberalism. As cultural sociologist Nick Stevenson (2010) argues:

> Only when public spaces become participatory and democratic spaces can we say that the project for an autonomous society has come to fruition. These are essential requirements, given that modern citizens are capable through cultural and democratic engagements of reformulating their current identities and reimagining their sense of self and of connection to others. (p. 276)

The relationship between belonging, culture, and public space has implications for multicultural planning. For example, the spaces of the city provide important locations where cross-cultural exchange occurs, and it is through place construction that cultural awareness, understandings, and meanings begin to take shape. The point here is that place-making practices, although not the only source of identity formation, are an important means for social groups to negotiate different imaginaries of cultural belonging, authorship, and power. Moreover, place ultimately reflects, and is the result of, these various negotiations – belonging to a community, participating in local decisions, and navigating geometries of power at multiple scales. A focus on cultural negotiation provides both an analytical and a practical framework for cross-cultural planning that revolves around three interrelated areas:

- *Negotiations of belonging* that centre on the exertion of citizens' rights to both be different and command a sense of belonging in the nation state, while also having the ability to participate in democratic processes;
- *Negotiations of authorship* that concern how individuals claim the specific cultural context in which they participate; and ultimately
- *Negotiations of power* that draw from the right to be visible, to be recognized, and to advance a group's cultural understanding towards economic, legal, and material outcomes.

Negotiations of Belonging

The exertion of citizens' rights both to be different and to command a sense of belonging in the nation state, while also having the ability to participate in the democratic process, requires negotiating one's position with respect to the different polities (Benmayor, Torruellas, & Juarbe, 1997). Much has been written about the "politics of recognition" and the acceptance of cultural differences in the public sphere (Taylor, 1994). However, to envision a sense of belonging, cross-cultural exchange necessarily involves the creation of new social and spatial imaginaries – not acceptance of the dominant culture's value system or holding on to resistant identities. Thus, a key task for planning is to challenge representations of marginalized places that essentialize identity vis-à-vis place. A sense of belonging in an increasingly multiethnic society does not necessarily translate into defending cultural differences but is constituted by the transactions of diverse cultural traditions that occupy the same geographic space.

Negotiations of belonging through creative acts of place making serve a vital role in engendering community while keeping open possibilities that can change social relations. For example, in the Market Creek Plaza neighbourhood in San Diego, California, the process began with creating a vision for a multicultural village where local citizens engaged in deep dialogue about past and present conditions to secure the right to envision a place that could belong to everyone. The story of the Village at Market Creek begins in the mid-1990s when residents and a handful of organizations participated in a community visioning exercise sponsored by the California Energy Commission, an unlikely convener owing to its mandated focus on statewide energy policy. The commission was interested in ways local communities could physically develop to reduce energy consumption. For the commission and its consultants, this intent included transit-oriented development strategies that would increase housing densities around a trolley station that was built in 1986 near the corner of Market Street and Euclid Avenue.

Building on this early effort, the Jacobs Family Foundation made an initial commitment to focus its mission on neighbourhood revitalization after funding several projects in the area. At the time, the area was known as the Four Corners of Death, as it was home to 42 gangs (Antoniadis, 2009). In 1997, the Jacobs Center for Neighborhood Innovation, the newly formed non-profit community development arm of the Jacobs Family Foundation, bought a 20-acre lot and hired Roque Barros, a seasoned community organizer, to engage residents in a participatory planning process. The hiring was a strategic move as there was significant mistrust in the community, much of it divided along ethnic and racial lines. To survey residents about their needs and desires for the site, Roque helped form international outreach teams of residents.

An explicit strategy was the identification of cultural diversity in the surrounding area and the creation of a safe and respectful space in which different world views could be expressed and shared. Beginning in 2010, the Jacobs Center for Neighborhood Innovation staff invited approximately 40 community members to form the VOCAL network, which stands for Voices of the Community at All Levels. For six months, weekly evening workshops were held to build the group's capacity and sense of belonging. A core focus of the weekly gatherings was a series of cultural workshops in which VOCAL members brought in cultural artefacts, described these items, shared their cultural histories, and entered into a deeper cross-cultural dialogue. One commonality that emerged from these learning sessions was the recognition that the different cultural groups experienced oppression in one form or another as a result of colonialism and imperialism, and that it was important not to lose sight of this common experience. During one workshop, an older African American male commented about blues music and the persistence of culture through music despite tragedy. He spoke about the persistence of pan-African culture even though more than two million Africans perished in the Atlantic Ocean during the slave trade and hundreds of thousands of families were separated. Similarly, other participants made reference to the violent histories of colonialism, whether it was French occupation in Laos or British rule in Sudan. A Kumiai Indian artist shared a photographic art piece depicting the mission wall where indigenous babies were buried, commenting, "We were also slaves," making reference to genocide inflicted by Franciscan missionaries at San Diego de Alcala.

The result of the various cultural exchanges (such as festivals, workshops, visioning forums, etc.) was a deep respect for cultural difference, which enabled participating groups to facilitate cultural communication after identifying common struggles of oppression that many of these groups had faced historically. This shared experience became a foundation from which to build solidarity and a dynamic multiethnic polity that is still evolving today. From this newly formed political space, different cultural groups made claims about their right to live in coexistence with other cultures and flourish together.

Negotiations of Authorship

Another characteristic of cultural negotiation is how individuals claim authorship of the specific cultural context in which they participate and which is inclusive of the right to challenge cultural consensus and hegemony (Boele van Hensbroek, 2010). This is demonstrated by participation in cross-cultural spheres of decision making and negotiating symbolic rights therein and can be described as a practising of cultural citizenship inasmuch as groups and individuals draw on their cultural resources to give voice to their concerns and navigate the procedural aspects of civic life (Nauta, 1992). Social learning and the "capacity for action" take place through participation in cultural and political discourse and the negotiations between the self and other (Delanty, 2003). This relational knowledge enables cross-cultural exchange and active engagement between participants, leading to mutual respect and equality rather than cultural passivity and dominance (Hart, 2009). However, this collective sense of association should not be conflated with a unitary idea of culture or flimsy multiculturalism, but rather should be rooted in solidarity and commitment to a dynamic political community, and the reciprocal acceptance and recognition of cultural differences (Parekh, 2000). In negotiating authorship through participation, cultural forms of citizenship provide the societal tools necessary to conceive and act according to one's self-identity within a capitalist economy (Isin & Wood, 1999). Performance is one way to create a social space of collective practice, identity, and authorship, as is the case of *Los Pastores*, a Mexican shepherds' play of the nativity in San Antonio, Texas, which combines the public display of traditional cultural performances with political acts and serves the purpose of forging solidarity, renewing identity, and providing a space of collective practice expressive of cultural power (Flores, 1997).

Insurgent cultural practices can transform negative perceptions of difference and impart to others openness and acceptance of different cultural norms that have consequences in the political realm (Müller & Hermes, 2010; Rocco, 1999). Within planning, negotiations of authorship challenge the normative spaces of decision making, including town hall meetings, charrettes, and other planning forums that convey particular cultural signifiers, norms, and styles of acceptable communication and behaviour but which often silence marginalized groups. By contrast, the creation of culturally alternative spaces allows for the promotion of different symbols and practices, which challenge planners to consider alternative methods and tools of communication and participation (see the discussion in chapter 12 by Main and Rojas). For example, the Mission Anti-displacement Coalition in San Francisco asserted an insurgent voice in response to the city's pro-development and market-driven agenda for the Mission District, a predominately working-class, immigrant, and multiethnic neighbourhood (Marti et al., 2012). The coalition's strategy was to institutionalize democratic participation to mitigate the effects of gentrification and zoning changes through a series of community benefit agreements. The coalition was a convener for a number of community-based organizations that had worked together in support of residents and businesses that were being displaced. Drawing from the tradition of Alinksy-style organizing and advocacy planning, the Mission Anti-displacement Coalition engaged directly in electoral politics through the promotion of citywide voter propositions aimed at closing loopholes in the planning process and imposing standard regulations and fees for live/work development projects. The coalition also borrowed from the radicalized experiences of many residents, some of whom came out of the Chicano movement or participated in the revolutionary struggles in Chiapas, El Salvador, and Nicaragua. These and other experiences led to the support from a majority of Mission District residents and a shift in the balance of power in a new board of supervisors.

With this momentum, the Mission Anti-displacement Coalition moved from protest to formulate policies and controls to prevent displacement, including the initiation of a community-driven planning process leading to the development of *el Plan Popular*, or the People's Plan. Parts of the community-driven plan included preserving a local industrial district and ensuring protection for blue-collar businesses, both of which were later adopted into policy by San Francisco's board of supervisors. Critical to the success of the plan was the engagement with a small number of city planning staff who aligned with the coalition's equity-planning framework; most of these were people of colour. For example, differing from workshops in other city neighbourhoods, the coalition coordinated efforts with city planning staff to incorporate popular education techniques imported from Latin America as well as Spanish translation for residents and child care for families. To educate residents on land use and zoning, coalition members developed a curriculum and trained a team of local residents as *promotoras* to conduct outreach, education, and organizing efforts. The *promatoras* model was developed in Mexico during the 1960s to reach out to marginalized communities with the goal of improving access to health care. A storefront in the Mission District neighbourhood was set up to showcase a large model used as part of a community-mapping process, the Barrio Uses Mapping Project. Residents conducted door-to-door surveys of the neighbourhood as well.

While the People's Plan was not fully adopted by the city, its recommendations were carried forward in the city's official plan and led to a series of resolutions passed by the board of supervisors that any rezoning must "conform to the city's affordable housing production goals, comprehensive needs and impacts, and coordination of relevant city agencies to implement the plan" (Marti et al., 2012, p. 211). A community advisory committee was also set up to ensure proper implementation and monitoring of the city-approved plan.

Negotiations of Power

Cultural negotiation and political rights are not mutually exclusive but intersect in ways that support one another. However, negotiated claims do not necessarily guarantee legal entitlements (Pakulski, 1997). It is from this understanding that cultural forms of citizenship are a means towards an end to negotiated power within a hegemonic system. The prospects of place making and, by extension, planning lie in the ability of cross-cultural coalitions to reimagine the production of space towards political ends that have economic, legal, and material benefits. By contrast, structures of belonging are also defined through hegemonic relationships and play an important role in how marginalized groups lay claim to space and how these claims are interpreted by the larger society (Diaz-Barriga, 2008). Dominant groups often use the claims of marginalized groups for radically different purposes. State and civil society groups (such as redevelopment authorities, planning agencies, community development corporations, and so on) often distinguish between particular ethnic groups for marketing and consumption purposes, while glossing over the claims made by these same groups. The result is the reproduction of racism and racial inequality (Ong, 1999). In this vein, Aiwa Ong describes cultural citizenship as "cultural practices and beliefs produced out of negotiating the often ambivalent and contested relations with the state and its hegemonic forms that establish the criteria of belonging within a national population and territory" (Ong, 1996, p. 738).

In reality, the two contrasting sides of cultural citizenship are part and parcel of place making, which suggests a dire need for an "active process of claiming space and rights rather than the passive acquisition of an arbitrary and limited set of rights" (Flores, 2003, p. 295). The building of cross-cultural coalitions among marginalized groups and allies within positions of power can create a broader sense of belonging and struggle for rights in the larger polity. The symbolic power inherent in cultural forms of citizenship is important, but it must be linked to formal rights if cultural citizenship is to serve as a vehicle to change social relations (Stevenson, 2003). In the Fruitvale and San Antonio districts of Oakland, California, the process has involved years of struggle by community-based organizations that fought for the rights of local citizens in the city's most culturally diverse neighbourhoods. Coordinated cultural festivals, commercial revitalization, transportation and infrastructure improvements, and new housing and parks to improve the quality of place were some of the outcomes resulting from a multiracial and multiethnic coalition that emerged out of necessity and capitalized on cultural diversity as a vital community asset (Maly, 2005).

One example of such a coalition activity was the creation of a partnership between non-profit organizations and community groups to increase the amount of parks and open space for the Fruitvale and San Antonio districts, an area of the city that had the smallest amount (0.68 acres per 1,000 residents), but also had some of the highest concentrations of youth under the age of 18. The Unity Council, a local community development corporation, led a campaign to request that Union Point, a nine-acre site of waterfront property owned by the Port of Oakland, be dedicated for a public park. The campaign educated the residents in the Fruitvale and San Antonio districts about the lack of public open space and the Union Point site. The Unity Council collected 3,000 signatures and 100 letters of support from citizens and organizations representing the multiethnic and multiracial community. As a result of the multiethnic coalition and demonstration of public support, the city and port of Oakland agreed to work in partnership with the Unity Council and its allies to develop a park at Union Point, despite the fact that open space would minimally benefit the port, which owned the land. A formal partnership was formed with the Unity Council and several other organizations including the California Coastal Conservancy, the Trust for Public Land, and the University of California. After the site was dedicated for park use, the Unity Council received a grant from the Coastal Conservancy to lead a process to design, and eventually build,

Union Point Park. In 2005, Union Point Park was completed and now provides the only major waterfront access to the Oakland Estuary for over 80,000 residents.

Negotiating Culture: Implications for Practice

Much of the focus on culture in contemporary planning practice is driven by a desire to capitalize on the authenticity of cultural heritage in specific places. However, a focus on authenticity is problematic, as it glosses over the diversity of subjectivities that make up cultural communities, leading to the essentialization of identity and the aestheticizing of marginalized groups as "other." This includes the thematization and commodification of multiethnic and multiracial places, whether by romanticizing a neo-traditional social imaginary of the past, promoting a pastiche postmodern image, or glorifying the ever-present postcolonial condition. In contrast to authenticity, sincerity sets up a different relationship entirely. It aims to understand how people think and feel about their identity in relationship to their everyday experiences, which are situated in a physical and social context that is ever changing due to so many causal factors beyond their immediate control. Drawing from Lionel Trilling's book *Sincerity and Authenticity*, cultural anthropologist John L. Jackson, Jr (2005) makes a further distinction in describing the etymology of the two terms. *Sincerity* comes from the Latin term *sincerus* (originally applied to things, not people), meaning without wax, unadulterated, not doctored. *Authenticity*, however, derives from Greek *authenteo*: to dominate or have authority over, even to kill. Jackson states, "sincerity was once about things, and authenticity about relations between people" (p. 15). However, in the present, he argues that their connotations have been reversed. Sincerity does not define relationships as subject-object,which authenticity presumes, but rather as a subject-subject interaction. This distinction between the two terms is more than a semantic difference and poses a challenge to the planning profession to move beyond the superficial elements of "authentic" culture towards sincere understanding and engagement.

Negotiations of belonging, authorship, and power are important inasmuch as negotiations also establish what groups can expect of one another. Some communication scholars have argued that negotiation is the basis for identity construction, the bridging of cultural difference, and the precondition for successful intercultural communication (Berger & Calabrese, 1975; Gudykunst & Ting-Toomey, 1988; R. Jackson, 1999; Ting-Toomey, 1999).

Negotiations are also the basis for agreements and provide shared experiences that maintain relationships into the future. These agreements, or "cultural contracts," measure the degree to which cultural values and commitments are exchanged between groups (R. Jackson, 2002). Cultural contracts require the adaptation of cultural values through negotiations between marginalized groups and institutions such as local organizations and planning agencies, as well as the co-creation of mutually valued agreements between marginalized groups. While cultural contracts often focus on exchanges between traditionally marginalized groups and majority groups in power, these agreements can measure the quality of cross-cultural planning that involves conflictive and non-conflictive, equal and unequal, social relationships, and relationships between individuals and groups from different backgrounds.

Key to these various forms of negotiation is the brokering between various identities. Brokering is a transactive process in which someone acts as an agent for others. In cross-cultural exchanges an agent facilitates the communication and transaction of customs and values towards a shared goal that is codified (explicitly or implicitly) in an agreement, whether short- or long-term. Important to effective cross-cultural planning are individuals who can bridge between different cultural groups and with various institutions. Planners need to be adept in facilitating dialogue and agreement between participants, as well as finding common ground between technical understandings of planning and the specific values of cultural communities.

Without the skilled brokering between various identities, it is doubtful that agreements can be reached that begin with cultural differences. The ability to broker between diverse groups will require instilling greater cultural competency in planning practitioners, students, and faculty alike. However, competency is not a skill to be obtained, mastered; or an attribute to be measured in quantitative terms. Rather, it is aspirational as cultures are constantly evolving and changing. From a linguistic perspective, competency can be defined as "the implicit, internalized knowledge of a language that a speaker possesses and that enables the speaker to produce and understand the language" (*Random House Webster's Unabridged Dictionary*, 2nd ed., s.v. "competency"). The "cultural" in cultural competency is the language of everyday life – the social practices and meanings produced in place by shared histories and present conditions of people who identify with a particular group. Cultural competency is experienced and understood by the level of cross-cultural communication between individuals and among different social groups. Cultural competency makes it possible for planners to engage in the kind of leadership and facilitation that transform perception and action across group identities. The primary task is then to emplace ourselves with others and continually discover how to speak (verbally and non-verbally) and act between and among different cultures with the goal of enabling greater exchange, understanding, and, ultimately, respect.

REFERENCES

Agnew, J. (1987). *Place and politics: The geographical mediation of state and society*. London: Allen and Unwin.

Amin, A. (2004). Regions unbound: Towards a new politics of place. *Geografiska Annaler, Series B: Human Geography, 86*(1): 33–44.

Antoniadis, G. (2009, Dec. 18). Jennifer Vanica '76 is transforming the national conversation about community development. *Wittenberg Magazine*. http://www9.wittenberg.edu/magazine/features/leading-change/

Benmayor, R., Torruellas, R., & Juarbe, A. (1997). Claiming cultural citizenship in East Harlem: "Si esto puede ayudar a la comunidad mia." In W.V. Flores & R. Benmayor (Eds.), *Latino cultural citizenship: Claiming identity, space and rights* (pp. 152–209). Boston: Beacon Press.

Berger, C., & Calabrese, R. (1975). Some explorations in initial interaction and beyond: Toward a developmental theory of interpersonal communication. *Human Communication Research, 1*(2), 99–112. http://dx.doi.org/10.1111/j.1468-2958.1975.tb00258.x

Blunt, A., & Dowling, R. (2006). *Home*. London, New York: Routledge.

Boele van Hensbroek, P. (2010). Cultural citizenship as a normative notion for activist practices. *Citizenship Studies, 14*(3), 317–330. http://dx.doi.org/10.1080/13621021003731880

Brah, A. (1996). *Cartographies of diaspora*. London, New York: Routledge.

Castells, M. (1997). *The power of identity, the information age: Economy, society and culture*. Cambridge, MA: Blackwell.

Delanty, G. (2003). Citizenship as a learning process: Disciplinary citizenship versus cultural citizenship. *International Journal of Lifelong Education, 22*(6), 597–605. http://dx.doi.org/10.1080/0260137032000138158

Diaz-Barriga, M. (2008). Distracción: Notes on cultural citizenship, visual ethnography, and Mexican migration to Pennsylvania. *Visual Anthropology Review, 24*(2), 133–147.

Flores, R.R. (1997). Aesthetic process and cultural citizenship: The membering of a social body in San Antonio. In W.V. Flores & R. Benmayor (Eds.), *Latino cultural citizenship: Claiming identity, space and rights* (pp. 124–151). Boston: Beacon Press.

Flores, W.V. 2003. New citizens, new rights: Undocumented immigrants and Latino cultural citizenship. *Latin American Perspectives*, *30*(2), 295–308. http://dx.doi.org/10.1177/0094582X02250630

Fraser, N. (1992). Rethinking the public sphere: A contribution to the critique of actually existing democracy. In F. Barker, P. Hulme, & M. Iverson (Eds.), *Postmodernism and the rereading of modernity* (pp. 197–231). New York: Manchester University Press.

Gudykunst, W., & Ting-Toomey, S. (1988). *Cultural and interpersonal communication*. Newbury Park, CA: Sage.

Hajer, M.A. 2005. Rebuilding Ground Zero: The politics of performance. *Planning Theory & Practice*, *6*(4), 445–464. http://dx.doi.org/10.1080/14649350500349623

Hart, S. 2009. The problem with youth: Young people, citizenship and the community. *Citizenship Studies*, *13*(6), 641–657. http://dx.doi.org/10.1080/13621020903309656

Harvey, D. (1996). *Justice, nature, and the geography of difference*. Oxford: Blackwell.

Hillier, J. 2003. "Agon"izing over consensus: Why Habermasian ideals cannot be "Real." *Planning Theory*, *2*(1), 37–59.

Isin, E.F., & Wood, P.K. (1999). *Citizenship and identity*. London, Thousand Oaks, New Delhi: Sage.

Jackson, J.L., Jr. (2005). *Real Black: Adventures in racial sincerity*. Chicago, London: University of Chicago Press.

Jackson, R. (1999). *The negotiation of cultural identity*. Westport, CT: Praeger.

Jackson, R. (2002). Cultural contracts theory: Toward an understanding of identity negotiation. *Communication Quarterly*, *50*(3), 359–367. http://dx.doi.org/10.1080/01463370209385672

Kalra, V., Kaur, R., & Hutnyk, J. (2005). *Diaspora and hybridity*. London, Thousand Oaks, New Delhi: Sage Publications.

Laguerre, M.S. (2005). Homeland political crisis, the virtual diasporic public sphere, and diasporic politics. *Journal of Latin American Anthropology*, *10*(1), 206–255. http://dx.doi.org/10.1525/jlat.2005.10.1.206

Laguerre, M.S. (2006). *Diaspora, politics, and globalization*. New York: Palgrave Macmillan. http://dx.doi.org/10.1057/9781403983329

Lake, R.W. (1994). Negotiating local autonomy. *Political Geography*, *13*(5), 423–442.

Maly, M. (2005). *Beyond segregation: Multiracial and multiethnic neighborhoods in the United States*. Philadelphia: Temple University Press.

Marshall, T.H. (1977). *Class, citizenship and social development*. Chicago, London: University of Chicago Press.

Marti, F., Selig, C., Arreola, L., Diaz, A., Fishman, A., & Pagoulatos, N. (2012). Planning against displacement: A decade of progressive community-based planning in San Francisco's Mission District. In M. Rios & L. Vazquez (Eds.), *Diálogos: Placemaking in Latino communities* (pp. 126–140). Abingdon, New York: Routledge.

Massey, D. (1991). A global sense of place. *Marxism Today*, no. 38, 24–29.

Massey, D. (2005). *For space*. London: Sage.

Mouffe, C. (2000). *The democratic paradox*. London: Verso.

Müller, F., & Hermes, J. (2010). The performance of cultural citizenship: Audiences and the politics of multicultural television drama. *Critical Studies in Media Communication*, *27*(2), 193–208. http://dx.doi.org/10.1080/15295030903550993

Nauta, L. (1992). Changing conceptions of cultural citizenship. *Praxis International*, *12*(1), 20–34.

Ong, A. (1996). Cultural citizenship as subject-making: Immigrants negotiate racial and cultural boundaries in the United States. *Current Anthropology*, *37*(5), 737. http://dx.doi.org/10.1086/204560

Ong, A. (1999). *Flexible citizenship: The cultural logics of transnationality*. Berkeley: University of California Press.

Pakulski, J. (1997). Cultural citizenship. *Citizenship Studies*, *1*(1), 73–86.

Parekh, B. (2000). *Rethinking multiculturalism: Cultural diversity and political theory*. London: Macmillan Press.

Rocco, R. 1999. The formation of Latino citizenship in southeast Los Angeles. *Citizenship Studies*, *3*(2), 253–266. http://dx.doi.org/10.1080/13621029908420713

Rosaldo, R. (1989). *Culture and truth: The remaking of social analysis*. Boston: Beacon Press.

Sandercock, L. (1998). *Towards cosmopolis: Planning for multicultural cities*. Chichester: John Wiley and Sons.

Sandercock, L. (2003). Integrating immigrants: The challenge for cities, city governments, and the city-building professions. Faculty Working Paper 03–20, Vancouver Centre of Excellence, Simon Fraser University.

Sanders, L.M. 1997. Against deliberation. *Political Theory*, *25*(3), 347–276.

Sanoff, H. (2000). *Community participation methods in design and planning*. New York: John Wiley and Sons.

Silvestrini, B. (1997). The world we enter when claiming rights. In W. Flores & R. Benmayor (Eds.), *Latino cultural citizenship: Claiming identity, rights, and space* (pp. 39–56). Boston: Beacon Press.

Stevenson, N. (2003). Cultural citizenship in the "cultural" society: A cosmopolitan approach. *Citizenship Studies*, *7*(3), 331–348.

Stevenson, N. (2010). Cultural citizenship, education and democracy: Redefining the good society. *Citizenship Studies*, *14*(3), 275–291. http://dx.doi.org/10.1080/13621021003731823

Taylor, C. (1994). The politics of recognition. In A. Gutman (Ed.), *Multiculturalism: Examining the politics of recognition* (pp. 25–74). Princeton: Princeton University Press.

Ting-Toomey, S. (1999). *Communicating across cultures*. New York: Guilford Press.

Young, I.M. (1990). *Justice and the politics of difference*. Princeton: Princeton University Press.

Young, I.M. (1996). *Intersecting voices: Dilemmas of gender, political philosophy, and policy*. Princeton: Princeton University Press.

Young, I.M. (2000). *Inclusion and democracy*. Oxford, New York: Oxford University Press.

Yuval-Davis, N. (1999). The multi-layered citizen. *International Feminist Journal of Politics*, *1*(1), 119–136.

Yuval-Davis, N. (2003). Citizenship, territoriality and the gendered construction of difference. In N. Brenner, B. Jessop, M. Jones, & G. MacLeod (Eds.), *State/space: A reader* (pp. 309–325). Malden, MA: Blackwell Publishers. http://dx.doi.org/10.1002/9780470755686.ch18

15 Educating Planners for a Cosmopolitan Society: A Selective Case Study of Historically Black Colleges and Universities

SIDDHARTHA SEN, MUKESH KUMAR,
AND SHERI L. SMITH

Introduction

Over the past 20 years, planning educators and planning practitioners in the United States have recognized the need to incorporate diversity into planning education. The term *diversity* encompasses more than multiculturalism. As pointed out by Agyeman and Erickson (2012) and in chapter 6 by Doan, multicultural discourses generally emphasize race and ethnic diversity and neglect differences in gender, age, class, sexual orientation, cultural heterogeneity, and disability. So a broader definition of pedagogical diversity can be defined as integrating issues of difference (race, gender, class, ethnicity, nationality, sexuality, physical disability, world views, religion, and age) in planning education. Many believe that such changes are necessary in order to transform planning education for the twenty-first century (Rodriguez, 1993; Thomas, 1996; Burayidi, 2000, 2003; Wolfe, 2003; Sandercock, 2003; Milroy, 2004; Rahder & Milgrom, 2004; Sen, 2000, 2005; Sweet & Etienne, 2011; Agyeman & Erickson, 2012). Although US cities have long been multicultural and suburbs are becoming more so, our discourse on diversity is still rooted in slavery, civil rights, affirmative action, and turbulent racial relations. There is a need to incorporate broader issues of difference in how we practise and teach planning (Sen, 2005, Agyeman & Erickson, 2012).

To meet the challenges facing our changing cities and suburbs, planning has to embrace multiculturalism and incorporate difference as the norm, regardless of the geographical location. Planning educators need to bring fuller treatment of and sensitivity to issues of diverse world views, values, cultures, race, ethnicity, and nationality, and these issues of redistributive social justice need to be articulated in planners' education (see, for example, Thomas, 1996; Burayidi, 2000; Gonnewardena et al., 2004; Milroy, 2004, Rahder & Milgrom, 2004; Sen, 2000, 2005; Sweet & Etienne, 2011, Agyeman & Erickson, 2012). To achieve this goal, one strategy will require active recruitment and retention of a diverse faculty and student body and a supportive academic environment to nurture them. It also requires curriculum changes to accomplish this goal (Forsyth, 1995; Thomas, 1996; Goonewardena et al., 2004; Sen, 2000; 2005; Sweet & Etienne, 2011; Wubneh, 2011; Agyeman & Erickson, 2012).

Programs at HBCUs, or historically Black colleges and universities, are historically and ideally suited to address these issues, as they are located predominantly in urban areas and serve a student body and a community that are directly involved and concerned with multicultural and diversity issues. HBCU planning programs are in fact teaching the interculturalism that Burayidi refers to in chapter 1. Such programs actively seek out different voices in their curricula and take a holistic approach to multiculturalism. HBCUs also have a history of embracing difference by accepting Whites, Hispanics, immigrants, and other disenfranchised populations. In this chapter we explore how accredited planning programs at three HBCUs – Morgan State, Texas Southern, and Jackson State – incorporate multiculturalism into their curricula using Banks's (1995a, 2001) guidelines as the analytic framework. We first briefly review multiculturalism and multicultural education and the methodological framework we use for the analysis and conclude with a discussion of the role of HBCUs in multicultural planning education.

Multiculturalism, Multicultural Education, and Diversity in Urban Planning

The word *multicultural* means "of many cultures." It is the existence or presence of people of different racial or ethnic backgrounds within a single municipality (Citrin, Sears, Muste, & Wong, 2001). Initially in the US, people of different cultures were expected to assimilate into one dominant culture that would contribute to a politically unified and hopefully harmonious society (Hartmann & Gerteis, 2005). Through social unrest, this expectation has changed so that now we view multiculturalism as a means to create a balance between unity and diversity, not to the benefit of one group, but for the enrichment of the entire society (Citrin et al., 2001). Chapter 1 provides a detailed definition of the term and explains its range of meanings, and this serves as a basis for our discussion.

The term *multicultural education* also has a spectrum of meanings. At its minimum or limited scope, it is the teaching of tolerance and recognition of "others" and their right to reside within a given municipality. At its ultimate, multicultural education results in the transformation of the educational system to empower students and a reform of the pedagogy to support diversity education (Banks, 1995b). Chapman (2004) and Banks (1995b) state that achieving the ultimate in multicultural education is a process, which when realized produces worthwhile goals for the individual, the university, and the societies our universities serve. As a guide to achieving true multicultural education, Banks (1995a, 2001) presented five dimensions as the acceptable direction for educational reform. These dimensions are 1) content integration, 2) knowledge construction, 3) prejudice reduction, 4) equity pedagogy, and 5) an empowering school culture. In this chapter we employ these five dimensions as an analytical framework for examining how these concepts are operationalized in planning education in HBCUs. Although Banks (1995a, 2001) developed these dimensions in the context of elementary school education, we contend that they are applicable to planning education in higher education as well.

Content integration consists of integrating examples, data, and information from diverse cultures in illustrating key concepts, generalizations, principles, and theories. Of concern here is the type of information that is included in the curriculum, how and where it is included, and its audience. Knowledge construction consists of making students cognizant of biases within a discipline and how race, ethnicity, and class influence knowledge creation. Prejudice reduction involves the use of strategies to develop more democratic attitudes and values among students. This is based on the understanding that significant prejudice reduction as well as improvement in the academic achievements of minority students can be gained through cooperative learning. Equity pedagogy consists of teaching techniques and methods that facilitate academic achievement among students from diverse ethnic, social, and racial groups. Such pedagogy should recognize differences in learning styles among the various ethnic groups, races, and social classes. An empowering school culture requires restructuring the culture of the organization so that students from diverse ethnic groups, races, and classes experience "educational equality and cultural empowerment" (Banks, 1995a, p. 5; Banks, 2001).

HBCUs and Their Role in Multicultural Planning Education

HBCU is a designation conferred by the US Congress through Title III of the Higher Education Act of 1965 for accredited, public and private colleges/universities that existed before 1964, with a historical and contemporary mission of educating Blacks while being open to all (Higher Education Act, 1965, Knight et al., 2012, United States Commission of Civil Rights, 2010, Brown & Ricard, 2007). While the commonly used term HBCU originated in the mid-1960s, the presence of African American or minority-serving universities has deeper roots in US history. As early as 1856, churches, benevolent societies, and Freedman Bureaus offered high school and postsecondary education to African Americans and anyone interested in pursuing an education regardless of their race or ethnicity (Higher Education Act, 1965; Knight, Davenport, Green-Powell, & Hildon, 2012, United States Commission on Civil Rights, 2010, Brown & Ricard, 2007, Fryer & Greenstone, 2007). The Morrill Act of 1890, also known as the Second Morrill Act, mandated that funds for education for land grant institutions created under the First Morrill Act of 1862 be extended to institutions that educated Black Americans. States had the opportunity to open the doors of the existing land grant institutions to African Americans and other minorities but instead developed a two-tiered system by creating 16 separate minority-serving institutions (Brown & Ricard, 2007; Fryer & Greenstone, 2007).

Eighty-five years later, the Civil Rights movement and subsequent legislation legally integrated and opened the doors of education to all (Title VI of the Civil Rights Act, 1964). African Americans and minority groups were officially allowed to apply to and enrol in traditionally White universities and receive degrees that had once been denied to them. For more than a century, HBCUs have been leaders in the secondary education of African Americans, minorities, and other disadvantaged individuals who have been denied education because of either colour, ethnicity, or a perceived inability to graduate with an undergraduate degree or higher (United States Commission on Civil Rights, 2010; Ashley et al., 2009). HBCUs were charged with a mission that stands today: promote universal access and offer a curriculum designed to meet the needs of both the institution and the community (Brown & Ricard, 2007). In doing so, they perpetuate the tradition of providing their students with a culturally, socially, economically, and politically relevant and diverse education (Brown & Ricard, 2007).

Despite their historical beginnings and laudable charge, HBCUs are besieged by controversy, surrounded by critics, faced with limited resources, and challenged about their educational relevance. Yet they continue to admit, educate, and graduate students who not only contribute to the well-being of African Americans but also make contributions to society at the local, national, and international levels (United States Commission of Civil Rights, 2010; Brown & Ricard, 2007, Fryer & Greenstone, 2007).

There are several ways to assess the contributions of HBCUs. From a quantitative approach, one could compare HBCUs to all degree-conferring colleges and universities in the United States and/or look at how HBCUs are serving African Americans, minorities, and disadvantaged populations (United States Commission of Civil Rights, 2010, Ashley, Gasman, Mason, Sias, & Wright, 2009). From a qualitative approach, one could evaluate HBCUs on their own merit and on how they have performed over time, reviewing the missions of their programs, the educational attainment of their students, the supportive learning atmosphere, or the calibre and achievements of their graduates.

How does this previous discussion address the role of HBCUs in multicultural planning education? With the exception of Qadeer, in chapter 3, who gives planning in the US some credit for responding to ethnocultural diversity and minority representation, most of the chapters in this book (see, for example, Burayidi's Introduction, Doan's chapter 6; Zaferatos's chapter 7, Nguyen, Gill, and Steephen's chapter 9, and Harwood and Lee's chapter 10) illustrate that, despite the progress made in bringing diversity to the attention of planners, the notion is not embedded in the everyday practice of planning. HBCUs are increasing this consciousness by not only producing planners who can integrate diversity into everyday practice but also diversifying the profession by increasing the supply of people who have been largely excluded from it. However, just as is experienced across the larger field of academia, the presence, value, and contribution of HBCUs to the planning profession is at best questioned and at worst overlooked. The contribution of HBCUs may be evaluated quantitatively as a valuable source for graduating numbers of African American planners. A review of the numbers provided by the Planning Accreditation Board (2013b) shows that, in 2012, 386 full- and part-time African Americans were enrolled in a master's program. Of those enrolled, 88, or 22.8 per cent, were enrolled at HBCUs. That percentage remained relatively constant over the four years between 2009 and 2012. As the number of

African Americans pursuing master's degrees in planning has vacillated over time, HBCUs were responsible for 23 per cent of African American planning students in 2009, 24 per cent in 2010, and 24 per cent in 2011. The four HBCUs that offer accredited master's degrees in planning – Alabama A&M, Jackson State, Morgan State, and Texas Southern – contributed between 22 and 24 per cent of African American students enrolled in urban planning programs annually over these four years (Planning Accreditation Board, 2013a). The magnitude of the contribution of HBCUs in increasing the representation of African Americans in the planning profession can be further understood from the fact that there were only four HBCUs among a total of 69 master's programs in US universities as of January 1, 2013 (Planning Accreditation Board, 2013b). The HBCU contribution to urban planning should therefore not be overlooked.

The Case Studies

The following three case studies use Banks's five dimensions of multiculturalism to discuss the role of HBCUs in educating planners for multicultural practice. Although there are similarities among the cases, there are also minor differences in the approach to the integration of multiculturalism in the curriculum in each school.

Morgan State University

The seeds of the planning program at Morgan were sown in 1956 with the arrival of a young African American faculty member, Homer E. Favour, who became actively involved in community planning activities in Baltimore neighbourhoods.[1] Favour's connections in the Baltimore area and a grant that the then-president of Morgan State College, Martin D. Jenkins, received from the state of Maryland to carry out extension work in Baltimore's neighbourhoods led to the establishment of the Urban Studies Institute in 1963. The institute blossomed into the Center for Urban Affairs in 1970, funded jointly by the state of Maryland and the Ford Foundation.

The MA in Urban Planning and Policy Analysis was initiated in 1970 as one of the instructional programs at the centre. In 1975, the Center for Urban Affairs became the School of Urban Affairs and Human Development. Institutional changes at Morgan sustained the development of the Built Environment Studies program within the School of Urban

Affairs and Human Development. The granting of university status in 1975 came with a new mandate: the responsibility for addressing and resolving urban problems. One interpretation of the mandate resulted in the creation of five new urban-oriented programs including Built Environment Studies, which in turn led to a change in the degree from an MA in Urban Planning and Policy Analysis to a Master of City and Regional Planning (MCRP). The Institute of Architecture and Planning was created in 1991 and housed the planning program (see Sen, 1997 for a detailed discussion). As discussed below, the program has been dedicated to incorporating multiculturalism since its inception because of its unique institutional history and location.

The presence of a large number of African Americans in the program makes the discussion on race and justice an integral part of all classes and contributes to prejudice reduction. Students are often concerned with community development by virtue of their residence in the inner city, ethnic background, social concern, or even employment. Even faculty members that come from traditional backgrounds or are of a different race are transformed by Morgan's student body and institutional culture and get engaged in issues of diversity through their teaching. Cooperative learning is employed in many of the classes by bringing together students of diverse backgrounds for the discussion of racial and cultural issues in the classroom.

Those who were instrumental in the establishment and growth of the program were dedicated to working with students and community groups to address issues of the inner-city neighbourhoods. This has left a legacy of serving the underprivileged and integrating diversity issues into the curriculum. For example, Homer E. Favour not only wrote a dissertation on property values and race but also, as stated above, became actively involved in community planning activities in Baltimore neighbourhoods. Martin D. Jenkins was also instrumental in cultivating a culture of community service for the underprivileged at the university. Jenkins was interested in undertaking extension work in the community because he was sensitive to what was then referred to as the "urban crisis," broadly defined as racial polarization, class alienation, physical decay, and deterioration of race relations in the inner cities. He was also aware of and concerned about the war on poverty programs and the university's role in resolving the urban crisis. He strongly felt that as a Black institution of higher learning, Morgan should be a pioneer in alleviating the urban crisis (Sen, 1997).

When the Urban Studies Institute was established in 1963, it focused on research and extension in the community. Institutional changes that were taking place at Morgan State also influenced the administration to undertake the initiative to transform the Urban Studies Institute into the Center for Urban Affairs. When the proposal for the centre was finalized in 1970, the Board of Trustees of the State Colleges, the Maryland Council for Higher Education, and the governor of Maryland were considering a recommendation to develop Morgan State College as *"a racially integrated, urban oriented"* (emphasis added) university. This expanded "urban thrust" prompted the administration to nurture the idea of a Center for Urban Affairs, which could act as a catalyst in propagating this new urban mission. There was a consensus among its administration, faculty, student body, alumni, and supporters that Morgan should play a greater role in alleviating the urban problems of the Baltimore metropolitan area in particular, and the state of Maryland in general (Sen, 1997). So even before the program was established, Morgan had a commitment and institutional culture of attending to planning problems of inner-city neighbourhoods. We must also remember that as an HBCU, Morgan was devoted to educating disadvantaged African Americans by reaching out to underprepared students from inner cities. A program that was born in this context and institutional climate was bound to incorporate diversity in its curriculum from the beginning.

An institutional change that further facilitated the program's commitment to the needs of the inner cities was the 1975 legislation that transformed Morgan State College into Morgan State University. The legislation stated in part that Morgan State should emphasize urban-oriented education. The State Board of Higher Education emphasized the need for developing programs addressing *"specific social problems of cities"* (State Board of Higher Education as cited in Morgan State University Catalog, 1984–1986, p. 13). While other programs were formed because of this mandate, the planning program made this mandate its central mission.

The university's self-study reports from the late 1980s, the 1990s, and the 2000s also reiterate this mission. These documents emphasize that the university should be an integral part of the resource base for the development of Baltimore and inculcate in its student body an understanding of urban America and a sense of social responsibility for improving the quality of life in urban areas. This is also true for the university's current mission, which includes "preparing high-quality, *diverse graduates*" and "priority to addressing *societal problems, particularly those prevalent in urban communities*" (Morgan State University, 2015a; emphasis added).

Although other programs have attempted to meet Morgan's historical emphasis on inner-city Baltimore and the underprivileged in various ways, the planning program has always been dedicated to meeting this mandate in its program mission. To achieve this, the planning program has focused on Baltimore from its early days. In contributing to the university's mission to address and help alleviate the problems of the inner city, the planning program at Morgan used Baltimore as a laboratory for most of its courses. Thus, multicultural content integration at Morgan has been taking place in Morgan's planning program since its inception.

Projects on disadvantaged sections of Baltimore have been an integral part of the curriculum. Such an approach has also led to prejudice reduction as it has developed more democratic attitudes and values among students by exposing them to the problems of the disadvantaged population. Most recently the program has been involved in research projects, studios, studies, and service learning on the Morgan Community Mile, which was initiated to fulfil one of the goals of Morgan's Strategic Plan: Goal 5: Engaging with the Community (Morgan State University, 2011). "The Morgan Community Mile partners with Northeast Baltimore neighborhoods and the private, public, and nonprofit sectors. This Initiative engages community stakeholders and university students, faculty, and staff in inclusive, democratic, and participatory processes that result in mutually defined community plans and projects, measurable outcomes, and positive community impact" (Morgan State University, 2015b).

Most faculty in the urban planning program at Morgan make students cognizant of biases in knowledge construction within urban planning as well as how race, ethnicity, and class influence how knowledge is created. Some courses not only include alternative reading material but ask students to critically reflect on the readings, given their racial and ethnic origins. Field trips and site visits encourage students to construct firsthand alternative knowledge about urban spaces and urban planning. Faculty have also used film and documentaries to inculcate alternative knowledge construction. Here too, it can be observed that even faculty members who may not be directly involved in alternative knowledge construction may have to deal with it because of the presence of students who are likely to question mainstream knowledge construction.

This usually happens because the students come from diverse places or ethnic backgrounds and have social concerns.

To meet the HBCU's role and the institution's mission of reaching out to disadvantaged population groups, the program at Morgan intentionally reaches out to and admits many African Americans with somewhat incomplete academic preparation and non-standard backgrounds but with high motivation and strong promise. Once admitted, these students are expected to meet the same standards as other students. This is achieved through small class sizes and nurturing by the faculty, and cooperative learning. The program has also reached out to a large number of part-time students, women, and non-traditional students of all races who may otherwise not have access to planning education. Classes are intentionally held in the evenings to accommodate working students. All this is done because the program embraces the HBCU and the university mission of providing access to graduate education for non-traditional students. Students are allowed to attain their master's over a period of seven years and even granted an extension in cases of exceptional hardships. Morgan also offers one of the most affordable accredited planning programs. Clearly, Morgan has an empowering culture where students from various ethnic groups, races, and class experience educational equality and cultural empowerment.

Texas Southern University

Texas Southern University is an urban university by location and mission. It is located in the inner city of Houston, which, according the US census, is the fourth largest city in the United States. Houston is also considered one of the most racially and ethnically diverse cities in America (Gates, 2012). The university became a four-year college in 1934 and transitioned from Houston's College for Negroes (1934–47) to Texas State University for Negroes (1947–51), and to Texas Southern (Texas Southern University, 2013). In 1951, after students petitioned the state legislature to remove the phrase "for Negroes," the school's name was changed to Texas Southern University (Texas Southern University, 2013). While the university has always been charged with educating minorities within the urban environment, it was not until 1972 that then-President Granville Sawyer stated officially that Texas Southern University was an "urban university" and that "everything we do, everything we project, all that we anticipate is to be evaluated in terms of what any given consideration offers toward the resolution of present problems in the urban community" (Sawyer, 1973, p. 41). He recognized that the university was a city within the city (Sawyer, 1976) and in his convocation address and subsequent writings, he acknowledged the complexities that many people living together can present and the university's responsibility to allocate resources to address these issues, "for blacks, the poor and their neighbors, whatever their race or condition" (Sawyer, 1973, p. 44).

In that same vein, the faculty and the student body of the Urban Planning and Environmental Policy Program at Texas Southern University represent the varied mix of races, ethnicities, and ages found not only in the US but across the world. Students and faculty from Asia, Africa, and South America study and work with African American, Caucasian, and Hispanic students and faculty from the US. All come together to Houston and Texas Southern University with a primary interest in the urban setting and its people and the natural and built environments. The program's evolving mission is to train policy-oriented planners and environmental policy analysts for leadership positions in planning and environmental policy-related organizations. The program's goal is to equip future professionals with analytical and policy formulation skills that will enable them to address, with vision and foresight, the current and future environmental problems caused by human impact on the environment (Urban Planning, 2011). The primary learning laboratory is the Houston region, expanding into the Gulf Coast communities. The dynamics, opportunities, and challenges that take place as a result of the area's diversity provide a rich learning environment for the next generation of planners, whose job it will be to represent and work with these individuals and neighbourhoods.

As is the case with other HBCUs, Texas Southern University faces resource challenges and must also overcome negative perceptions. Faculty are challenged with offering the requisite accreditation courses or subject matter, expanding the students' knowledge base, preparing students to be successful in their chosen profession and community, and creatively holding to their core values of promoting justice and equity.

At Texas Southern University, content integration starts by expanding the sources of information provided to the students in the form of required or supplemental readings. Students are expected to seek out works not cited as seminal texts in the planning profession but reflective of their interest or their perspective on a specific issue or time frame. For example, the history course requires readings by W.E.B. Dubois and John Hope Franklin. Discussions on the reform movements include both the traditional Jane Adams's Hull House and the Garveyism Movement with the United Negro Improvement Association. Both readings expose students to those efforts that impacted the immigrant and African American populations, the urban reform movement, and city development.

Students have asked questions about and are challenged to go deeper in finding their own understanding of the role of immigrants, minorities, and women at the Chicago World's Fair. Wells (n.d.) explains the presence of exhibits of peoples in an atmosphere where those same peoples were not allowed admittance into the fair. The impact of "ethnography" on the African American way of life is also examined. The end assignment is to reconstruct history going beyond the planning historical timeline. Students select a city to explore and expand on the peoples, events, and writings that contributed to the city's development. The result is recognition of the significant contribution that diverse people, especially minorities, have made to society.

As the years pass, the student body continually represents a younger cohort and more nationalities from outside the US, requiring videos and documentaries to assist the students in visualizing and grasping pivotal events and issues in US history. The most effective are those that illustrate the civil unrest of the 1960s from the vantage points of women, Hispanics, and African Americans, played out on topics such as the environment and public housing. All efforts are designed to assist students to effectively tie theory to context and understand that all groups had a role to play in the development of cities.

Knowledge construction and equity pedagogy go hand in hand at Texas Southern. Knowledge construction and equity pedagogy recognize that people learn differently and that teaching styles must acknowledge and adapt to these differences to facilitate the students' grasp and application of information. Added to this dynamic, in any given year, approximately 50 per cent of the students are non-traditional in that they are older than the typical 21- to 24-year-old graduate student. Often, they are employed full time and have children. In addition, their backgrounds vary by income, culture, and life exposure. Their experiences shaped their perspectives and are therefore not discounted. The faculty recognize that they teach adults (andragogy), who have history and a unique and insider's perspective of the communities in which they live and the challenges many of these communities face. Current students and alumni are able to effectively convey examples to younger students through storytelling. Storytelling is an art in minority communities and is used at Texas Southern as a way to provide tangible examples of concepts so that students are able to crystallize the ideas in their minds and are then better able to see the potential for application.

In reflection, faculty facilitate knowledge construction in various ways through an equity pedagogy that represents a combination of field-independent and field-dependent approaches (Bennett, 2003). In addition to storytelling, students are required, in various courses, to meet and interview individuals who are salient to the project or subject they are studying. The goal is to move them away from media and its interpretation to acquire first-hand knowledge and information by creating the opportunity to ask questions and probe deeper into a problem. Students must collect and analyse their own field data through surveys or observations and compare their information to existing reports, theories, and concepts. This is important not only to grasp critical skills but to also understand that data can be misrepresented and misunderstood and to examine how groups are advantaged or disadvantaged depending on how numbers, statistics, or facts are presented.

Not all students accurately express their grasp of the information in written papers or exams. And, while the ability to express oneself in written form and successfully pass exams such as the AICP certification exam is important, the path to achievement is varied. Students can demonstrate their grasp of the concepts or produce their evidence of learning in ways that reflect the subject matter. For example, history students participate in a debate using *Ten Successes That Shaped the 20th Century American City* (Gerckens, 2000) as a base. They can choose from Gerckens's selected list or introduce one of their own, but must substantiate their argument from the course material and beyond. In the Community Development course, students host public meetings with city officials on a topic covered in class. Hosting the meeting includes planning, communicating to the public, conducting the meeting, and providing follow-up analysis. This includes reflection on who was represented, who was not, and what efforts were made to include the traditionally disenfranchised.

In essence, a large part of knowledge construction and what is emphasized at TSU is critical thinking. As information is presented in the media, students are asked who is affected, how they are affected, and what other information exists that we do not know. These questions are posed to challenge their perspectives, their way of thinking, and what they believe they know to be true. For minority and lower-income students, this can be especially challenging if their life exposure has been limited prior to entering the field. Letting go of long-held beliefs, even if one is not sure where these beliefs originated, is in itself a process. The result has been a call to action and the recognition of the power of involvement such as the starting of a support group for transgendering young adults, gaining membership to park boards, and organizing community clean-ups.

Alumni working in the planning or related professions are the most effective way to reduce existing prejudices. Graduates from the Urban Planning and Environmental Policy Department are prominent and effective within the various planning, housing and community development, and public works departments in Houston. In addition, they head non-profit efforts throughout the city. Alumni have a strong sense of giving back to the program; having faced similar challenges when pursuing their degree, alumni work with faculty and students to create internship opportunities that complement students' work schedules. These internships expose students to the planning profession and assist them in understanding the policy-making process, seeing how diverse peoples must work through their differences and come together in order for progress to be made.

Current students recognize the benefits of gaining exposure as part of the educational process. Students are an integral part of the $33 million revitalization effort of Emancipation Park, a historical park in Houston's Third Ward, heading neighbourhood clean-ups, volunteering for homeless shelters, and assisting with community sustainability efforts. These efforts have resulted in, among other things, the naming of a student to the Houston Clean City Commission Board of Directors. Students, past and present, understand that racial barriers continue to exist; however, through alumni support and their own efforts, they work to recognize and overcome such barriers.

Embracing multicultural education encompasses more than the willingness of the faculty. Administration and accreditation entities must also recognize that "one size does not fit all," which Texas Southern does. Conversations regarding graduation rates, matriculation rates, percentage part-time or full-time enrolment, and external funding have an impact on the context or atmosphere under which students matriculate. It is clear that graduate students must progress to graduation. However, the program maintains that a 48-hour master's program may not be attainable within two years. Courses are offered in the evening, which allows students to obtain or maintain employment. The average semester load is three courses or nine credit hours. Students are able to complete their 48 credit hours with a combination of fall and spring semester and summer session courses. This allows the majority of students to graduate in two and a half to three years.

Dedicated alumni of the department are integral to an empowering of the school's culture. Alumni contribute their resources to ensure that the department as a whole is in tune with the skills and experience students need to be competitive in the professional market. In addition, alumni willingly serve on department, college, and city boards and committees, where they advocate for the student body. When the opportunities present themselves, alumni craft junior committee positions, allowing students to experience the committee or boardroom culture and thus be better prepared upon graduation.

The department realizes that for many of its students, some form of financial support is necessary, so financial assistance is provided. In addition, when they can be made available, departmental assistantships and book awards are provided to defray the cost of education. Pursuing a Master's of Urban Planning at Texas Southern is one of the more affordable programs in the United States. Therefore, students are able to pursue the program with a level of comfort that they will have the financial means to complete it once they start. Texas Southern creates an empowering school culture by providing students with unique exposure to the professional world, an opportunity to graduate over a longer period of time, evening classes, affordable tuition, financial aid, and other assistance.

Jackson State University

In January 1975, Jake Ayers, Sr, a factory worker and local civil rights activist from the small town of Glen Allen, Mississippi, filed a lawsuit on behalf of his son and other Black students in Mississippi. The lawsuit accused the state of operating a segregated system of higher education and systematically underfunding the state's three HBCUs. Joined by the US three months later with complaint-in-intervention, the lawsuit charged "that state officials had failed to satisfy their obligation under, inter alia, the Equal Protection Clause of the Fourteenth Amendment and Title VI of the Civil Rights Act of 1964 to dismantle the dual system" (*United States v. Fordice,* 1992) of maintaining five almost completely White and three almost completely Black universities. After protracted court battles, a $503 million settlement was reached that included funding for expanded programs at three HBCUs (Jackson State University, Alcorn State University, and Mississippi Valley State University). The Urban and Regional Planning (master's and PhD) program was one of the programs specifically mandated in the settlement.

The first accredited master's program in Urban and Regional Planning in the State of Mississippi was at the University of Mississippi from 1968 to 1986 (Planning Accreditation Board, 2013c). After 1986, for 13 years, no planning program existed at any of the universities. As part of the Ayers Settlement, a Department of Urban and Regional Planning was created at Jackson State University. The settlement called for initiating graduate programs in areas that were assessed to be lacking in the state. Hence, though not explicitly noted, these graduate programs in planning were initiated to fill a clear gap of planning education in Mississippi. Since its inception in 1999, the master's program has graduated over 80 students, about half of whom serve in a planning capacity with various local and state governments within the state. The master's program was accredited in 2009 for five years.

The two main racial groups in the state of Mississippi are African American and White, 37 per cent and 59 per cent respectively of the population in 2010. The largest city in the state, Jackson peaked in 1980 with a population of about 203,000. Since then the population has steadily declined to 197,000 in 1990, to 184,000 in 2000, and to 174,000 in 2010. The estimated population for 2012 showed an increase to almost 176,000. During the years of decline in the central city, the nearby suburbs and counties gained significant population. For 2010 the metropolitan area had a population of 539,000, of which 53 per cent were White and 45 per cent were African American. This is in sharp contrast to the central city with over 79 per cent Black and about 18 per cent White. While the metropolitan area is more reflective of the state's racial makeup, the city of Jackson is overwhelmingly Black. White flight from Jackson has occurred at a very high rate during the last three decades. As a result, many parts of the city are often overwhelmingly Black (over 95 per cent) and very poor. Since the inception of the planning program at Jackson State University, the program recognized that the change in the racial makeup of the city also paralleled the decline of middle-income Black neighbourhoods, and this provided a fertile opportunity to influence planning practices in the city.

The curriculum at Jackson State University was set up to orient students and faculty towards developing appropriate responses to the challenges that racial history posed for city planning in general and the city of Jackson in particular. Core courses in the program reflect this emphasis by integrating the history of early planning practices in the city that resulted in the development of segregated neighbourhoods. While core courses extensively prepare students to understand the historical progression of local planning that marginalized minority communities, studio courses offer opportunities to create plans that fit the neighbourhoods not in terms of values that are external to the communities but in terms of recognizing and highlighting the values that often emerge from active community engagement.

Knowledge construction at Jackson State University takes place within the context of planning practices in the city of Jackson. The first Black mayor of the city, Harvey Johnson, was singularly important in bringing about a very different approach to planning for the city. One of the achievements of Harvey Johnson's administration (1997–2001) was the production of Jackson's first comprehensive master plan, titled Fabric, an acronym for For a Better Revitalized Inclusive Community. The effort to prepare this comprehensive plan coincided with the university's inception of the Department of Urban and Regional Planning, and so several faculty members were directly involved with the public efforts to produce a comprehensive plan that reflected the value preferences of the wider citizenry. This was the first significant attempt to comprehensively address the issues of zoning, transportation, and infrastructure. However, few of the components of the comprehensive plan were implemented. More recently, another visioning process called Vision 2022 commenced in 2012 as an attempt to provide a comprehensive vision for the metropolitan area focused on the core city of Jackson. The process is led by the Greater Jackson Chamber of Commerce and is expected to produce a 10-year plan to address the issues of creating places, wealth, talent, and connections (Greater Jackson Chamber of Commerce, 2013). At present it is widely viewed as a business-supported process, and is expected to utilize a participatory visioning process to produce a strategic plan that is embraced by both economic and political leaders.

The Department of Urban and Regional Planning is similarly directly involved with components of the visioning process that deal with the city of Jackson. Faculty members and several students serve on committees that are expected to generate components of the long-term vision for the city.

Jackson State's success in incorporating prejudice reduction in its curriculum is evidenced from its alumni's contribution to transforming planning in the city. Until very recently, as has been typical of most southern cities, planning in Jackson was led by a growth coalition, composed of large developers, engineering firms, architecture firms, and other business and finance elites. Sixty-five per cent of all businesses in Jackson are owned or controlled by Whites, and almost no large businesses are owned or controlled by Blacks. Until recently, it meant that the growth coalition was largely White and operated with a highly prejudiced view of minority business owners, minority business leaders, and poorer neighbourhoods in general. Over the last decade there has been a remarkable shift, and this coalition that easily found allies among most political elites and planners has gradually lost power to influence planning. This growth coalition had hitherto neglected all areas of the city with the exception of downtown, since this was where its direct business interests were concentrated. There was no concern for planning in any of the inner-city neighbourhoods, as most of the planning occurred either for downtown revitalization (mostly business-led developments) or developments at the suburban fringe.

The growth coalition came head to head with politically empowered minority communities during the 2013 mayoral election, when the candidate favoured by the coalition was defeated by a populist progressive candidate. Although the effects of this election result will be felt over the years, there is evidence of the formation of progressive coalitions that are forcing a fundamental shift in thinking about planning in Jackson.

There is a highly inclusive master plan effort underway for West Jackson, the poorest part of the city. The current planning director (an alumnus of the URP program at Jackson State University) has been leading several planning efforts that aim more directly at bringing the community interests to the table, unlike the earlier efforts at introducing the typical mainstream approach of promoting large, isolated individual projects in the name of growth and development. Several of these efforts include involvement of churches in the neighbourhood association activities, strategic use of overlay districts to enhance the immediate needs of communities, and rethinking transportation corridors as integral to diverse community needs, to name a few.

Jackson State's success in incorporating equity pedagogy in its curriculum and empowering school culture is evidenced through its alumni's contribution to transforming planning in the city. As graduates of the programs, many of whom come from underprivileged backgrounds, have taken on influential positions among the city and business leaders, there has been a significant change in how issues of infrastructure investment, regional planning, and economic development projects are strategized and implemented. Until recently, the distribution of costs and benefits was almost never questioned. Often costs were borne by the central city residents (majority Black), while the benefits accrued to either the privileged few inside the city or the largely White population in the suburban areas. Examples include development of the airport and a reservoir outside the city limits. In both cases, the city of Jackson provided significant financial support for the projects and yet has no influence on the consequent developments or control of direct financial benefits accruing from the developments. Lately, however, as new large projects are proposed, a vocal group of concerned citizens repeatedly demand a more equitable distribution of costs and benefits on the lines of race, income, and political representation.

Although the Urban and Regional Planning program at Jackson State University cannot take all the credit for this fundamental shift in planning in Jackson, it is no mere coincidence that the city is witnessing a more progressive planning movement as more and more of the program's graduates join the ranks of city administration. Several of the program's alumni, who either work for the city or lead neighbourhood planning associations, and faculty of the program have pushed the city towards adopting a more progressive planning agenda. In alumni surveys, one of the more common statements relates to how their education at Jackson State University influenced their thinking on considerations of inner-city neighbourhoods through a more participatory and bottom-up, as opposed to top-down, approach. All of the studio courses in the program deal directly with one or more of the severely declining inner-city neighbourhoods. Through a highly active internship program with the city, the program maintains a close relationship with the city's planning department that mutually benefits the program and the city by disseminating newer progressive approaches to planning and preparing graduates with a deep understanding of the planning processes influencing inner-city neighbourhoods.

The transformation observed over the last decade in the city of Jackson is quite eye-opening in terms of how urban planning has taken a significant turn towards progressive views on economic development, environmental concerns, neighbourhood designs, and affordable housing. In the first major effort to prepare a comprehensive plan for the city, the section on transportation simply endorsed the transportation plans proposed by the local MPO with little awareness that this did not address the most pertinent issues of the minority city residents. Lately, however, transportation is being discussed more as an enabler of community economic development than as a mere traffic management plan that calls for building more roads. There is a planning process underway that is looking into a regional trails system, complete streets, bike paths, and enhanced public transportation, and many of these efforts are led by alumni, current students, and the faculty.

Conclusion

The case studies of the three HBCU planning programs demonstrate how they operationalize and incorporate the five dimensions of educational reform for multicultural planning education developed by Banks (1995a & b, 2001). Their contribution to increasing minority representation in the planning profession is also significant. The importance of their contribution to increasing minority representation in planning can be further understood from the fact that four HBCUs are providing a quarter of the supply of African American planners from a total of 69 master's programs in the United States. Given the makeup of their student bodies, faculty interest, and extensive utilization of local inner-city communities in studio courses and projects, all of the HBCUs directly contribute to diverse and multicultural planning practices in their respective geographic areas. A clear emphasis on issues of equity and justice in their curricula not only prepares their graduates for a multicultural society but also affords them better opportunities to bring minority concerns to mainstream planning practices.

Although these planning programs have made highly valuable contributions to changing the planning practices in their respective cities, much work is yet to be done. Cities throughout the country continue to work with a paternalistic process in which minority concerns are treated as public input to a somewhat participatory process without adequate institutional mechanisms to incorporate such concerns into the initial phases of planning (also see Burayidi's Introduction, Doan's chapter 6, Zaferatos's chapter 7, Nguyen, Gill, and Steephen's chapter 9, and Harwood and Lee's chapter 10 in this context). Planning programs at HBCUs will require institutional support from both their respective institutions and our professional bodies – the Planning Accreditation Board, the American Planning Association, and the Association of Collegiate Schools of Planning – to accomplish the objective of inclusive planning practices in their respective cities in particular and the nation in general.

While the Planning Accreditation Board (PAB) has introduced new language to include diversity in planning curricula, it is a paradox that it is undercutting HBCU planning programs that have done such an outstanding job not only in increasing diversity in the profession but also in teaching diversity. The biggest threats to HBCUs' ability to continue their tradition of imparting a multicultural education and serve a diverse population are the new PAB criteria for accreditation: criteria 2C: Size of the student body, 6C: Student retention and graduation rates, and 6D: Outcomes (Planning Accreditation Board, 2015). The requirement of a minimum student population as stated in criterion 2C may be the biggest impediment to planning programs at HBCUs, since many such programs are usually small with a large number of part-time students, minorities, women, and non-traditional students who may otherwise not have access to planning education. Such students often do not have the resources to attend graduate school full-time. The new accreditation criteria are likely to undermine access to planning education for the above-mentioned population groups and HBCUs' mission of providing access to graduate education for such groups. Because of disproportionate funding, most HBCUs have limited financial aid for recruiting students. Program degree productivity as specified by the new criterion 6C could also be a concern for small HBCU programs with part-time students. Such students take longer to graduate due to work and family obligations. Furthermore, the HBCU tradition of admitting students who may be somewhat underprepared for graduate work but have high motivation may also extend students' time to graduate and therefore affect graduation rates. Alternative means for assessing student and program outcomes such as a cohort-dependent graduation rate can accurately reflect the true graduation rates for such programs. Graduation rates should also take into account the context of overall HBCU graduation rates, which are generally low.

Finally, the new criterion 6D to quantify all outcomes data may pose an additional burden on HBCU programs. Most HBCU planning programs do not have sufficient resources to collect the "outcomes" data that is now required. While we should know what happens to our graduates, we need to examine the broader trends and build analyses around data that is obtainable. There should be some flexibility in letting small HBCU programs compile anecdotal and qualitative data and determine what kind of data they can collect to document alumni success and satisfaction. Success in passing the AICP is also not an accurate measure for showing the success of alumni in HBCUs, as their graduates may choose alternative planning careers. Letting HBCUs decide what a successful planning career is and what are the indicators of success will be more meaningful. All these reform measures will help HBCUs to continue to serve as institutional models for increasing diversity in the profession and support their role for multiculturalism planning education.

The PAB can better integrate multiculturalism in planning education by introducing an additional criterion under the "Programs Assessment" (Accreditation Standards and Criteria 6) (Planning Accreditation Board, 2015). This new criterion should require all programs to demonstrate how they are increasing diversity in planning education. This would change planning education in general, by encouraging all programs to make a serious effort to integrate diversity in their programs in terms of the student and faculty composition as well as the curriculum.

The discussion presented in this chapter has serious implications for planning theory and practice. As pointed out by Qadeer, in chapter 3, multicultural planning theories are normative and prescriptive. By teaching students how to deal with diversity, HBCUs are contributing towards the development of normative theories of multicultural planning. As we have demonstrated, HBCUs are training planners that have the skills and cultural competency to broker among various identities and cultures to reach the agreements that Rios calls for in chapter 14. They are also well versed in the communicative planning methods that Talen argues for in chapter 11 to plan for the sustainability of diverse communities. In this way, HBCUs are contributing to practice, by training planners who are better equipped to deal with multicultural cities. Finally, HBCU programs are also contributing to planning practice by increasing minority representation in the ranks of planners. Certainly, planning practice can deal with diversity in better ways than what is discussed in several chapters of this book (see Burayidi's Introduction, Doan's chapter 6, Zaferatos's chapter 7, Nguyen, Gill, and Steephen's chapter 9, and Harwood and Lee's chapter 10) if the ranks of practising planners become more diverse.

NOTE

1 The section on Morgan State is based on numerous interviews and reports. For a detailed discussion of these reports and interviews, see Sen (1997).

The data has been updated since then through participant observation and other reports that are referred to in this chapter.

REFERENCES

Agyeman, J., & Erickson, J.S. (2012). Culture, recognition, and the negotiation of difference: Some thoughts on cultural competency in planning education. *Journal of Planning Education and Research*, *32*(3), 358–366. http://dx.doi.org/10.1177/0739456X12441213

Ashley, D., Gasman, M., Mason, R., Sias, M., & Wright, D. (2009). *Making the grade: Improving degree attainment at historically Black colleges and universities (HBCUs)*. New York: Thurgood Marshall College Fund.

Banks, J.A. (1995a). Multicultural education: Historical development, dimensions, and practice. In J.A. Banks & C.A. Banks-McGee (Eds.), *Handbook of Research on Multicultural Education*. New York: Simon and Schuster Macmillan.

Banks, J.A. (1995b). Multicultural education and curriculum transformation. *Journal of Negro Education*, *64*(4), 390–400. http://dx.doi.org/10.2307/2967262

Banks, J.A. (2001). Multicultural education: Historical development, dimensions, and practice. In J.A. Banks & C.A. Banks-McGee (Eds.), *Handbook of Research on Multicultural Education*. San Francisco: Jossey-Bass.

Bennett, C.I. (2003). *Comprehensive multicultural education: Theory and practice* (5th ed.). Boston: Pearson Education.

Brown, M.C., & Ricard, R.B. (2007). The honorable past and uncertain future of the nation's HBCU. *Thought and Action*, special issue, 117–130.

Burayidi, M.A. (2000). Tracking the planning profession: From monoistic planning to holistic planning for a multicultural society. In M.A. Burayidi (Ed.), *Urban planning in a multicultural society* (pp. 37–51). Westport, CT: Praeger.

Burayidi, M.A. (2003). The multicultural city as planners' enigma. *Planning Theory & Practice*, *4*(3), 259–273. http://dx.doi.org/10.1080/1464935032000118634

Chapman, T.K. (2004). Foundations of multicultural education: Marcus Garvey and the United Negro Improvement Association. *Journal of Negro Education*, *73*(4), 424–434. http://dx.doi.org/10.2307/4129626

Citrin, J., Sears, D.O., Muste, C., & Wong, C. (2001). Multiculturalism in American public opinion. *British Journal of Political Science*, *31*(02), 247–275. http://dx.doi.org/10.1017/S0007123401000102

Forsyth, A. (1995). Diversity issues in a professional curriculum: Four stories and suggestions for a change. *Journal of Planning Education and Research, 15*(1), 58–63.

Fryer, R.G., & Greenstone, M. (2007). The causes and consequences of attending historically Black colleges and universities. *National Bureau of Economic Research*, working paper 13036.

Gates, S. (2012). Houston surpasses New York and Los Angeles as the "most diverse in nation." Retrieved May 20, 2015 from *http://www.huffingtonpost.com/2012/03/05/houston-most-diverse_n_1321089.html.*

Gerckens, L. (2000) Ten successes that shaped the 20th century American city. *Planning Commissioner's Journal*, no. 38 (Spring), 3–11.

Goonewardena, K., Rankin, K.N., & Weinstock, S. (2004). Diversity and planning education: A Canadian perspective. *Canadian Journal of Urban Research, 13*(1), supplement, 1–21.

Greater Jackson Chamber of Commerce. (2013). Vision 2022. Retrieved August 20, 2012 from http://www.greaterjacksonpartnership.com/files/3103.pdf.

Hartmann, D., & Gerteis, J. (2005). Dealing with diversity: Mapping multiculturalism in sociological terms. *Sociological Theory, 23*(2), 218–240.

Higher Education Act. (1965). Pub.L. 89–329. 79 Stat.1219. Retrieved February 10, 2013 from http://c.ymcdn.com/sites/www.ncher.us/resource/collection/90515964-F9A5-45E4-83E5-06C2A26E3125/00004C57%28OriginalHEAof1965%29.pdf.

Knight, L., Davenport, E., Green-Powell, P. and Hildon, A.A. (2012). The role of historically Black colleges or universities in today's higher education landscape. *International Journal of Education*, 4(2), 223–235.

Milroy, B.M. (2004). Diversity and difference: A comment. *Canadian Journal of Urban and Regional Research, 12*(1) supplement, 47–49.

Morgan State University. (2011). *Growing the future, leading the world: The strategic plan for Morgan State University, 2011 – 2021*. Baltimore: Morgan State University.

Morgan State University. (2015a). Mission statement. Retrieved from http://www.morgan.edu/about_msu/mission_and_vision.html.

Morgan State University. (2015b). The Morgan community mile. About us. Retrieved from http://communitymile.morgan.edu/?page_id=587.

Planning Accreditation Board. (2013a). Student and faculty data. Retrieved from http://www.planningaccreditationboard.org/index.php?s=DataLibrary.

Planning Accreditation Board. (2013b). Accredited planning programs. Retrieved from http://www.planningaccreditationboard.org/index.php?id=30.

Planning Accreditation Board. (2013c). History of accredited programs. Retrieved from http://www.planningaccreditationboard.org/index.php?id=29.

Planning Accreditation Board. (2015). Accreditation standards. Retrieved from http://www.planningaccreditationboard.org/index.php?s=file_download&id=212.

Rahder, B., & Milgrom, R. (2004). The uncertain city: Making space(s) for difference. *Canadian Journal of Urban and Regional Research*, *12*(1), supplement, 27–45.

Rodriguez, S. (1993). Schools for today, graduates for tomorrow. *Journal of the American Planning Association*, *59*(2), 152–155. http://dx.doi.org/10.1080/01944369308975864

Sandercock, L. (2003). Planning in the ethno-culturally diverse city: A comment. *Planning Theory & Practice*, *4*(3), 319–323. http://dx.doi.org/10.1080/1464935032000118661

Sawyer, G.M. (1973, Sept.). One university's urban commitment. Revised article of paper presented at the opening convocation, Texas Southern University, Houston, Texas.

Sawyer, G.M. (1976). *The cities within the city*. Unpublished manuscript. Texas Southern University, Houston, Texas.

Sen, S. (1997). The status of planning education at historically Black colleges and universities: The case of Morgan State University. In J.M. Thomas & M. Ritzdorf (Eds.), *Urban planning and the African American community: In the shadows* (pp. 239–257). Thousand Oaks, CA: Sage.

Sen, S. (2000). Some thoughts on incorporating multiculturalism in urban design education. In M.A. Burayidi (Ed.), *Urban planning in a multicultural society* (pp. 207–224). Westport, CT: Praeger.

Sen, S. (2005). Diversity and North American planning curricula: The need for reform. *Canadian Journal of Urban and Regional Research*, *14*(1) supplement, 121–139.

State Board of Higher Education as cited in Morgan State University Catalog. (1984–1986). Baltimore: Morgan State University.

Sweet, E.L., & Etienne, H.F. (2011). Commentary: Diversity in urban planning education and practice. *Journal of Planning Education and Research*, *31*(3), 332–339.

Texas Southern University. (2013). History. Retrieved from http://www.tsu.edu/About /History.php.

Thomas, J.M. (1996). Educating planners: Unified diversity for social action. *Journal of Planning Education and Research*, *15*(3), 171–182. http://dx.doi.org/10.1177/0739456X9601500302

United States Commission on Civil Rights. (2010). The educational effectiveness of historically Black colleges and universities: A briefing before the United States Commission on Civil Rights in Washington DC. Retrieved from www.usccr.gov/pubs/HBCU_webversion2.pdf.

United States v. Fordice. (1992). 505 U.S. 717. Majority opinion by Justice White. Retrieved from https://www.law.cornell.edu/supremecourt/text/505/717.

Urban Planning and Environmental Policy. (2011). Program overview. Retrieved May 20, 2015 from http://bjmlspa.tsu.edu/departments/urban-planning-environmental-policy/.

Wells, I.B. (Ed.). (n.d). *The reason why the colored American is not in the world's Columbian exposition:* The Afro-American's contribution to Columbian literature. Retrieved May 20, 2015 from http://www.digital.library.upenn.edu/women/wells/exposition/exposition.html.

Wolfe, J.M. (2003). Politics and planning schools. *Plan Canada*, *43*(3), 15–17.

Wubneh, M. (2011). Commentary: Diversity and minority faculty perception of institutional climate of planning schools – results from the climate survey. *Journal of Planning Education and Research*, *31*(3), 340–350. http://dx.doi.org/10.1177/0739456X11402089

16 Moving the Diversity Agenda Forward

MICHAEL A. BURAYIDI

Introduction

The contributors to this book have discussed how cities and urban planners are attending to the politics of difference in cities of the twenty-first century and provided practical ways to manage this difference. They have also provided pragmatic ways to make planning more effective in a pluralist society.

Learning to get along is one of the greatest challenges of the twenty-first century, and planners can help create living environments that promote social cohesion rather than fragmentation. This is especially so in those developed countries where an unprecedented number of immigrants from non-European countries are transforming the demographic composition. Planners are accomplices in creating just outcomes, since spatial and physical development policies have a direct bearing on where people live, whom they encounter on a day-to-day basis, and whom they interact with in their lives. The native and immigrant multiculturalism policies discussed throughout this book support enhancing the social contract between host countries and their multiple publics.

The social contract has been a long-standing political philosophy in Western democracies since at least the seventeenth century. It is a mutual covenant between the state and the governed, who freely give up their rights to the state in return for the creation of a civil society and for the protection of the rights of all. Without this, Hobbes contends, life will be "solitary, poor, nasty, brutish and short." A social contract then is an unwritten compact between individuals and society about the expectations of relationship between the state and the individuals who live in it. It is from this social contract that government is obligated to care for the poor and infirm, to protect the elderly and children, and to defend the rights of its citizens. Fair housing laws, building codes, even zoning ordinances all spring from this well of the social contract.

Myers (2008) views the social contract "less as a statement of political philosophy than as an expression of shared understanding of a unified purpose" (p. 153). This unified purpose is the glue that holds society together. With respect to managing diversity in a pluralist society, the social contract protects the rights of minority groups and in turn obligates them to abide by the principles of a civil society. According to Kymlicka (2004), in Canada the social contract motivated the national government's policy of multiculturalism, which was initially conceived to recognize the duality of English and French as the "Founding Nations" of the country and eventually broadened to include all ethnic and minority groups.

In the US the social contract was invoked by African Americans in seeking the extension of civil rights to all persons in the country. As in the case of Canada, the social contract also protects the rights of women, the elderly, children, and ethnic minorities. In turn, society expects that the values that bind and provide common ground in a civil society are upheld by everyone. Myers (2008) argues that the social contract in the US is composed of three strands: 1) cultural cohesion and the American creed; 2) unrestrained upward mobility; and 3) collective protections and services (p. 156). Cultural cohesion does not imply cultural homogeneity but refers to shared belief systems, particularly that of "civic unity through liberal democratic and egalitarian principles" (p. 161). American society also provides the opportunity for everyone, regardless of their family background, history, or place of origin, to become whoever they want to be as long as they work hard and play by the rules that govern a civil society. In turn, society provides the social net and protection for those in need. Planners have a role to play in advancing each of these strands.

The social contract obliges both native residents and recent immigrants and their host country. By agreeing to migrate to a new country, immigrants are implicitly accepting the rules and regulations as well as the values of the host country. Thus, it is important that immigrants understand the principles embedded in the social contract when they migrate to Western countries. In many cases, however, immigrants do not learn about the host country's history and values until much later, after they have already settled in the country. As Harper and Stein observed in chapter 2, liberal democratic societies cannot assume the resilience of their core values with increased migration and lapping with new cultures. Thus, these core values, such as tolerance, democracy, liberty, and freedom of the individual, have to be persistently communicated to the citizenry, especially new immigrants. As a precondition for a visa, immigrants must understand and accept these core liberal democratic values of the country to which they migrate.

It is important that these values are communicated to immigrants early in the immigration process, perhaps starting at the embassies of the host countries when an application for a visa is made. In the US, for example, the history, values, and governing principles of the country are not taught to immigrants until they apply for citizenship. At this time, the person must have lived in the country and been a "permanent resident" for at least five years. This requirement comes far too late in the process. As part of the responsibilities for immigrating and living in the host country, immigrants should first go through an orientation process that teaches them about the values of the host country. Agreement to uphold these values while visiting or living in the country should be required at this time.

In *Assimilation American Style,* Peter Salins (1997) prefers the metaphor of "religious conversion" to that of the "melting pot" to describe his view of assimilation. This is because, while new immigrants to the US are converted to the values of liberal democracy, they are free to discard as little or as much of the culture of their homeland as they wish and do not need to change their behaviour in other respects. Planners can assist the host country in upholding its part of the social contract and "religious conversion" of recent immigrants, a task to which we now turn.

Steps for Delivering the Planning-Related Social Contract

The purpose of planning is to help communities make the transition from where they are to where they want to be. This requires that planners develop an intimate knowledge of the community for and with whom they plan. In a pluralist society, effective planning requires that planners: 1) acknowledge that diversity and difference matter in planning; 2) plan with difference in mind; and 3) enhance social cohesion in communities with diversity. We discuss each of these requirements below.

Acknowledge Diversity and Multiculturalism

The just city must recognize and embrace difference in all its forms. Despite the demographic shifts that are taking place in the West, some planners still contend that difference does not matter to planning practice. Such planners remain committed to treating everyone the same, using universal planning principles. The contributors to this book provide convincing evidence of why acknowledging cultural pluralism may be the first principle of effective planning.

Qadeer's discussion in chapter 3 reiterates the value of diversity and inclusivity in planning but points to the limits of urban planning in creating the just city. He argues that the jurisdiction of planners is limited to helping frame policies that advantage the poor and those that are marginalized in society. Nonetheless, planners can inject diversity concerns (race, ethnicity, and immigrant status) into policy making to complement those of social class in the allocation of resources. Qadeer believes that the urban landscape already manifests planners' attention to diversity. As examples, he points to the development of ethnic neighbourhoods and enclaves and ethnic economies and places of worship that now dot cities in North America as testimony of planning practice's successful response to diversity.

In chapter 4, Laws and Forester discussed the conflict that arose between native Dutch residents and Moroccan immigrants in the West borough in Amsterdam and how a resolution of the case required an appreciation of the differences in cultures between the two groups. Whereas native Dutch readily apologized for a wrong they committed, Moroccan immigrants saw an apology as an affront to their honour, and although they may acknowledge the wrongdoing, would never openly apologize because they are constrained by their cultural practices. Planners would do well to understand these cultural norms to help resolve conflicts that are likely to arise in the increasingly diverse multicultural milieu of the twenty-first-century city.

Doan in chapter 6 showed how pursuing universal planning practices worked to the detriment of Atlanta's non-normative population. The attitudes of planners and Atlanta's civic leaders to the non-normative population in the city's Midtown neighbourhood were due to ignorance of the gay culture and fear of the "other." As a result, the LGBT residents in the neighbourhood were not recognized and given a voice in the redevelopment process. The situation cried out for advocacy from the planning profession but was met with a resounding silence.

When a group feels the danger and ostracism of the outside world, it recoils into safe places and neighbourhoods. In doing so, it gains the support of its members. This is how new immigrants responded to perceived and real threats from the larger society in Western Europe and how the non-normative population in Atlanta responded to the cruelty of the outside world when it sought safety in Atlanta's Midtown neighbourhood.

However, the redevelopment of the neighbourhood did not value the contributions of the LGBT residents. The displacement of gays who initiated the gentrification of the Midtown neighbourhood accorded a privileged position to business and favoured urban renewal and "order" over incremental and organic redevelopment of the neighbourhood. Thus Atlanta's "governing coalition," driven by business interest and xenophobic attitudes, saw the LGBT residents as a "problem" to be fixed. This not only destroyed a safe haven for the group but also made its members more vulnerable to the larger society.

The observations made by Stone several years ago about Atlanta's growth coalition remain instructive: "business excels in getting strategically positioned people to act together, thereby expanding its realm of allies and imposing opportunity costs on those who decline to go along" (Stone, 1989, p. xi). In the course of redeveloping the Midtown neighbourhood, it was the non-normative population on whom the opportunity cost was imposed. A more recent assessment of the redevelopment of the Midtown neighbourhood by Fowler reached similar conclusions: "A close analysis of recent planning efforts in Midtown Atlanta reveals that while a process that takes steps to include residents does exist, an overall domination on the part of the neighborhood's business community can be detected" (Fowler, 2001, p. 2).

All of the professional planning organizations, including the Canadian Institute of Planners and the American Planning Association, entreat planners to consider the effects of their actions on marginalized groups and to seek just outcomes for all. The charter of the Royal Town Planning Institute in the UK states that planners "shall not discriminate on the grounds of race, sex, sexual orientation, creed, religion, disability or age and shall seek to eliminate such discrimination by others and to promote equality of opportunity." The Australian Code of Professional Conduct for planners states in part that: "2) members shall uphold and promote the elimination of discrimination on the grounds of race, creed, gender, age, location, social status or disability. 3) Members shall seek to ensure that all persons who may be affected by planning decisions have the opportunity to participate in a meaningful way in the decision-making process" (Planning Institute of Australia, 2002, p. 2).

That the neighbourhood residents, and the gay residents in particular, were not recognized and given a voice in the redevelopment of the Midtown neighbourhood is a shortcoming of the planning process. It is also a clear indication of the utilitarian values that drove the redevelopment of the neighbourhood, one that undervalued "place" in the accounting of development.

Zaferatos recounted in chapter 7 how protracted conflicts between Native American tribes and state governments were resolved through alternative conflict resolution strategies that appeal to the culture of Native Americans. His account of tribal planning in the state of Washington shows how recognition of difference works to the mutual benefit of all. By recognizing tribal rights and the use of alternative dispute resolution approaches rather than litigation, the state and tribal leaders were able to resolve long-simmering conflicts in land use and resource management. The Comprehensive Cooperative Resource Management program used negotiation rather than litigation to resolve conflicts between Native American tribes and state agencies involving the use of water resources, in comprehensive and economic development, and in transportation and wetland disputes.

In my previous writing (Burayidi, 2000), I pointed out six ways that misunderstandings can arise between planners and ethnocultural groups. I noted with respect to dispute resolution in particular that non-Western cultures may prefer other approaches to resolving dispute than the use of litigation: "While Western democracies regard confrontational dialogue as necessary and desirable for resolving disputes, Eastern cultures generally regard this as demeaning and embarrassing" (Burayidi, 2000, p. 6). After recognizing the cost and time it took to go through litigation, the tribes and the state government were able to reach an amicable solution to their land use disputes by creating an institutional arrangement that is less confrontational and more conciliatory of tribal interests. The Centennial Accord between Washington and Native American tribes ended the 100-year conflict in the management of natural resources. The success of the program was in large part due to recognition of difference; that there is a more effective way to resolve disputes with Native Americans than litigation. Without this recognition, successful resolution of the disputes would have been futile.

Burayidi and Wiles in chapter 8 provided examples of majority-minority cities that changed their planning procedures and practices to recognize diversity. In Los Angeles, for example, recognition of the growth of the Armenian population led then Mayor Villaraigosa to appoint an Armenian resident to the city's planning commission. Other cities such as Honolulu, Las Vegas, and Philadelphia rewrote their vision and proclamation statements to include recognition of the diversity of the cities' populations. Though these vision statements may be symbolic in nature, they provide the first efforts in the recognition of diversity, following which more substantive policies may be implemented to advance social cohesion. Nguyen, Gill, and Steephen (chapter 9) and Harwood and Lee (chapter 10) also provided examples of cities that recognized and acknowledged the diversity of their population and therefore enacted proactive policies and programs to support, acculturate, and integrate their new residents into the community's fabric. Such policies have included establishing a multicultural affairs office, creating an immigrant advisory board, enforcing fair housing laws, reducing language barriers, encouraging cross-cultural understanding, and providing a citizenship course, among others. These programs speed up the integration of immigrants and abate social fragmentation, which ultimately help these cities build socially cohesive communities.

Plan with Difference in Mind

When planners plan with difference in mind, difference is not viewed as a deviation from the norm; it is the norm. With such a view, planners understand that procedures and approaches that work well in one community may not work in all communities. When difference is the norm, the standardized approaches to planning that involve data collection through surveys, the holding of public hearings to obtain community input, and the assumption that all groups and persons will assert their rights equally in the public sphere have limited application with ethnocultural groups. In such cases, planners take the time to understand new and diverse groups that they encounter, explore alternative approaches for gaining information from multiethnic groups, seek unconventional ways to solicit their views and opinions with respect to plans, and gain the competence to work with these groups and to handle conflicts when they arise.

Sen, Kumar, and Smith in chapter 15 showed how HBCUs inculcate a critical and social justice orientation in the pedagogy of their planning programs. In these schools, students are encouraged to question received knowledge, delve further into sources of data and information, and come up with their own and alternative interpretation of the stories they hear and read about in the popular media. The curriculum in HBCUs weaves together five dimensions of education that include content integration, knowledge construction, prejudice reduction, equity pedagogy, and an empowering school culture. The schools make a valuable contribution to the profession by increasing the diversity of practising planners and by educating planners with a knack for social justice and inclusiveness in planning practices.

Planning with difference in mind is inclusive planning. This is both a process and an outcome. Procedurally, inclusive planning requires that planners provide meaningful and effective ways for all those who are impacted by plans (stakeholders) to be heard in the planning process. The substance of inclusive planning ensures:

> that everyone in society can access work, good schools, quality health care, and affordable healthy foods; as well as live in a safe environment … Good planning makes a positive difference to people's lives and helps to deliver homes, jobs, and better opportunities for all. It is important to ensure that people are able to play a full and independent role in society, and are not precluded in doing so by the inaccessibility of land, buildings, transport and other facilities. (Reading Borough Council, 1998)

Rios's discussion in chapter 14 shows how planners can assist marginalized groups to become a part of the planning process and assist diverse cultural groups to negotiate place making in the urban sphere. The cultural contracts that develop through negotiations of belonging, authorship, and power not only enable the different groups to establish what groups can expect of one another but also empower them to push planning and political institutions to help them meet their felt needs.

Planning with difference as the norm needs articulation at all levels of government. In the case of Canada, the federal government officially adopted multiculturalism as a national policy in 1971. The goal was "to recognize all Canadians as full and equal participants in Canadian society" and to protect the rights of ethnic and cultural minorities. Prime Minister Pierre Trudeau contended that "a policy of multiculturalism within a bilingual framework is basically the conscious support of individual freedom of choice. We are free to be ourselves" (Canada House of Commons, 1971, p. 8546).

Such an inclusive ideology at the national level trickles down to the local level as well. For example, many organizations and municipal governments in Canada adopt similar inclusive policies in their work. The Toronto Board of Trade noted the significance of immigrants to the city's development, observing in particular that the city's labour market growth is primarily due to immigration. So the board committed itself to inclusive planning because:

> Engaging the entire population broadens our base of skills and expands our networks to better prepare us to compete in a global marketplace. Further, successful investments in social cohesion and economic inclusion have been shown to reduce public expenses in health, education and justice. In some instances, the returns on these investments are so strong that they are offset within the first year (with the cost of savings continuing in the future). (Toronto Board of Trade, 2010, p. 4)

When communities plan with difference in mind, they modify planning procedures and regulations to accommodate the changing needs of their residents. For example, cities such as Seattle and Charlotte have modified their zoning codes to allow accessory dwelling units in some predominantly single-family residential neighbourhoods. This is due to the realization that a growing number of their residents live in multigenerational households.

Public participation processes that use innovative and alternative approaches to soliciting ideas from residents such as photovoice and Place It! give voice to marginalized groups in the planning process. As Main and Rojas explained in chapter 12, residents use photovoice to take pictures of their city and place of residence and use them to tell stories about what they like and dislike about their community. With Place It! residents who otherwise would find it difficult to participate in public forums are able to use random objects and recycled material to model their ideal neighbourhood. In the process, planners are able to extract information from these groups and take it into consideration in the design and development of the community.

It is worth acknowledging, however, that planners work within a political and institutional environment that may constrain what they do. In some cases, even where there is a will on the part of planners, as Gale and Thomas pointed out in chapter 5 with respect to England, statutory regulations may limit their ability to act. The authors suggest that these constraints explain why UK planners remain procedurally focused and less involved with social issues than they are with managing the statutory planning system.

Enhance Social Cohesion through Planning

The fraying of community in the West has been attributed to several causes, chief among which is increased immigration (Council of Europe, 2004), isolation (Home Office, 2001), and religious extremism and the rise of identity politics (Reich, 2001). Some have argued that the increased ethnic and cultural diversity of Western democracies will undermine social cohesion (Huntington, 2004), although these concerns have not been borne out in other studies (see, for example, Johnson & Soroka, 1999, and Commission on Integration and Cohesion, 2007). Nonetheless, planners must ensure that planning practices do not work to the detriment of the social structure and the glue that binds communities together.

Social cohesion refers to the prevalence or absence of shared values in a society or community, to feelings of belonging and place identity, trust, and cooperation as well as to equity, liberty, and justice. The Council of Europe defines social cohesion as:

> society's capacity to ensure the well-being of all its members by minimising disparities and avoiding marginalisation; to manage differences and divisions and to ensure the means of achieving welfare for all. Social cohesion is a dynamic process and is essential for achieving social justice, democratic security and sustainable development. Divided and unequal societies are not only unjust, they also cannot guarantee stability in the long term. (Council of Europe, 2010)

Embedded in this definition is the need to decrease inequality, to embrace all members of society as productive citizens, and to ensure social stability. Planners can contribute to social cohesion and community identity because planning practices affect the quality of the public realm where social interaction takes place.

The physical environment also impacts a person's sense of belonging and identity. In this sense, planners can help create socially cohesive places through design and regulation. Litman (2012), for example, noted that "transportation and land use planning decisions can affect community cohesion by influencing the location of activities and the quality of the public realm (places where people naturally interact, such as sidewalks, local parks and public transportation) and therefore the ease with which neighbors meet and build positive relations" (p. 2).

In chapter 13 Agrawal showed that the spatial agglomeration of religious institutions can have the effect of increasing inter-group contact and promoting inter-religious understanding. Because the religious institutions were located in close proximity, people of different religious persuasions were able to work together to promote their mutual interests and common social goals such as that of providing disaster relief for those affected by hurricanes in Asia. Over time, and although people clung to their individual religions, they began to see the humanity in everyone despite their religious differences. This is a starting point for cooperation or at least altering views about the "other" and for reducing inter-religious conflict.

Forrest and Kearns (2001) identified five domains of social cohesion as follows: 1) common values and civic culture, 2) social order and control, 3) equity and wealth disparity, 4) social capital, and 5) place identity and attachment. Let's consider how planners' actions affect these attributes in a community.

COMMON VALUES AND A CIVIC CULTURE

This criterion relates to the moral principles and codes of behaviour of people in a community and to the level of support for political institutions. A socially cohesive community is one in which residents share the same goals in life, abide by similar moral principles and codes of behaviour, and are politically active. It is precisely for this reason that Harper and Stein (chapter 2) called for adherence to the core liberal democratic principles to guide behaviour and expectations. In their contribution, Nguyen, Gill, and Steephen (chapter 9) identified the two main immigrant integration strategies of cities to be those of developing greater cross-cultural interaction between immigrants and the receiving community, and promoting greater civic participation. These strategies ensure that new immigrants and the receiving community develop common understanding and values that are necessary for the development of a socially cohesive community.

The design of the physical environment also impacts these attributes. McMillan and Chavis (1986) found that emotional connection is fostered in communities that provide opportunities for people to "interact, important ways to share ... opportunities to honor members, opportunities to invest in the community, and opportunities to experience a spiritual bond among members" (p. 14).

Similarly, Cochrun (1994) observed that:

> The components of sense of community – membership, influence, fulfillment of needs, and shared emotional connection – work in concert with each other to produce observed individual and group behaviors, such as political participation and other community action ... An awareness of the shared needs among neighborhood residents can foster concerns about political allocation of resources, and shared emotional connections can motivate residents to take political action on behalf of other people in the neighborhood. (p. 95)

Creating a sense of community is one of the goals of planning. For example, Miami Beach, Florida, states, "The goals and objectives of the Neighborhood Services Department are to build neighborhood capacity, redesign public services *and create a sense of community*" (City of Miami Beach, 2013; emphasis added). The document goes on to state that "Neighborhood revitalization ultimately depends on a sense of neighborhood identity and a commitment by all residents to make their neighborhood a better place to live, work, learn and play."

Planners can enhance a community's common purpose and shared values by providing ample public spaces and a quality public realm that encourage chance encounters of neighbourhood residents. Such interactions can lead to discussions about the welfare of the neighbourhood and galvanize residents to act to promote the common good. This is because "residents living in walkable, mixed-use neighborhoods are more likely to know their neighbors, to participate politically, to trust others, and to be involved socially" (Leyden, 2003, p. 1550).

Talen argued convincingly in chapter 11 for using urban form to support stable, socially diverse neighbourhoods because such neighbourhoods have a high level of vitality and strong economies, and are socially just, resilient, and sustainable. To achieve this goal she advocates a targeted neighbourhood development process that ties together three leading views on the city: a communicative and inclusive planning discourse that acknowledges difference and reaches out to marginalized groups in society, New Urbanist ideals of physical form that include mixed housing types and lot sizes to cater to different socio-economic groups, and the just city ideals that support strategic public investment in infrastructure to create quality public spaces and accessible neighbourhood amenities such as parks and places of worship. Such neighbourhoods facilitate walking and thus increase opportunities for spontaneous social interaction along sidewalks, neighbourhood parks, and shops. As people get to know each other in the neighbourhood, a sense of community and belonging is created.

SOCIAL ORDER AND CONTROL

Communities vary in their level of civility, informal social control, cooperation, tolerance, and respect for difference. Some observers such as Oscar Newman attribute high crime rates in some neighbourhoods to the lack of social control, which is engendered by the nature of the built environment. It is for this reason that in outlining the principles of "defensible space" Newman states that the goal is to:

> ... restructure the physical layout of communities to allow residents to control the areas around their homes. This includes the streets and grounds outside their buildings and the lobbies and corridors within them. The programs help people preserve those areas in which they can realize their commonly held values and lifestyles. (Newman, 1996, p. 9)

The design of the physical environment, particularly the placement of buildings in relation to the street and public areas, can influence property owners' sense of control of their neighbourhoods. For example, when residents consider the street and sidewalk to be an extension of their property, they are more likely to take better care of it and to monitor activities that take place there.

Because compact neighbourhoods facilitate walking, such neighbourhoods have significant pedestrian activity and so help to keep "eyes on the streets." Residents are thus able to identify intruders and to report crime to law enforcement. These desirable qualities have ripple effects in the neighbourhood. Low crime rates make neighbourhoods desirable places to live and so there is less residential transience. Residential stability increases attachment to the neighbourhood, a necessary ingredient for social order and control. By contrast, a high degree of transience in a neighbourhood means fewer long-term residents remain and anonymity increases. In such neighbourhoods weaker social ties bind residents, leading to a high degree of instability and social fragmentation. Such neighbourhoods can quickly disintegrate into chaos and economic disinvestment. Adopting the strategies for creating socially diverse, sustainable, and just neighbourhoods suggested by Talen in chapter 11 also supports this goal of social order and control.

SOCIAL SOLIDARITY AND EQUITY IN WEALTH

Significant disparities in wealth and access to public services create tension and cynicism, as some members of the community may feel left out. Planners are involved in deciding the service needs of communities and where such services should be located. The seminal article by Norman Krumholz (1982) on planning practice in Cleveland from 1969 to 1979 leaves no doubt that planners can assist in providing a wider range of choices for those who have few options and in so doing contribute to social equity.

Equity and advocacy planning assist in redressing the needs of the disadvantaged. The contributing reasons for the riots in Western Europe at the turn of the century were high unemployment rates and the disengagement of the immigrant population from the host society. That is why the advocacy role of the planner is important to sustaining social cohesion. As Johnson (2008) noted, the problem is not diversity – it is poor housing, poor education, fear of difference, and the perception that particular groups are granted better access to services.

Furthermore, place prosperity programs such as tax increment financing and enterprise zone programs can help decrease income disparity by drawing businesses to locations of economic distress and providing employment for people who live in such places. By decreasing poverty and enhancing job prospects, such programs help decrease economic disparity and social unrest. As discussed by Harwood and Lee (chapter 10), Chicago's New Americans Plan has a goal to fully integrate immigrants into the city through 27 initiatives, including support for immigrant business and the development of immigrants' human capital. In unveiling the plan in December 2012, Mayor Rahm Emanuel stated:

> These immigrants are small business owners, teachers, parents, clergymen, elected officials, and leaders in our communities. This is why I am committed to making Chicago the most immigrant friendly city in the nation. I thank the [Office of New Americans] Advisory Committee for assisting us in developing a plan that will ensure Chicago continues to thrive and grow and attract the world's leading human capital to compete in the 21st century global economy and beyond. (City of Chicago, 2012)

Other cities such as Dayton, Ohio, have similar immigrant integration programs, all aimed at unleashing the creativity of their new residents and fully engaging them in the civic culture of the cities.

SOCIAL NETWORKS AND SOCIAL CAPITAL

Social networks are the mutually reinforcing formal and informal relations and associations of people in a community. These social networks strengthen cooperation between people and the community at large. The French writer Alexis De Tocqueville linked the higher level of social networks in the US to a stronger democracy and to equality of conditions. Social networks are developed through social interaction, civic engagement, and associational activity and are particularly important to minority groups, in part because the networks enable them to obtain information about job prospects that they otherwise may not know of.

Where strong social networks exist, there are positive relationships between people in the neighbourhood, workplaces, and third spaces. Planners can provide community residents opportunities that increase social networks through physical design, as it has been shown that interaction between different ethnic groups helps to decrease fear of the "other." This is the basis of Talen's call in chapter 11 for the creation of diverse socially sustainable neighbourhoods.

Neighbourhoods that isolate minority groups make it difficult for them to encounter people who are different from them. This lack of ethnic group intermingling breeds mistrust and makes it difficult to have a cohesive community. Since 1994, the Moving to Opportunity program in the US has enabled poor families to move to higher-income neighbourhoods. When the Department of Housing and Urban Development evaluated the program in 2011, it found that families that participated in the program increased their social connections, had close friends, and increased their social networks. The report observed that:

> [Moving to Opportunity] also helped families move into neighborhoods where neighbors were more willing to work together to support shared norms, a measure of informal social control that previous research suggests may be particularly important in improving the lives of neighborhood residents. (US Department of Housing and Urban Development, 2011, p. 57)

These findings support the use of planning to promote social inclusion through mixed-income neighbourhoods and inclusionary zoning practices. Zoning and land use regulations that support mixed-use development increase the social mix of neighbourhoods and provide opportunities for people of different incomes, abilities, and racial groups to get to know each other.

PLACE ATTACHMENT AND IDENTITY

People who feel connected to a place have a strong emotional attachment to it and are therefore willing to dedicate their time and effort to preserve and/or enhance its quality. Places that have no identity do not cultivate this sense of belonging and place attachment. Places with identity are places that have a special character, in either their architecture or design, and provide "the affective link that people establish with specific settings, where they tend to remain and where they feel comfortable and safe" (Hernandez et al., 2007, p. 310). People attach to places that have meaning to them and where they have a sense of belonging.

Place attachment can be enhanced where there is a feeling of authorship in its creation. Rios in chapter 14 discussed the "cultural contracts" that different groups negotiate with each other and with planning institutions to come to negotiated settlements in the creation of place. Such negotiations usually take three forms: negotiations of belonging (the right to be different and still be a part of the larger polity), negotiations of authorship (providing room for marginalized groups in particular to contribute ideas to the development of place in ways that may be different from the conventional approaches that are typically used to elicit information and participation from the public), and negotiations of power (the extraction of legal rights to the city by cultural groups). Rios gives examples of how such negotiation has led to the enhancement in quality of place for neighbourhood residents in the Fruitvale and San Antonio neighbourhoods in Oakland, California, through transportation and infrastructure improvements and the development of housing and neighbourhood parks. Such negotiated outcomes increase a cultural/ethnic group's sense of belonging and neighbourhood satisfaction.

Place attachment matters in creating social cohesion because residents in such places develop a unifying bond with others in the neighbourhood. Freeman's (2001) study of adults in Atlanta, Boston, and Los Angeles led him to conclude that there is an inverse relationship between neighbouring and people's reliance on their automobiles. He also noted that "the proportion of residents who drive to and from work ... has a strong and statistically significant relationship to whether or not an individual has a neighborhood social tie" (p. 74). This suggests that walkable neighbourhoods are better able to create place identity and attachment than those that are car oriented, and is also supported by Rogers's (Rogers, Halstead, Gardner, & Carlson, 2010) study in New

Hampshire, which led him to conclude that social capital is higher in walkable neighbourhoods than car-dependent ones.

Similarly, Leyden (2003) found the level of social capital to be highly and positively related to mixed-use and walkable neighbourhoods, and residents in such neighbourhoods were more likely to be involved in the community, to know their neighbours, and to be socially engaged. Where there is extreme segregation of residential neighbourhoods by class, income, and ethnicity, it is easy for conflict to arise between the different groups since they don't know each other. Physical form is important to creating place identity.

Conclusion

In conclusion, to move the diversity agenda in planning forward, we need to cast a wider view of what it means to be different. In the past, planning viewed diversity only through a narrow racial and ethnic lens. As we've seen in the essays in this book by Doan (chapter 6) and Zaferatos (chapter 7), there is more to diversity than this conception. By casting a wider net, we appeal to the broader spectrum of the human condition and hence are more inclusive of all persons in society. As a first step to effective planning in a pluralist society, planners need to acknowledge that diversity matters and plan with difference in mind. Only then can we assist in creating socially cohesive places that are just and fair and where everyone has the potential to achieve their ultimate best in life. The contributors to this volume tell us how this can be done!

REFERENCES

Burayidi, M.A. (2000). *Urban planning in a multicultural society*. New York: Greenwood Publishing Group.

Canada House of Commons. (1971). *Statement to the House of Commons on Multiculturalism*, House of Commons, Official Report of Debates, 28th Parliament, 3rd Session, 8 October, 8545–8546.

City of Chicago. (2012). Press release: "Mayor Emanuel unveils first-ever Chicago New Americans Plan." Retrieved November 21, 2013 from http://www.cityofchicago.org/content/dam/city/depts/mayor/Press%20Room/Press%20Releases/2012/December/12.4.12NewAmericans.pdf.

City of Miami Beach. (2013). Work plan. Retrieved October 6, 2013 from http://web.miamibeachfl.gov/housingcommdev/scroll.aspx?id=17286.

Cochrun, S.E. (1994). Understanding and enhancing neighborhood sense of community. *Journal of Planning Literature*, *9*(1), 92–99. http://dx.doi.org/10.1177/088541229400900105

Commission on Integration and Cohesion. (2007). *Our shared future*. Wetherby, UK: Commission on Integration and Cohesion.

Council of Europe. (2004). A new strategy for social cohesion. European Committee for Social Cohesion. Approved by Committee of Ministers of Council of Europe on March, 31, 2004. Council of Europe Workshops.

Council of Europe. (2010). The action plan for social cohesion. Retrieved October 1, 2013 from http://spiral.cws.coe.int/tiki-index.php?page=action+plan+social+cohesion.

Forrest, R., and Kearns, A. (2001). Social cohesion, social capital and the neighbourhood. *Urban Studies*, 38(12), 2125–2143.

Fowler, D.P. (2001). *Midtown Atlanta: Privatized planning in an urban neighborhood*. Major paper submitted to Virginia Polytechnic Institute and State University in partial fulfilment of the requirements for the degree of Master of Urban and Regional Planning, Blacksburg, Virginia.

Freeman, L. (2001). The effects of sprawl on neighborhood social ties: An explanatory analysis. *Journal of the American Planning Association*, *67*(1), 69–77.

Hernandez, B., Hidalgo, M. C., Salazar-Laplace, M. E., Hess, S. (2007). Place attachment and place identity in Natives and Non-natives. *Journal of Environmental Psychology*, 27(4), 310–319.

Home Office (2001). *The Cantle Report: Community Cohesion*. Coventry, UK.

Huntington, S. (2004). *Who are we? America's great debate*. London: Simon and Schuster.

Johnson, N. (2008). Diversity and social cohesion. Bridging Social Capital Seminar Series, Seminar No. 3: Diversity and Social Cohesion, Carnegie UK Trust, September 29, 2008.

Johnson, R., & Soroka, S. (1999). Social capital in a multicultural society: The case of Canada. Paper presented at the Annual Meeting of the Political Science Association, Sherbrooke, QC.

Krumholz, N. (1982). A retrospective view of equity planning: Cleveland 1969–1979. *APA Journal*, *48*(2), 163–174.

Kymlicka, W. (2004). The Canadian Model of Diversity in Comparative Perspective. Eight Standard Life Visiting Lecture, University of Edinburgh, April 29, 2004.

Leyden, K.M. (2003, Sept.). Social capital and the built environment: The importance of walkable neighborhoods. *American Journal of Public Health*, *93*(9), 1546–1551. http://dx.doi.org/10.2105/AJPH.93.9.1546

Litman, T. (2012). *Community cohesion as a transport planning objective*. Victoria, BC: Victoria Transport Policy Institute.

McMillan, D.W., & Chavis, D.M. (1986). Sense of community: A definition and theory. *American Journal of Community Psychology, 14*(1), 6–23. http://dx.doi.org/10.1002/1520-6629(198601)14:1<6::AID-JCOP2290140103>3.0.CO;2-I
Myers, D. (2008). *Immigrants and boomers: Forging a new social contract for the future of America.* New York: Russell Sage Foundation.
Newman, O. (1996). Creating defensible space. US Department of Housing and Urban Development, Office of Policy and Research. Retrieved May 20, 2015 from http://www.huduser.org/publications/pdf/def.pdf.
Planning Institute of Australia. (2002). Code of Conduct. Retrieved May 20, 2015 from https://www.planning.org.au/membershipinformation/code-of-conduct.
Reading Borough Council. (1998). Reading Borough local plan.
Reich, R. (2001). *The future of success.* New York: Alfred A. Knopf.
Rogers, S.H., Halstead, J.M., Gardner, K.H., & Carlson, C.H. (2010). Examining walkability and social capital as indicators of quality of life at the municipal and neighborhood scales. *Applied Research in Quality of Life, 6*(2), 201–213.
Royal Town Planning Institute. Professional Standards. Retrieved October 2, 2012 from http://www.rtpi.org.uk/membership/professional-standards/.
Salins, P.D. (1997). *Assimilation, American style.* New York: Basic Books.
Stone, C.N. (1989). *Regime politics: Governing Atlanta, 1946–1988.* Lawrence: University Press of Kansas.
Toronto Board of Trade. (2010). Lifting all boats: Promoting social cohesion and economic inclusion in the Toronto region. Votetoronto2010.com.
US Department of Housing and Urban Development. (2011). *Moving to Opportunity for fair housing demonstration program.* Washington, DC: Office of Policy Development and Research.